W9-CFU-250

Whole House Remodeling Guide

Whole House Remodeling Guide

S. Blackwell Duncan

TAB BOOKS Inc.
Blue Ridge Summit, PA

FIRST EDITION
FIRST PRINTING

Library of Congress Cataloging-in-Publication Data

Duncan, S. Blackwell.
   Whole house remodeling guide / by S. Blackwell Duncan.
      p. cm.
   ISBN 0-8306-9281-9   ISBN 0-8306-3281-6 (pbk.)
    1. Dwellings—Remodeling—Amateurs' manuals. I. Title.
TH4816.D86   1989
643.7—dc20
                                       89-39624
                                           CIP

TAB BOOKS Inc. offers software for sale. For information and a catalog, please contact TAB
Software Department, Blue Ridge Summit, PA 17294-0850.

Questions regarding the content of this book should be addressed to:

    Reader Inquiry Branch
    TAB BOOKS Inc.
    Blue Ridge Summit, PA 17294-0214

Acquisitions Editor: Kimberly Tabor
Book Editor: Cherie R. Blazer
Cover Design: Lori E. Schlosser

Contents

3
CHAPTER

Walls   **130**

4
CHAPTER

Windows and Doors   **236**

5 **Ceilings**   **348**

CHAPTER

Index   **446**

Introduction

Few homeowners today have escaped facing the high (and continually increasing) costs of home maintenance, repairs, and general fix-up work. When it comes to redecorating and remodeling, the expense is so substantial that many people decide the work they would like to have done is simply unaffordable. Furthermore, in many areas of the country finding competent tradespeople to do a decent job in a timely fashion for a price that is at least within reach is at best a struggle and at worst impossible. To top it all off, the cost of both existing and new homes has gone so high that qualifying for the purchase of either one has become very difficult. For many people, "trading up" into a more suitable and attractive home, even of the same or smaller size, has become an unattainable goal.

Difficulties such as these leave only one practical course open for the homeowner who wants to live in more amenable, attractive, functional surroundings: hang onto the present house and remodel and redecorate it as extensively as need be through a do-it-yourself program. Although economics are perhaps the most common reasons for undertaking do-it-yourself work of this sort, they are by no means the only ones. A recent nationwide survey reveals that another reason for the popularity of owner-remodeling and assorted do-it-yourself home-improvement projects is that an increasing number of homeowners are discovering the simple enjoyment of doing their own work. They find great satisfaction in the completion of refurbishing and upgrading projects and take great pride in what they have accomplished. There are other good reasons for following this course. The do-it-yourselfer has complete control over many important aspects of each job, such as the timing of various phases of work; the time span for each project; the quality of the materials, workmanship, and the completed job; the overall design or appearance or effect of the project; the ability to easily make changes as the work goes along; the convenience of carrying on the work at the most appropriate times; and the ease of suiting the expenditures to fluctuating budgetary concerns without any pending contractual obligations to others. Also, many (if not most) do-it-yourselfers find that they can end up with better results and far fewer hassles than if they hire out the work to be done.

The fact that you have never done any of your own remodeling or redecorating work is really no reason that you shouldn't, or can't. By using the *Whole House Remodeling Guide,* you will find that there is much you can do yourself even if you have no experience at all, and much more you will be able to do as you gain experience. And as the book explains, there is always outside help available for whichever aspects of your projects are simply too complex or time-consuming for you to undertake.

This book has two main thrusts. The first is to expose you to the many ways in which you can remodel or redecorate the interior of your house. (The text does not deal with house exteriors or structural remodeling, such as adding on rooms or second stories, which is a whole other subject.) There are literally hundreds of ways to change not only the entire appearance but also the livability and functional practicality of any house interior—ranging from simple to quite complex—and the combinations are limitless. You might find some ideas or materials or decorating possibilities that immediately appeal and that you can put right to work. Some of the material discussed might engender other ideas or schemes and set your imagination percolating. Because so much ground is covered, you will probably discover some remodeling or redecorating possibilities you never thought of, or even knew existed.

The second thrust of the book is to explain the nuts-and-bolts aspects of numerous specific projects. Many of the more popular ones are covered in detail, with step-by-step instructions for completing them successfully. You will learn, for example, various ways to repair a floor or install a wood floor over a concrete one, how to lay wood parquet flooring, the details of wainscoting, how windows and doors are made and how to install them, how to hang a suspended tile ceiling, and how to go about preparing for and making many other installations. For some projects, such as installing wall-to-wall carpeting, texturing ceilings, and putting up wall trimwork, the basics are covered but the minor details are left to you. The skill and knowledge levels required for these installations range from zero to advanced.

A few possibilities—more unusual or difficult or requiring professional installation (terrazzo flooring, for example)—are merely described as potential options, so you will at least be aware of them and be able to take them into consideration. Although the text is separated into many short pieces on specific subjects, by going through the entire book you will end up with a well-developed picture of the way most houses are typically put together and the many ways in which they can be refurbished and redecorated.

You don't have to live with your house the way it is, nor do you have to place yourself at the mercy of the real estate agents, the lending instititutions, or the remodeling contractors (though all those folks are helpful at times) if you don't want to. You can make changes yourself. Check it out.

1

Do-It-Yourself Interior Remodeling

FOR MUCH OF OUR LIVES WE ARE ENCLOSED IN A PERSONAL ENVIRONMENT WE CALL home. We arrange and decorate this environment to suit our own tastes, interests, life-styles, whims, and perceptional mirrorings of who we are (or could be, or should be). For similar reasons we frequently rearrange, redecorate, and remodel our environment. When you come to this point, you must make some choices: Should you hire someone to undertake these tasks, do them yourself, or do them partly yourself? When choices exist there are decisions to be made, and that can't be done without information. This chapter will deal briefly with the generalities of do-it-yourself interior redecorating and remodeling and give you an overview of what is involved. The following chapters will explore specific options and projects in more detail. Hopefully you will acquire enough information to assess your choices and select those best suited to your remodeling needs.

THE FIVE ELEMENTS OF REMODELING

Our artificial interior environments are comprised of five principal elements: floors, walls, windows, doors, and ceilings. These are also the five elements of interior redecorating and remodeling because each can be given a wide variety of different treatments.

Each element can be *redecorated*; this is the simplest and usually the least expensive course because it is a cosmetic process, a facelift that involves changing patterns, colors, or coverings without making changes in fundamental structure, form, or shape. Each element also can be *remodeled*, which means that changes are also made in some of the components of the structure itself. Often the two processes go hand in hand, and they can be carried out in any combinations of the five elements, or parts thereof. There is infinite flexibility.

For example, if you have a varnished hardwood floor that is getting tired-looking, you can revarnish it for a better appearance: this is simple renovation. You could cover it with wall-to-wall carpeting or sheet vinyl to change the appearance altogether: this is redecorating of a more complex nature. Or you could tear up the old hardwood, beef up the floor frame, and lay a quarry or marble tile floor, which changes the character and construction of the floor: this is a major remodeling project. Many possibilities exist, from simple to complex, for redoing floors of all kinds.

The same is true of the walls. Changing their appearance by redecorating is easy. You can repaint, repaper, or add decorative moldings, and put a fresh coat of paint on the wood trimwork. You can change the trimwork entirely—rip it out and put up something new and different. Building in a cove is another possibility. You might add a wainscot or tear an old one off. Don't like the shabby wall surface? Cover it up with paneling. Or, though the job is a little more difficult, take down the sheathing and start over. Where there isn't a wall you can always put one up. Where there is one, you can often tear it down—or shorten it, lengthen it, or change its direction or location.

You don't have to live with your existing windows. For one thing, you can retrim them. Add moldings to the existing trimwork to change the appearance. Take off the old casings and put up another style, add cornices or sill shelves or put in false muntins. You can take out the old windows and replace them with modern, more efficient, more presentable units, and at the same time change the shape, or the size, or the style. Taking a window out and blocking off the opening is not terribly difficult. Cutting a new one into a new location is always a viable option, although it's also a bigger job. And how about a skylight or two? Making major changes in an existing window installation can make a major difference in how livable a house is and can increase its value.

Your doors don't have to stay the same, either. A fresh coat of paint is easy enough and might help, of course, unless there are already many layers on the doors. But there are a lot of other options that rarely seem to be considered. You can strip painted doors and redo them in a natural finish, or cover them with another material, or add decorative moldings to them. You can replace them with different kinds or patterns of doors. You can add to the trimwork, or replace it with something different. If a doorway is too narrow, you can usually make it wider; if too wide, you can always make it narrower. Got a door in an awkward place? Block the doorway off and cut a new one through in a different spot, perhaps after moving a wall. Need another door somewhere for greater convenience? Put one in. Again, there are plenty of possibilities here for not only changing appearance but also improving home safety, livability, and value.

Then there is that often-ignored element always hanging over our heads: the ceiling. Is there any decent redecorating/remodeling hope there? Absolutely! A ceiling in poor shape can always be fixed, or covered, or both. Sags, dips, droops, slants, cracking plaster, popping nails, splitting seams, dingy finish, blank appearance—these can all be improved. And once the ceiling is fixed, or if it is already in sound condition, there is a raft of options and an endless number of redecorating schemes from which you can select.

Another coat of paint appears to be the most common choice for ceilings, which is perhaps a good reason to select something else. At least you could go to a textured finish for some added character. How about wallpapering it? Not as hard as it sounds. Flat ceilings can be perked up in dozens of ways by adding decorative moldings, which is a fairly simple job. You might put up wood planking, or even ceramic tile—don't worry, they'll all stay up there. Acoustic tile is another option; there are many handsome new patterns that work nicely in any kind of room. If your ceiling is too high, you could install a dropped ceiling or hang a suspended ceiling; new lighting fixtures are easy to mount in either one. One way or another, you can make the ceilings an important part of your interior decor.

SKILLS AND KNOWLEDGE

If all of this sounds overwhelming and well beyond your abilities, and you are wondering what kind of a superworker you have to be to put these ideas into effect—relax. Begin by realizing that, as with any other endeavor, there will be a certain amount that you can do, or learn to do, and a certain amount that you can't, or that will hold no interest for you. Bear in mind, too, that most of the tradespeople who do all these redecorating and remodeling chores for a living are no more dexterous or intelligent than you are—a lot of them probably less so.

There are two distinct types of skills involved in redecorating and remodeling. The first is the artistic, imaginative skill coupled with the aesthetic sensibilities needed to plan the colors, patterns, forms, perspectives, and associated subtleties that make up the overall interior decor of a house. The other is the manual skill required to make the plans a reality.

Artistic skill involves both interior decorating skill, and for some projects, architectural design skill. Both can be learned, although it often seems these talents are inborn. You need not be an expert, however; that is necessary only for professionals in those fields. This subject is not specifically addressed in this book, but there is an endless supply of interior decorating and design ideas readily available to you in the many books and magazines that deal with home decorating, interior design, and improvements. Other sources of ideas and information are interior decorating shops and finish building materials suppliers. The latter often have display rooms, as well as knowledgeable personnel with whom you can discuss ideas. As to architectural design and engineering—both complex fields—your best bet is to consult with a professional if and when the need arises.

Manual skills are a different matter. You can get the gist of a procedure or a process from a book or magazine, but you cannot actually learn how to do them this way. If you have skills at the outset, you can relate them to a written exposition, but that neither gives you skills nor improves your own. The same is true of watching a demonstration, or a video cassette. The only way to acquire and develop a manual skill to whatever level you are mentally and physically capable is by doing, and then doing some more. Hands-on experience is what counts, and no amount of watching, listening, or reading is a substitute for that. Remember, however: Don't bite off more than you can chew. Proceed at a rate that is right for your skill level.

Manual skill also has two distinct areas. First, the mechanical aspect, the part that requires a reasonable degree of manual dexterity, gross and fine motor skills, muscular and hand/eye coordination, and general physical well-being. If you can perform the normal run of workday and household tasks, that is all that's needed. The second aspect, developing what is generally called a "feel" for the task at hand, is less tangible and can only be gained through experience. It is a combination of combined sensory preceptions and learned and stored data that you have on tap. For example, when you are using a certain kind of paint on a certain kind of wood and the brush starts to drag, you know the paint is too thick, or perhaps the temperature is too high or the humidity too low. You know when the glue skims over too soon that the working conditions are wrong. You know when the mortar reaches the right consistency and plasticity for effective workability.

In many instances this "feel" comes from recognizable sensory signals that prompt a response. You know the board is not sanded smooth enough because you can feel slight ripples with your fingertips (tactile). You know you are using too much heat when lifting paint with a torch because you can smell the scorching material (olfactory). You know when your table saw blade is starting to bind because you can hear the change in the pitch of the blade noise (aural). You know when a joint does not fit quite right, you can see it (visual). It takes practice and experience to recognize, understand, and act upon these sensory perceptions.

Knowledge is power, and this certainly is applicable to the do-it-yourselfer. Not only does knowledge of the many aspects of redecorating, remodeling, and the building trades allow you to know and understand what you are doing and why, it also allows you to understand what someone else is doing or talking about. Before you start your project you can gain basic knowledge about tools, materials, supplies, construction methods and practices, building codes, and all associated elements from books, magazines, technical manuals, conversations, and observations. More complex and detailed comprehension will come to you, growing and expanding constantly, as you actually undertake project after project and become more involved in and familiar with whatever your special interest areas might be. Some knowledge will be gained almost automatically. Much more can be absorbed by conscious study. You can never have too much knowledge.

Now, how much do you have to know about carpentry, painting, masonry work, etc., in order to do your own redecorating and remodeling with creditable results? That is an open question. At the outset, none. We all start from scratch in every new subject. Gain the initial knowledge that you need for some particular project by investigation and study beforehand, more by doing the work and by continued reading or observation. You need to know as much as is required to successfully put up an acoustic tile ceiling in your rec room, but you don't have to know everything there is to know about acoustic tiles, ceilings, and every possible installation practice and variation to get the job done. You need to know enough about windows to be able to talk intelligently with a supplier and select a good product. And you need to know enough about window installation to realize the difference between a good job and a shoddy one when a subcontractor installs the window.

You need to be able to specify how the installation is to be made, and then to tell whether or not you are getting what you pay for.

In short, the more you know, the better. But none of us can know everything. Expect some gaps, expect some mistakes or omissions stemming from those gaps, expect to not have all the answers. However, don't hold back from going ahead with your plans and projects just because you think you might not know enough about a subject—at least, not if it's something you really want to do.

TOOLS AND EQUIPMENT

When you think about the whole gamut of redecorating and remodeling possibilities, there is an enormous array of tools and equipment that is either necessary or useful to make the job go faster and easier. However, you certainly don't need everything listed in a master tool catalog. In fact, for many projects you need very few tools, most of them common and inexpensive and often already on hand around the house. Exactly what you will need depends on the job to be done. Most of the projects discussed in this book, and most redecorating and interior remodeling work in general, can be done with a relatively small selection of hand and hand power tools, along with some more specialized but inexpensive equipment. For the few exceptions where high-priced equipment is needed, try to rent it from a tool rental outlet.

For example, the gear you might need to paint some walls amounts to a decent brush or a roller and pad set, a few plastic drop cloths, and maybe a putty knife. A ten-dollar bill will put you in business. If you want to hang some wallpaper, that takes a few special tools. These come in a kit for less than $10, too. Add a bucket and a big wooden spoon for mixing paste, a pair of long, good-quality shears, a framing square and a good straightedge, a pack of razor blades, maybe a small X-acto knife, an 8- or 10-foot measuring tape, a short stepladder, and a pencil—all of which comes to another $30 or $35. Only the pencil and the blades will wear out, so for around $50 you can wallpaper for as long as you keep the equipment.

The cost goes up a little if you want to remove and replace your door and window casings, or perhaps add ceiling perimeter moldings or put up some decorative moldings. You will need a claw hammer and a nail set, a small pry bar and perhaps a cat's claw, a measuring tape, maybe a coping saw, and a miter box. You could buy a cheap miter box for $20 or less, but you probably wouldn't be very happy with it. A good box runs about $125 and will give excellent results. Or, you can go up to around $200 and get a motorized compound-angle chop saw, which is ideal. You can also rent miter boxes, so figure from $50 to $250 for the project.

Some of the major jobs require surprisingly few tools. For example, if you want to sand a floor you don't need anything—rent all the necessary equipment, including a shop vacuum if you need it. To tear up a wood floor or take down wall sheathing, a claw hammer and a wrecking or pry bar is all you need. Sometimes a hand power saw is also useful for the flooring job. To install a new wood floor you can often get by with a hammer and nail set, while for certain floors you will have to rent a flooring nailer.

If you want to tackle a broad range of renovation projects, then you will find use for a fairly wide variety of carpentry tools, and even a few mechanic's tools. Hammers, screwdrivers, chisels, pry bars of different sizes, nail sets, planes, several kinds of handsaws, measuring tapes, plumb bob, chalkline, pliers, wire cutters, putty knives, scrapers or shave hooks, utility knife, squares, socket, adjustable wrenches, clamps, straightedge, files, and levels are all standard fare. A few power tools are handy, too, such as a $3/8$-inch drill and bit assortment, jigsaw, pad sander, circular saw, plane, and a table or radial arm saw. Other equipment, such as a heat gun, torch, electric paint scraper, soldering iron, or stud locater, will be needed for some projects.

There is no point in assembling a whole assortment of tools before starting to do your own remodeling and redecorating work. You can pick them up gradually as the need arises. That doesn't seem to hurt the pocketbook as much, and eventually you will find that you have enough gear to deal with just about anything. As you purchase tools, whether power or hand, stay away from the cheap, poorly made items. The so-called "homeowner" quality, common in lines of hand power tools, is not a good choice if you expect long life and good service from the equipment. Buy top-quality, name-brand tools that you can rely on, because there is nothing more frustrating than trying to do a decent job with equipment that continually lets you down. Good tools are expensive, but they will perform properly and last a long time if you care for and don't abuse them.

ECONOMICS

How often have you heard that if you do it yourself you'll save a whole lot of money? Sometimes it's true and sometimes it isn't, so regard the statement with mild skepticism. And be aware that there is more to the subject than simply saving money.

Assuming that you have the time, patience, and skills, you can do most redecorating and interior remodeling jobs yourself and it is likely that you will save money. That is, your cash outlay will be less than if you hired a professional. In many cases you can purchase not only all the required materials and supplies of the job but also all the tools and equipment, and still come out ahead. Take wallpapering as an example. You can buy the equipment to hand-strip the old and hang the new paper, plus the wallpaper and supplies for an average-sized room, for approximately half what it would cost to have a professional do the job. And you will still have the equipment to use again.

As another example, consider laying 200 square feet of parquet tile on plywood underlayment. Tile at $3 a square foot totals $600, adhesive for that area would be about $80, the necessary hand tools might run another $50, and if you also buy a moderately priced table saw for the trimming you would invest about another $350. Total outlay is about $1000 to $1100, and you have your new floor, plus the equipment is yours for further use. To have the same tile professionally laid costs about $1200 to $1500, depending upon your local pricing structure. If the professional cost were about the same as your own, would you still be ahead? Yes, because part of your investment is in reusable equipment that presumably will continue to be of value to you.

However, you must also consider the economics of time, in several respects. First, what is your time worth? Not to you, but to someone else. If a contractor is going to charge you $30 an hour to do a job that you are equally capable of doing, and your average earnings are $15 dollars an hour and you have the spare time anyway, you are better off doing the job yourself. But if you make $40 an hour and don't have much spare time, let the contractor do it.

Second, what is your time worth to you? Not necessarily in dollars, but in the overall sense. You might feel that any time you put in working on your house is time well spent, just as enjoyable or more so than golfing or skiing or watching a movie. On the other hand, you might be delighted to meet a contractor's stiff price just to get out of painting the living room ceiling.

Third, what is your time worth to your family? Dollars are not involved here, just feelings. Are you better off spending that time with your friends and family, who perhaps don't see enough of you as it is, or doesn't it matter?

Fourth, you have to make a reasonable assessment of how much time your project will actually take, from start to finish—including adequate planning before and cleaning up afterward. Then you need to decide whether or not you really do have that much time available to spend on the project, and whether or not you will be able to steal some time from other activities, if necessary. If so, what will the impact of that theft be on yourself, your work, or your family? Understand that one of the most troublesome problems in do-it-yourselfing, and one that can lead to all kinds of additional problems, is underestimating the time required for a project. It could drag on forever or even be left unfinished because of a lack of worktime. When you make your worktime estimate, double it, or at least add half again.

Finally, consider the elapsed-time factor. How many hours or days will the job take? When you divide that by the work periods that you can schedule for evenings or Saturdays or vacations, how long will the job go on? If too long, you might lose your enthusiasm for the project or other unexpected calls on your time might crop up, so bringing the job to a successful conclusion could turn into a real struggle.

When considering how much you can save by doing your own remodeling and redecorating work, versus how much you will have to spend to hire the work done, don't neglect to make a fair appraisal of not only the time elements but also your capabilities. The other big stumbling block in do-it-yourselfing is the old problem of getting in too deep, overestimating what or how much you can really do, and then getting stuck or terminally discouraged partway through a project. When the job is half done—and maybe fouled up as well—or you have no more time or you don't know what to do next, you have to call professionals in to finish the job. That can be discouraging, embarrassing, *and* expensive. There go all the savings, plus some.

However, don't neglect to consider the bright side of do-it-yourselfing. Some folks consider any kind of project around the house sheer drudgery, but others thoroughly enjoy the work. To be able to say they did all the wallpapering, or put up the suspended ceiling in the kitchen, or laid the pecan flooring in the dining room, means a great deal. A tremendous amount of satisfaction is involved in seeing these projects to a successful conclusion. What is the value you place on this self-satisfaction, of pride on a job well done, of the pleasure and enjoyment of working on your own unique projects, of the self-confidence engendered by achievement? All are part of the equation.

CODES AND PERMITS

There are numerous building, safety, and health codes, regulations and covenants, and permit processes involved in all construction, rebuilding, and remodeling. They differ greatly from place to place. In some areas all building activity is strictly regulated and the rules are rigidly enforced, while in others there are few if any rules in effect. It is up to you to investigate and determine how, or if, you might be affected by any rules and regulations in your locale.

Most interior remodeling work that is undertaken by the owner of the property does not require adherence to any regulations or require any permits. You should have a free hand with such projects as painting, papering, redoing moldings and trimwork, laying finish flooring materials, and similar ordinary refurbishing chores. In most places you can also make major interior changes, such as moving a wall, tearing down a ceiling, or converting an unfinished basement or attic. Whatever you do inside, that cannot be detected from outside, is usually nobody's business but yours. If you remodel in such a way as to change the exterior appearance, as by cutting in new doors or windows, however, you might be required to follow certain regulations and/or get a permit to do the work.

You will find exceptions. In many places, particularly cities, all electrical or plumbing work must be done by a licensed contractor, under permit and inspection. There might be restrictions only on certain kinds of work. For example, you might be able to paint and paper to your heart's content, but not be permitted to take down a wall or build an office in your attic without submitting plans, purchasing a building permit, and undergoing specified inspections. As you are probably aware, many people regard this as an invasion of their privacy, which can be summed up in the question: "Who's to know?" Ignoring the rules is not recommended, but you must consult your own conscience and make your own decisions. There are many shades of gray here, because in many places where regulations are in effect, compliance and enforcement are loose, variable, or nonexistent.

If you rent your home, you will certainly be restricted in what you can or cannot do in the way of remodeling and redecorating. Your rental agreement may or may not spell this out in detail. If you rent a single-family residence or a duplex from a private owner, there is usually a considerable amount of leeway and flexibility in what you can do inside your home. Often an owner will allow a renter to take care of ordinary redecorating like painting and papering, and perhaps the owner will even pay for the materials. Sometimes more involved remodeling projects are possible. If you rent an apartment from a commercial organization, however, you probably won't be able to do much on your own.

Whether you own or rent your home, before you begin a project, check to see if you might run afoul of any local restrictions that could cause noncompliance problems or compel a change in plans, materials, or methods. Information can be obtained at town, city, or county government offices. Local libraries also usually have copies of whatever regulations apply in the general area. Following are the items to consider.

Zoning Laws. Zoning laws primarily govern new construction and determine what can be built where. They do not exist everywhere, and when imposed, they do not in themselves affect redecorating and remodeling. If zoning laws are in effect, however, they are usually accompanied by building codes.

Building Codes. Building codes are legally enforceable regulations that spell out in detail certain minimum standards of design, construction, and quality of materials for buildings and structures. They apply not only to new construction, but to alteration and repair as well. Thus, where they are in effect they sometimes cover some kinds of remodeling projects. A wide variation exists throughout the country as to just what is covered and what is not.

Building Permits. Building permits go hand in hand with building codes. If a building code specifies that a certain remodeling project must comply with its provisions, a building permit will also be required to carry out the project. In this case it would be a remodeling permit, entailing submission of plans and specifications (which usually can be informal), payment of a fee, and subjection to one or more inspections during the work and/or after completion.

Plumbing Permits. Plumbing permits are required in many places if any part of the drain-waste-vent (DWV) system or the hot or cold domestic water supply system will be removed, altered, or added to in any way. Such permits are issuable only to licensed professionals. It is possible for a plumbing permit to be required even when a remodeling permit is not.

Electrical Permits. Electrical permits are often required if any part of the electrical system will be altered or extended, or even if part of the system needs only to be disconnected, removed, and directly replaced in order to allow completion of a remodeling job. Here again the permit is issuable only to a licensed contractor, and such a permit might be needed even if a remodeling permit is not. Also, it is possible for an electrical code to be in effect in areas where a building code is not.

Fireplace Permits. Fireplace permits are new on the scene. Many towns and cities now require them, with the intention of mitigating the health and safety hazards so commonly caused by improper installations and/or to reduce potential air pollution by allowing the installation of only certain approved clean-burning units. The regulations apply to both fireplaces and wood/coal stoves.

Covenants. Covenants are common in formal housing subdivisions. They consist of list of do's and don'ts applicable to all housing units within the subdivision, and can be very restrictive. They are used along with, and sometimes in lieu of, local building codes. Covenants are legally enforceable by local authority, and usually form a legal part of the property deed. They do not regulate what may or may not be done inside the house, but do often require approval for any changes in the exterior appearance.

Lease/Rental Agreements. These commonly contain clauses that specify what a lessee or renter can do in the way of redecorating or remodeling. In the absence of specific details, always check with the landlord before doing anything. In fact, it pays to check even if the agreement restricts you from doing anything, because individual agreements overriding the contract clauses can sometimes be reached. If you do get permission to undertake some project, get it in writing, and in detail.

CONTRACTING

If you do not want to do your own redecorating or remodeling work, you can hire a contractor to take care of the whole project. Or, you can plan the work and set up the specifications and act as your own general contractor—hiring and coordinating the various subcontractors needed to complete different parts of the job. You can do part of the work yourself and have the harder parts taken care of by professionals. You can hire a contractor to act as your advisor and work along with him to complete the job. There are numerous approaches, and there is no best way.

A general contractor, or a building contractor, will oversee an entire job from start to finish. He will take care of all the facets either with his own men, or by hiring independent workers who specialize in certain trades. In that case they are the subcontractors, such as painters, paperhangers, electricians, tile-setters, plumbers, drywall hangers, and sheet-metal workers. However, as independent companies they are also called contractors in their own right—electrical contractor, plumbing contractor, and so on.

For remodeling and redecorating purposes, you have three options. You can call in a general contractor to take care of the whole thing. Some contractors only work in new construction, but many do remodeling as well. You could also call in a remodeling contractor or individual trade contractors as needed.

If you are planning a project that is complex and difficult, or involves some highly specialized area that is unfamiliar to you or that you feel is beyond your capabilities, by all means call for help. For example, removing a large part of a loadbearing wall can be mighty tricky and requires equipment and know-how. If not properly done, serious structural damage can result. Fiddling around with the electrical system can be disastrous if you don't know what you are doing, and messing up the plumbing system can get you in trouble, too.

But when you do call for help, don't just pick a name out of the Yellow Pages. It pays to investigate first. Look for a company or a tradesperson with a good local reputation for being knowledgeable, fair, capable, trustworthy, and able to get the job done in timely fashion. Compare pricing structures and get bids, especially on the larger projects, but be aware that neither the lowest nor the highest may be either the best or the worst. For your own protection, and the contractor's, get all the specifications and plans—as well as the contract clauses—set down in detail so that all parties know exactly where they stand, what is to be done, and for how much money.

You might not be able to get firm prices, only estimates. Whether written or verbal, estimates have a way of often going sour. The same can be true of time-and-materials arrangements, where the contractor goes ahead on the basis of so much money per hour plus the cost of the materials, usually marked up a certain percentage. This is known as "cost plus" work. Both arrangements can work for or against you, which reinforces the need for a reputable and honest contractor, especially in cases where it is impossible to get a firm figure.

For more advice on contracting and subcontracting, as well as names of remodeling contractors in your area, contact the National Association of the Remodeling Industry, 1901 N. Moore St., Suite 808, Arlington, VA 22209. Books and articles on the subject can also be obtained at or through your local library.

2
CHAPTER

Floors

BECAUSE MOST PEOPLE TEND TO LOOK DOWNWARD WHEN THEY WALK ABOUT, THE FLOORS of a house—or any building, for that matter—are usually the most initially obvious part of the interior decorating scheme. When we are sitting, we readily notice the floors because we are close to them. If the finish or the floor covering is shabby and worn, we notice it immediately, and unfavorably. If it is in fine shape and complements the decorating scheme, we notice that favorably.

The floors also incur the greatest amount of abuse of any part of a house. Scuffs, scratches, general wear, stains, dents and chips, dirty trails along the traffic patterns, and overall shabbiness appear sooner than we like on the surface. Beneath the surface, the floor frame takes a pounding day after day from tramping feet, traffic vibration, running children and dogs, and furniture being dragged and boosted around, while at the same time supporting a huge weight of assorted possesions. Sometimes there are worse, slowly accumulating problems—moisture incursion between floor layers from leaking drains or pipes or toilet seals; rotting from ground moisture condensation in crawl spaces; separation of structural members because of foundation settling; gradual weakening of joists because of poor or undersized materials, improper construction, or just too much weight for too long a period of time; and weakening from insect attack, to name a few.

Obviously, then, the floors of a house need periodic refurbishing. This could be as simple as applying a new finish or installing a new covering to preserve or enhance the appearance of the floors or to change the decorating scheme of the rooms involved. However, the job also might entail partial rebuilding of the floor or even the floor frame for a satisfactory end result. After all, there is little point in spending a lot of time, effort, and money to put a new surface on a floor that is in bad shape to begin with. The following sections will cover the many complexities of floor system renovation and the various ways in which you can refinish or recover the floors in your house.

NEW/OLD FLOORING COMBINATIONS

Remodeling or redecorating often includes installing a new finish flooring, or perhaps refinishing an old one. There are literally hundreds of ways to install new over existing floorings, involving thousands of variable application details. Many combinations are easy and practical, others are more difficult. Some are possible but impractical, and some are theoretically feasible but actually impossible under the given job conditions. Direct installations can give good results with a minimum of effort, but often require one or two intermediate steps.

Table 2-1 lists the most common basic combinations of new over old flooring materials and indicates generally what must be done to apply or install them. To use the table, look to the left for the new flooring material you would like to have. Read across to the vertical column for the old flooring that you want to cover, and refer to the explanation of the indicated letter at the bottom of the table. This will tell you what intermediate steps, if any, you must take to accomplish your purpose. If no interface between the new material and the old surface is needed, you can continue with the instructions given in the following pages of text.

Be aware that the combinations listed here are only workable in the general sense and factors such as cost or technical difficulties or job conditions might make a particular combination unfeasible. There are usually at least several approaches and a lot of differing details that have to be considered for each installation. Exercise judgement and common sense in assessing the possibilities that might suit your needs—a little ingenuity sometimes helps, too.

The floorings listed on the left are standard materials, but note that there is often a wide variety of each product available, with differing characteristics and qualities. Make selections according to the requirements of particular job conditions, and heed manufacturers' recommendations.

The "old flooring" entries across the top of Table 2-1 are self explanatory, with two exceptions; dry concrete floors, and those with dampness or alkali problems. For flooring installation purposes, concrete floors are set in three categories, according to location: suspended (which means at least 18 inches above grade), on grade and below grade. Suspended concrete floors are presumed to be always dry, while the other two are presumed to be susceptible to dampness or moisture. Because this has a deleterious effect, most flooring products are not recommended for those locations. However, the location is not the problem, the moisture is. If moisture definitely is not present, or can be barred or eliminated, these floorings can be safely used.

For our purposes here, the entry "Concrete—Dry" refers to the absence of dampness and moisture or alkali and efflorescence problems; with on grade or below grade concrete slabs protected by effective waterproofing, vapor barriers, and/or foundation drainage systems. "Concrete—Damp and/or Alkaline" refers to situations where barriers, dampproofing, or drainage systems are not positively known to exist, and/or the concrete does have perhaps slight but demonstrable moisture or alkali problems. If you are unsure whether moisture or alkali exist in a concrete slab, you can have it tested. There is no entry for a

Table 2-1. Basic Combinations of New-Over-Old Floorings.

Old \ New	Applied Coatings	Wood Strip or Plank—1/2" or Less	Wood Strip or Plank—3/4" to 1 1/4"	Wood Strip or Plank—1 1/2" or more	Parquet—1/2" or Less	Parquet—3/4" or More	Sheet Vinyl—Vinyl-Backed	Sheet Vinyl—Felt-Backed	Vinyl Tile—Solid	Vinyl Tile—Felt-Backed	Vinyl Tile—Composition	Rubber Tile	Asphalt Tile	Carpeting—Glued
Subfloor-Board or Plywood 3/4" or less	NA	J	B	NA	J	B	J	J	J	J	J	J	J	J
Open Joists	NA	M	M	B	—	—	—	—	—	—	—	—	—	—
Earth	NA	—	L	NA	L	L	—	L	—	L	—	—	—	—
Soundboard Underlayment	NA	H	J	NA	H	H	H	H	H	H	H	H	H	H
Particleboard Underlayment	NA	B	B	NA	B	B	B	B	B	B	B	B	B	B
Hardboard Underlayment	NA	B	B	NA	B	B	B	B	B	B	B	B	B	B
Plywood Underlayment	NA	B	B	NA	B	B	B	B	B	B	B	B	B	B
Concrete-Damp and/or Alkaline	B	G	G	NA	G	G	—	G	—	G	—	—	—	—
Concrete-Dry	B	E	G	NA	E	E	E	E	E	E	E	E	E	E
Masonry and Brick	D	E	G	NA	E	E	E	E	E	E	E	E	E	E
Ceramic Tile	D	E	G	NA	E	E	E	E	E	E	E	E	E	E
Stone	D	E	G	NA	E	E	E	E	E	E	E	E	E	E
Terrazzo	D	E	G	NA	E	E	E	E	E	E	E	E	E	E
Resilient Tile-Self-Stick	C	F	B	NA	F	F	F	F	F	F	F	F	F	F
Carpeting-Glued	NA	F	K	NA	F	F	F	F	F	F	F	F	F	F
Resilient Tile	C	E	B	NA	E	E	E	E	E	E	E	E	E	E
Sheet Vinyl	C	E	B	NA	E	E	E	E	E	E	E	E	E	E
Parquet	B	E	B	NA	E	E	E	E	E	E	E	E	E	E
Finish Wood	B	E	B	NA	E	E	E	E	E	E	E	E	E	E
Applied Coatings	A	E	B	NA	E	E	E	E	E	E	E	E	E	E

Carpeting—Stretched	Carpet or Tile—Self Stick	Seamless Flooring	Terrazzo	Stone—Tile	Stone—Flag or Rubble	Ceramic Tile	Masonry and Brick	Concrete or Mortar	Plywood Underlayment	Hardboard Underlayment	Particleboard Underlayment	Soundboard Underlayment	Subflooring—Board or plywood
J	H	H	H	H	B	H	B	B	B	NR	B	B	NA
I	M	—	—	—	—	—	—	M	M	M	M	—	B
L	L	—	—	—	B	—	B	B	B	L	L	NA	L
H	H	H	H	H	K	H	K	K	B	NR	B	B	NA
B	H	H	H	B	B	H	B	B	B	B	B	B	NA
B	H	H	H	B	B	H	B	B	B	B	B	B	NA
B	B	P	P	B	B	B	B	B	B	B	B	B	NA
E	G	P	P	E	B	E	B	B	G	G	G	G	L
E	G	P	P	E	B	E	B	B	G	G	G	G	L
E	G	P	P	E	B	E	B	B	G	G	G	G	L
E	G	P	P	E	B	E	B	B	G	G	G	G	L
E	G	P	P	E	B	E	B	B	G	G	G	G	L
E	G	P	P	E	B	E	B	B	G	G	G	G	L
E	F	F	F	F	B	F	B	B	B	B	B	E	NA
E	F	F	F	F	K	F	K	K	K	K	K	K	NA
E	F	F	F	E	B	F	B	B	B	B	B	E	NA
E	E	F	F	E	B	F	B	B	B	B	B	E	NA
E	E	F	F	E	B	E	B	B	B	E	E	E	NA
E	P	P	P	E	B	E	B	B	B	E	E	E	NA
E	E	E	E	E	B	E	B	B	B	E	E	E	NA

Notes

A. Direct application, after preparation, depending upon the compatibility of the old and new coatings.
B. Direct application, after preparation, with usual applicable methods and according to usual practice and/or manufacturer's instructions.
C. Direct application only of manufacturer's recommended waxes, if any.
D. Direct application only of manufacturer's recommended sealers and/or waxes.
E. Direct application, after preparation, to a smooth, sound, even, dry substrate, according to usual practice and/or manufacturer's instructions. Uneven, rough, wide-jointed, or damaged substrates must be smoothed, filled, and leveled, or repaired, removed, and/or a new substrate provided.
F. Old covering must be removed. Adhesive must be scraped or sanded away to provide a clean, smooth surface, or a new substrate must be provided.
G. New wood subfloor on screeds or floor frame must be installed first.
H. Plywood underlayment must be installed first.
I. Concrete or wood substrate must be installed first.
J. Plywood or particleboard underlayment must be installed first.
K. Old floor covering must be removed first.
L. Floor frame or complete subfloor system must be installed first, as applicable, of pressure-treated Permanent Wood Foundation (PWF) material or wood over a concrete slab.
M. Plywood or board subfloor must be installed first, with underlayment added as applicable.
P. Professional installation required.
NA. Not applicable.
NR. Not recommended.

Table 2-1. *Continued.*

concrete slab that is damp to the point of wetness, or has standing water on it at any time of the year, because such a floor cannot be covered with a finish flooring until the condition is permanently corrected.

PREPARATION FOR NEW FLOORING MATERIAL

When you cover an old floor with a new flooring material, whether a finish flooring or a substrate, a certain amount of preparation is always needed—in some cases very little, in others quite a bit. Either way, preparation is usually the key to a successful result.

A certain amount of preparation work is common to virtually all flooring jobs, and the information outlined in Table 2-2 presupposes that such matters have already been taken care of. Remove furnishings and appliances from the room, and all base and base shoe moldings as necessary. This might also require the removal of electrical convenience outlets from mopboards, and floor outlets should be disconnected and removed as well. Take up floor heating registers and cold air return duct grilles and remove baseboard or other heating equipment as required. Remove thresholds or any other objects that might interfere with the job at hand. If necessary, provide protective covering for anything left in the room that might otherwise become damaged, dust laden, or soiled.

Table 2-2 contains the same headings as Table 2-1, and the same information given in the previous section holds true. It is assumed that the new flooring material is to be applied or installed directly on the the old flooring. To use this table, start with Table 2-1 to select a new/old flooring combination and its required interface, if any. Then find the new material, whether a finish or an interface, on the left of Table 2-2. Read across the row to find the old flooring over which the new material will be applied or installed, and refer to the indicated number at the bottom of the table. This will tell you the basic preparation steps that will be necessary, and refer you to other parts of this chapter for more information, if necessary. The preparation procedures follow common practice using standard equipment and materials, so the exact details vary—often substantially—with specific job conditions. Use your own judgement.

REMOVING OLD FINISH FLOORING

If a finish floor covering is in such bad shape that it cannot be satisfactorily refinished, it might be better to remove it. The same is true if the condition of the current flooring is not good enough that is can be covered with another flooring, or if the new covering is incompatible with the old, or if the addition of a compatible underlayment would cause problems with the levels of adjacent flooring or interfere with trimwork. If there are structural problems with the floor frame or its underpinnings, or the subfloor, the finish flooring must often be removed in order to make repairs. Whatever the case, the procedures are as follows.

Wood Strip or Plank

To take up a wood strip or plank finish floor, first remove the shoe molding around the perimiter of the room. If the flooring extends under the baseboard, remove that as well (Fig. 2-1). Number each piece, if salvageable, and key all of them to numbers on the wall for easier replacement later. Pry up any edging strips or threshholds, and remove registers, baseboard heating units, or other items that rest upon the flooring.

If there is a convenient crack between boards or an exposed end, start here. Pry up the first board with a pinch bar, wrecking bar, or a similar tool. From that point, take up the remaining boards or strips in whatever order seems logical. If there is no handy starting point, as might be the case with tight-fitting strip flooring, determine the thickness of the flooring at an edge or by drilling a hole down to the subfloor. Set the shoe of a portable electric circular saw to that measurement and make one or more plunge cuts at any handy spot, down the center of a board or strip. If the cut cannot be made from end to end of the board, make a crosscut at each end of the rip cut. Use the saw kerf as a starting point and pry up the pieces, then start working the adjacent pieces free. Using a combination of sawing, prying, and simply splitting the boards apart by force, they will all come up.

If the flooring extends under a partition, cut the wood away with the saw, as close to the wall as you can, then trim the stubs back flush to the wall. If the flooring continues through a doorway into another room, you can take that up as well. If you wish to leave it intact, use the saw to make a crosscut on a straight line across the middle of the doorway opening, coming as close to the door jambs as possible. Remove the cut pieces, then cut off the remaining chunks with a wood chisel (Fig. 2-2). The joint between the new and the

Table 2-2. Basic Preparation Required for Combinations of New-Over-Old Flooring.

Old \ New	Applied Coatings	Wood Strip or Plank—1/2" or Less	Wood Strip or Plank—3/4" to 1 1/4"	Wood Strip or Plank—1 1/2" or More	Parquet—1/2" or Less	Parquet 3/4" or More	Sheet Vinyl—Vinyl Backed	Sheet Vinyl—Felt-Backed	Vinyl Tile—Solid	Vinyl Tile—Felt-Backed	Vinyl Tile—Composition	Rubber Tile	Asphalt Tile
Subfloor—Board or Plywood 3/4" or Less	NA	NR	7	NA	NR	7	NR	NR	NR	NR	NR	NR	NR
Open Joists	NA	NA	NR	8	NA	NA	NA	NA	NA	NA	NA	NA	NA
Earth	NA	NA	NA	NA	NA	NA	NA	NA	NA	NA	NA	NA	NA
Soundboard Underlayment	NA	NR	7	NA	NR	NR	NA	NA	NA	NA	NA	NA	NA
Particleboard Underlayment	NA	7	7	NA	7	7	7	7	7	7	7	7	7
Hardboard Underlayment	NA	7	7	NA	7	7	7	7	7	7	7	7	7
Plywood Underlayment	NA	7	7	NA	7	7	7	7	7	7	7	7	7
Concrete—Damp and/or Alkaline	3	NR	NR	NR	NR	NR	5	NR	5	NR	5	5	5
Concrete—Dry	3	5	NR	NA	5	5	5	5	5	5	5	5	5
Masonry and Brick	2	NR	NR	NA	5	5	5	5	5	5	5	5	5
Ceramic Tile	2	5	NR	NA	5	5	5	5	5	5	5	5	5
Stone	2	NR	NR	NA	5	5	5	5	5	5	5	5	5
Terrazzo	2	5	NR	NA	5	5	5	5	5	5	5	5	5
Resilient Tile—Self-Stick	2	NR	7	NA	NR	NR	NR	NR	NR	NR	NR	NR	NR
Carpeting—Glued	NA	NA	NA	NA	NA	NA	NA	NA	NA	NA	NA	NA	NA
Resilient Tile	2	4	7	NA	4	4	6	6	6	6	6	6	6
Sheet Vinyl	2	4	7	NA	4	4	6	6	6	6	6	6	6
Parquet	1	4	7	NA	4	4	6	6	6	6	6	6	6
Finish Wood	1	4	7	NA	4	4	6	6	6	6	6	6	6
Applied Coatings	1	4	7	NA	4	4	6	6	6	6	6	6	6

Carpeting—Glued	9	9	9	9	NA	5	5	5	5	5	NR	7	7	7	NA	NA	NR
Carpeting—Stretched	10	10	10	10	11	10	NR	10	10	10	NR	7	7	7	NA	NA	NR
Carpet or Tile—Self-Stick	4	4	4	4	NR	5	5	5	5	5	NR	7	NR	NR	NA	NA	NR
Seamless Flooring	12	12	NR	NR	NR	12	12	12	12	12	12	12	NR	NR	NA	NA	NR
Terrazzo	12	12	NR	NR	NR	12	12	12	12	12	12	12	NR	NR	NA	NA	NR
Stone—Tile	4	4	4	4	NR	5	5	5	5	5	5	7	7	7	NA	NA	NR
Stone—Flag or Rubble	13	13	13	13	13	14	14	14	14	14	14	14	13	13	NA	15	14
Ceramic Tile	16	16	13	13	13	17	17	17	17	17	17	16	16	16	8	8	NR
Masonry and Brick	18	18	18	18	18	19	19	19	19	19	19	19	18	18	NA	15	NR
Concrete or Mortar	13	13	13	13	13	14	14	14	14	14	14	14	13	13	NA	NA	14
Plywood Underlayment	8	8	8	8	8	NA	NA	NA	NA	NA	NA	8	8	8	8	NA	8
Hardboard Underlayment	8	8	8	8	8	NA	NA	NA	NA	NA	NA	8	8	8	NR	NA	NR
Particleboard Underlayment	8	8	8	8	8	NA	NA	NA	NA	NA	NA	8	8	8	8	NA	8
Soundboard Underlayment	8	8	8	8	8	NA	NA	NA	NA	NA	NA	8	8	8	8	NA	8
Subfloor—Board or Plywood	NA	NA	NA	NA	NA	NA	NA	NA	NA	NA	NA	NA	NA	NA	NA	8	NA

Notes

1. Preparation depends upon the nature and condition of the floor surface and the coating to be applied.

2. Vacuum the floor, strip off old wax if necessary, and wash thoroughly. Allow the floor to dry completely before proceeding.

3. Preparation depends upon the nature and condition of the floor surface and the coating to be applied.

4. For a floor in good condition, vacuum thoroughly. In addition, if making an adhesive-bonded installation, clean the surface thoroughly of grease, oil, wax, dirt accumulations, loose old finish, dust, and any other contaminants. Nail down loose flooring. Lightly roughen shiny surfaces with fine sandpaper. Level worn-down or low spots with mastic or leveling compound. For old flooring in poor condition, provide suitable substrate.

5. For a floor in good condition, thoroughly clean the surface of grease, oil, wax, dirt accumulation, loose old finish, or other contaminants. If the surface is very smooth or slick, roughen it slightly by rubbing with an abrasive block or etching with muriatic acid. Fill and level defects, recessed grout joints, cracks, and low spots with a suitable patching or leveling compound. Vacuum the surface thoroughly. For flooring in poor condition, provide a new, smooth substrate.

6. Proceed as in Note 4. In addition, fill all holes, defects, cracks, and joints greater than $1/16$ inch in diameter or width with a suitable filler or leveling compound.

7. Renail any loose spots. Set all nail heads flush with the surface, as applicable. Sweep or vacuum up any loose debris before proceeding. For thin, flexible coverings, fill all holes, defects, cracks, and joints greater than $1/16$ inch in diameter or width with a suitable filler.

8. No preparation needed, except cleanup of loose dirt and debris.

9. Remove base shoe molding if necessary. Fill all holes, defects, cracks, and joints greater than $1/8$ inch in diameter or width with a suitable mastic or leveling compound. Level worn spots or depressions with an appropriate leveling compound. Nail down any loose spots. Roughen medium-to high-gloss surfaces with a medium-grit sandpaper. Strip wax, clean away dirt or grease deposits, or wash the floor, as necessary. Vacuum thoroughly just before proceeding. For flooring in poor condition, provide a suitable substrate.

10. Remove base shoe if necessary. Fill all holes, defects, damaged areas, cracks, and joints greater than $1/8$ inch in diameter or width with a suitable leveling compound. Nail down any loose flooring. Vacuum the floor thoroughly just before proceeding.

11. Refer to "Carpeting" in chapter 2 for details.

12. Professional preparation and application only.

13. Remove base moldings and provide protection for finished wall surfaces, as necessary. Sweep up loose dirt and debris and provide isolation membrane/moisture barrier for required isolated mortar bed installation.

14. For an isolated bed, see Note 13. For a bonded bed, remove base moldings and provide protection for finished wall surfaces, as necessary. Thoroughly clean the floor of loose debris, grease, oil, wax, dirt accumulations, efflorescence, and any other contaminants, and strip off old coatings. If the surface is very smooth or slick, roughen it by rubbing with an abrasive block or by etching with muriatic acid.

15. For direct laying, rough-level, compact, and thoroughly dampen the earth surface immediately before starting, and keep exposed portions damp as the job proceeds. For a moisture-isolated installation, rough-level and compact the earth, add a layer of 1 to 2 inches of sand raked or screeded smooth and level, and cover the sand with a lapped layer of 6-mil polyethylene sheet.

16. For a thin-set mortar installation, see Note 4. For an isolated thickbed installation, see Note 13.

17. For a thin-set or bonded thick-bed mortar installation, see Note 14. For an isolated thick-bed mortar installation, see Note 13.

18. For dry laying, sweep up loose debris. In addition, for laying with grout, moisture-cured sand/cement joints, or an isolated mortar bed, cover the floor with a lapped layer of 15- or 30-pound roofing felt or 6-mil polyethylene sheet.

19. For laying in a bonded mortar bed, see Note 14. Otherwise, see Note 18.

NA. Not applicable.

NR. Not recommended.

Table 2-2. *Continued.*

old flooring can be covered later with a threshold even if the two are not quite level. An alternative is to cut the old flooring flush with the faces of the door casing in the room where the old flooring will remain. Fashion a length of wood the same width as the door jambs and fit snugly between them to form a break between the new and old floors. (Fig. 2-3).

Wood Parquet

Wood parquet flooring is almost always glued down and might not come up easily. Remove the shoe molding; this should reveal the tile edges. Take up any registers or other items resting upon the tile. Sometimes a good starting point for taking up the tile is a doorway. If this is not feasible, check the tile thickness (probably about $5/16$ inch) and make a saw cut. With a hammer and a wood chisel, chop out enough tile to open up a sizeable space. Then shift to a broad-bladed mason's chisel or similar tool, and continue lifting the tiles by driving the chisel beneath them. Alternatively use a posthole bar, which is long and heavy and has a wide chisel-like blade. It can be rammed under the tiles from a stand-

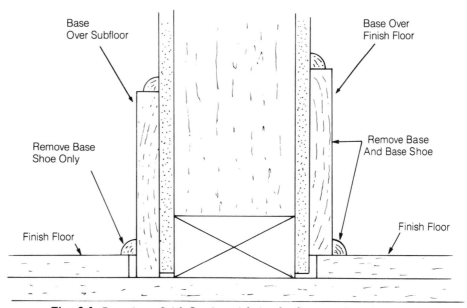

Fig. 2-1. *Removing a finish flooring might involve first removing a base shoe, or both the shoe and the base.*

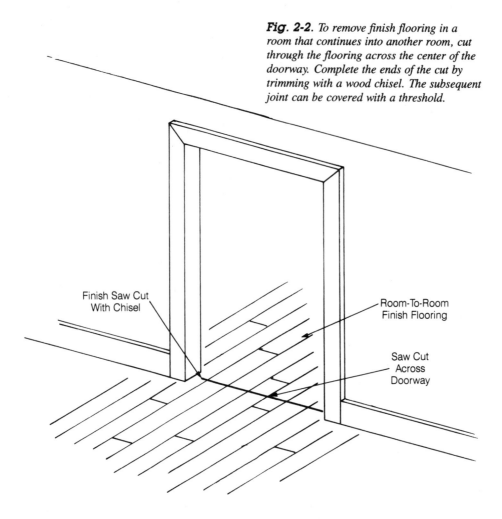

Fig. 2-2. To remove finish flooring in a room that continues into another room, cut through the flooring across the center of the doorway. Complete the ends of the cut by trimming with a wood chisel. The subsequent joint can be covered with a threshold.

Finish Saw Cut
With Chisel

Room-To-Room
Finish Flooring

Saw Cut
Across
Doorway

ing position. A heavy-bladed, square-edged ice chopper also works well, and if the adhesive is not too tough, a square-bladed shovel will do the job.

Vinyl Sheet and Resilient Tile

The procedure for removing these floor coverings, which are usually glued down, is much the same as for wood parquet tile. If the adhesive is fairly old, the flooring can be lifted off by driving an ice chopper or a shovel blade under it. Vinyl sheet can sometimes be peeled back by hand, forcibly tearing it up in chunks, once a corner or edge has been started. Tiles have to be chipped away at, bit by bit. Often, some judicious heating with a propane torch or a heat gun (which is safer), played along the adhesive line as the tile is peeled back will speed up the process. There are times, however, when these coverings are very difficult to remove cleanly. At this point you have two choices: If the covering has been laid on an underlayment, rather than directly on a subfloor, it might be easier and

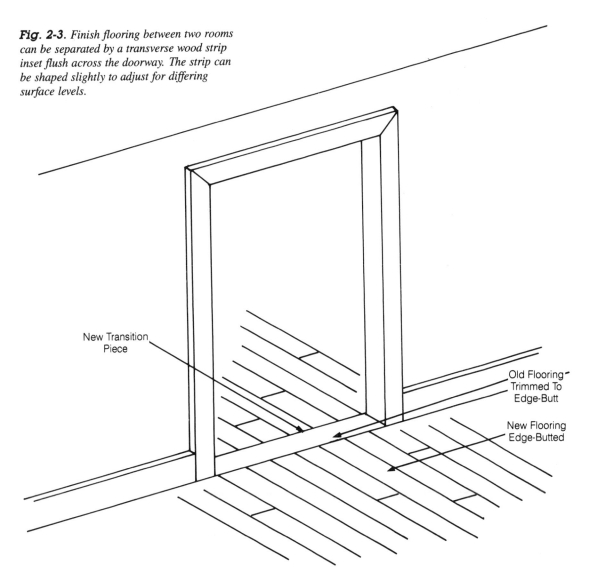

Fig. 2-3. Finish flooring between two rooms can be separated by a transverse wood strip inset flush across the doorway. The strip can be shaped slightly to adjust for differing surface levels.

New Transition Piece

Old Flooring Trimmed To Edge-Butt

New Flooring Edge-Butted

faster to take up the underlayment and the covering together. If the covering was laid directly on the subfloor and the flooring system as a whole is sound, lay underlayment over the old finish flooring to provide a clean base for a new covering. For both procedures, see the section entitled "Underlayment" in this chapter.

Masonry, Stone, and Ceramic Tile

Masonry or stone finish flooring, such as pavers, quarry tile, slate, marble, brick, and ceramic tile, might be glued directly to a wood underlayment or subfloor with mastic, laid in thin-set mortar on a wood subfloor, or bedded in mortar on concrete. If a thick mortar bed is bonded to a concrete subfloor, removal is not practical.

First remove any moldings or other items that might interfere. Look for a spot where you can insert a pry bar or the claws of a hammer, and break a piece free. If there isn't a handy spot, smash one of the units with a light hand sledge (wear safety goggles) or break out a grout line with a hammer and cold chisel. If the units are glued down with adhesive or set in mastic, clear away a large enough starting spot with a hammer and cold chisel so that you can continue with a posthole bar or an ice chopper. Drive the tool under the units to free them. If the units are bedded in mortar on concrete, they might not break loose easily and/or the mortar bed might not lift from the concrete. This requires working at both with a hand sledge and a broad mason's chisel, air chisel, or electric hammer. If the surface cannot be cleaned down smooth, you might have to go over it with a carborundum stone or take it down with a terrazzo grinder, which is somewhat like a floor sander. If the units are set in mortar on a wood subfloor, the whole affair will easily break up into pieces, and can often be taken up with just the claws of a hammer.

Carpet

Wall-to-wall carpet is secured in one of two ways: it is glued down, or stretched onto tackless strips nailed into the subflooring around the perimeter of the room.

First remove all carpet edgings. If the carpet continues under the base moldings (usually it does not), remove them as well. If the carpet is glued down, dig at an edge or a seam with the claws of a hammer until it begins to pull free from the adhesive. When you get a chunk loose that is large enough to grab with both hands, tear it free.

If the carpet is held by tackless strips, start by ripping it free at a corner, or where you have removed a piece of edging. Or, if you can find a seam, rip it open with a carpet knife. Once you begin to ease the tension on the carpet it will come off the tackless strips more easily. If it is a heavy carpet and resists coming off the spikes in the strip, work some of the strips free by prying them up with a pinch bar or claw hammer. Once there is no tension on the carpet, free it from the tackless strip. First use a screwdriver to pry up the folded-down carpet edge behind the tackless strip and next to the wall, then push or pull the edge toward the wall to slide it off the spikes. When the carpet is up, pry up all the tackless strip (unless a new carpet is to be laid and the strip is in good condition and of the correct kind).

Stretched-in carpet almost always has a padding underneath it. This might be a heavy felt, which needs only to be rolled up and removed. Various sorts of foam paddings are usually stapled down, and will rip free easily. Pull up remaining staples with pliers or drive them flush using a hammer.

Adhesive Removal

There is no easy way to remove the adhesive residue of floor coverings. Some old adhesives become dry and crumbly, and can be scraped away with little trouble. Those that are tough and rubbery can sometimes be scraped away or sanded off. Sanding, however, is often difficult because it clogs the paper. Solvents like mineral spirits or lacquer thinner, naptha, or benzene might work in conjunction with scraping, but most of these are dangerous to use and highly flammable. They must be used with plenty of ventilation, the electricity turned off, and no flame anywhere in the vicinity (including pilot lights). Another possibility is an electric paint scraper. It works with a heated blade, which will soften and loosen the adhesive so you can scrape it up.

If the adhesive cannot be satisfactorily removed, or the job turns out to be more work than it is worth, there are two other possibilities. If the subsurface is an underlayment on a subfloor, take up the underlayment and replace it. Or, if the subsurface is the subfloor, install a new layer of underlayment over it. See the section entitled "Underlayment."

SUBFLOOR/UNDERLAYMENT REPAIR

The subfloor is the first layer of flooring attached to the joists. Here it will also refer to any other layer between it and the finish floor covering. If subfloor damage from moisture and rot is widespread, or if the floor framing must be renewed, removal of an entire subfloor section might be necessary. See the section entitled "Subfloor Replacement."

To repair a relatively small section of subflooring, such as beneath a toilet, first determine the thickness of the entire flooring assembly and how many layers of materials are laid. If this is not visible, drill a 3/4-inch hole through the flooring close to the damaged area. Measure the thickness of the various layers. If the finish floor covering is to be entirely removed and later replaced, take that up. If it is to be patched in with a matching material, remove a section that is at least 1 foot wider all around (or back to a wall if necessary) than the planned subflooring repair area. If the damage is more widespread than anticipated, keep cutting the finish flooring back to about 1 foot over sound material. Pull any nails and sweep up the debris.

Next, cut away the underlayment, if necessary. Do this with a portable electric circular saw, with the shoe set to a cut depth the same thickness as the underlayment. Whatever cannot be cut with a circular saw can be finished with a reciprocating saw or saber saw fitted with an offset blade, or with a hammer and wood chisel. Start with a relatively small area, and as with the finish flooring, keep cutting back from the damaged subflooring area until about 6 inches to 1 foot back over sound material. Again, pull any nails and sweep up.

The subflooring itself is the final layer. Reset the saw cut depth to match the subflooring thickness, and cut away a small square of material so that you can see just where the floor joists lie. Then make a series of cuts to take up the remaining damaged portions. Make as many cuts as possible at the outside edges of the final opening, exactly along the centerlines of the floor joists, so that you will have bearing surfaces to which the new material can be nailed (Fig. 2-4). Before making these cuts, pry up the old subflooring nails. This can be done with a small pry bar, but a cat's claw works best.

Inspect the joists to make sure they are undamaged and still sound. If they are not, refer to the section entitled "Floor Frame Repair." If they are, cut a piece of plywood of matching thickness (or match in with lengths of board) and fit it to the opening with the face grain running at right angles to the joists. Nail the bearing edges down every 6 inches and nail intermediate supports every 10 inches with 6d ring-shank nails (Fig. 2-5). Simi-

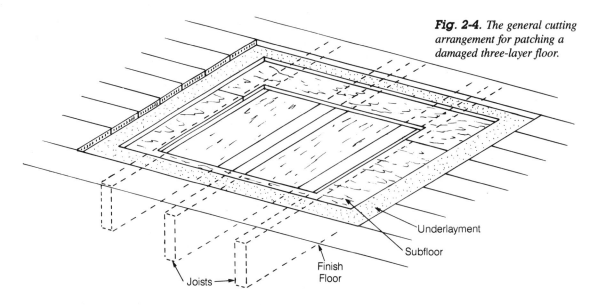

Fig. 2-4. *The general cutting arrangement for patching a damaged three-layer floor.*

Underlayment

Subfloor

Finish
Floor

Joists

Fig. 2-5. *A patch panel installed in the subfloor layer.*

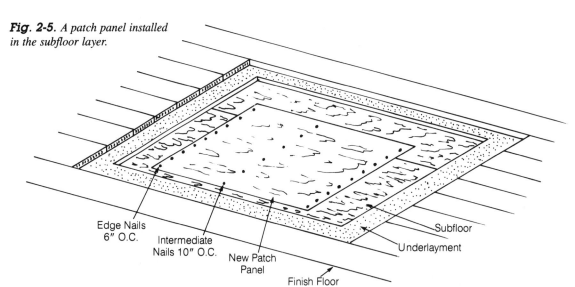

Edge Nails
6″ O.C.

Intermediate
Nails 10″ O.C.

New Patch
Panel

Finish Floor

Subfloor

Underlayment

larly, cut and fit a piece of new underlayment that matches the thickness of the old (it need not be the same material). Mark the location of the floor joists, then nail the underlayment in place through to the joists at every location possible. The full nailing pattern should have the nails about 6 inches apart around the perimeter and 10 inches apart in all directions in the field (Fig. 2-6), except for $1/4$-inch-thick hardboard. Hardboard should be nailed every 3 inches around the perimeter and at 6-inch intervals in the field. Use 4d ring-shank nails except when nailing through to the joists; use 6d nails there.

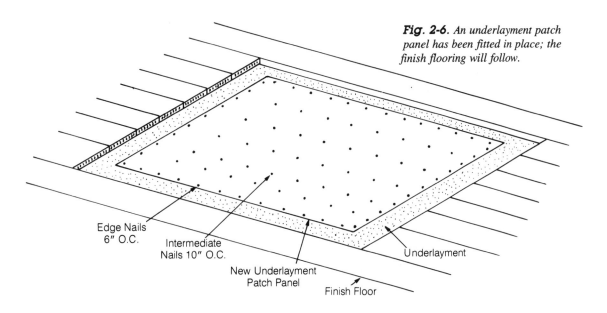

Fig. 2-6. *An underlayment patch panel has been fitted in place; the finish flooring will follow.*

Edge Nails
6″ O.C.

Intermediate
Nails 10″ O.C.

Underlayment

New Underlayment
Patch Panel

Finish Floor

Finally, replace the finish flooring in the manner best suited to the material. Wherever mechanical fasteners cannot be used on floorings that normally are so laid, construction adhesive is an adequate substitute. Weight the glued portion fairly heavily overnight for best results.

FLOOR INSULATION

In all climates—except those that are moderate all year and neither heating nor cooling is required—certain floors should be fitted with thermal insulation and vapor barriers. These include floors over open or enclosed crawl spaces (except those employed as heating plenums or solar airways); over unheated or partially heated basements; between areas that are heated or cooled to temperatures with a differential of about 10 degrees; between areas with different mechanical heating or cooling zones that might be closed off from one another, with one being left completely or partially unheated or uncooled at times; and floors of unoccupied attics, whether there is insulation in the roof or not. Thermal insulation is also sometimes installed in floors as part of a soundproofing procedure. If no insulation exists, a good time to install it is during a flooring renovation project. It is also a good time to add to a skimpy insulation, or replace any that might be damaged or in poor condition.

For the do-it-yourselfer the most convenient type of insulation to install is either fiberglass or mineral wool blanket insulation, which comes in batts or rolls. Or, if access can be gained to the floor frame cavities from above and there is a solidly built ceiling attached to the underside of the frame, fiberglass or mineral wool, perlite, vermiculite, or cellulose loose-fill insulation are good alternatives. (Note: Insulation should not be installed to lie directly upon an acoustic tile ceiling.)

In mild climates, the usual choice is batt or roll insulation 3½ inches thick with or without a foil or kraft paper facing, and a thermal resistance value of R-11. In colder climates, the thickness should be 6 inches with an R-value of 19. In very cold climates, insulation beneath the floor of an unoccupied attic with an uninsulated roof is considered cap insulation. It is usually installed in layers with a minimum R-value of 38. Loose-fill insulations can be poured to any appropriate thickness. The R-values are different for the different types, so the total thickness should be based on the characteristics of the insulations and the local weather conditions. For example, a value of R-11 for fiberglass loose-fill would require an insulation thickness of 5 inches.

Installation

To install loose-fill insulation from above, first lay out a couple of planks for walkways; they can be shifted around as needed. Check all around the floor frame to make sure there are no spaces where the insulation can drift down into wall cavities or into the area below. If there are, seal them off with duct tape for small cracks, or blocks of wood for larger spaces. Do *not*, however, block off the spaces between chimneys and adjacent wood framing (there should be a minimum 2-inch gap there). Stuff these openings with chunks of mineral wool or fiberglass blanket insulation with the facing removed. Other openings or spaces can be similarly stuffed full if blocking or taping is difficult.

Wear old clothes, button up the shirt sleeves and collar (elastics around the pants cuffs help, too), and put on a hat, safety goggles, and above all, a dust mask. Then just pour the loose-fill out of the bags into the spaces between the floor joists. With a narrow rake or a garden hoe, fill the joist bays to the desired level and even the material out without compacting it—it must remain loose and fluffy (Fig. 2-7).

Fig. 2-7. *Loose-fill thermal insulation is poured between the joists and raked out level to any desired depth. It should not, however, rest directly upon ceiling tile.*

When you install blanket insulation, remember the following: When you put in faced insulation (kraft or foil), the facing is a vapor barrier and should always be to the *inside* of a heated or cooled space. However, if the insulation is installed between two interior areas, if should have *no* facing. Thus, the facing is up in a floor over an unheated crawl space, down in a floor in an unheated attic without roof insulation, but lacking in a floor between two variably heated levels.

To install faced blanket insulation from above, over an unheated area (crawl space, for instance), set up for the job as for loose-fill insulation. Install the batts or lengths by pushing them, facing up, down into the joist spaces until the facing is about 1 inch below the joist tops. Run your hands between the edges of the insulation and the joist sides to make sure the material is not caught up anywhere and is fluffed out to full thickness. Fold the flanges along the edges of the facing upward and staple them to the joist sides at 10- to 12-inch intervals (Fig. 2-8). Staples about $^1/_4$ or $^5/_{16}$ inch long work well, and a power stapler makes the job easy. For odd-sized openings, cut the insulation with long shears, about 1 inch wider and $^1/_2$ inch longer than the opening size. Tuck the pieces in place and staple wherever possible. Fold the facing upward to make a flange on cut edges or ends.

To install blanket insulation from below, push the insulation upward between the joists with the facing up. Run your hands between the insulation edges and the joist sides, and push the material snugly against the floor. Leave the attaching flanges folded down. In a tightly enclosed (except for vents) crawl space or unheated basement, the insulation will remain in place by friction alone; no fasteners are needed. In an open or skirted crawl

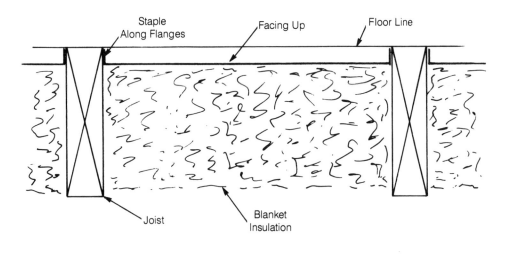

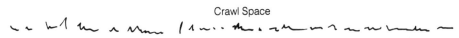

Fig. 2-8. *When installing blanket floor insulation from above, place faced insulation with the facing upward between the floor joists. Secure it by driving staples through the mounting flanges.*

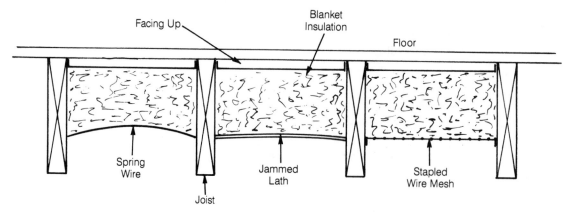

Fig. 2-9. *Blanket insulation can also be placed from below and held in place by any of the methods shown.*

space, however, retainers should be installed to keep the insulation in place. You can use springy wire lengths that are made for the purpose, jamming them between the joists about 18 inches apart. Or you can nail lath or furring strips across the joist bottoms, or staple up lengths of chicken wire (Fig. 2-9).

To install blanket insulation under an unheated attic floor where the roof is uninsulated, stuff the lengths down between the joists until the facing nearly touches the upper side of the ceiling. The facing should be down, and no fastening is needed. If a second layer is needed to provide the required R-value, lay the lengths out across the tops of the joists and at right angles to them, tucked snugly together but not jammed. If there is an airspace between the layers, so much the better (Fig. 2-10). To install the insulation from

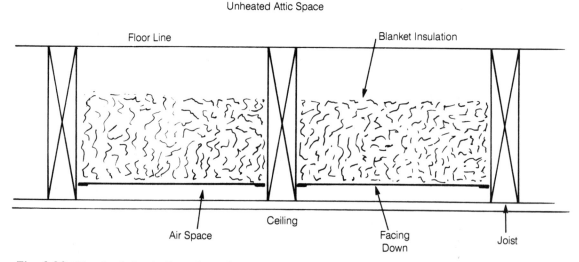

Fig. 2-10. *When insulating the floor of an unheated attic with faced blanket material, leave the mounting flanges folded and slip the material between the joists with the facing downward.*

below, follow the same procedure as for working upward from a crawl space, but with the facing downward. Staple the flanges to the joist sides.

To install blanket insulation under floors that separate interior areas heated or cooled to different temperatures, buy unfaced insulation or peel the facing off standard insulation. Push the material into place and smooth the edges and surfaces. Make sure the material is completely fluffed out, especially at the hidden back corners. Make cut pieces about 1 inch oversize; friction will hold it in place.

FLOOR VAPOR BARRIER

Any floor that separates a heated or cooled area from the exterior or from an unheated area such as a crawl space, should be fitted with a vapor barrier. The barrier is always placed to the inside of the insulation. Some insulations are provided with a built-in vapor barrier, but these are often ineffective because of tears, open seams between pieces, and gapped edges or flanges. For this reason, good construction practice calls for the installation of another, separate barrier.

The most common, least expensive, and most effective vapor barrier material used today is polyethylene sheet in either 4-mil or 6-mil thicknesses (Fig. 2-11). It is available

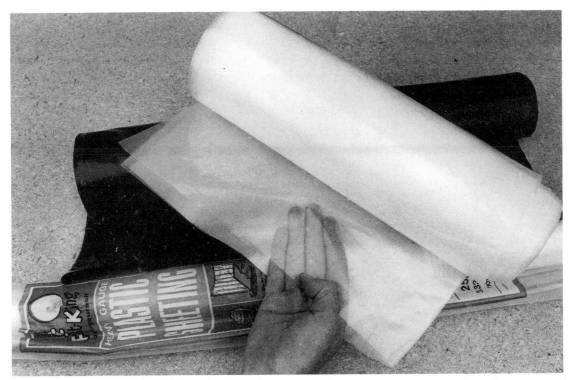

Fig. 2-11. *Polyethylene sheet makes an excellent, inexpensive vapor barrier. Black is commonly used to cover crawl space floors, but the clear is easier to use when it must be stapled to joists or studs.*

at any building supply house in numerous lengths and widths. The barrier can be placed between the joists and the subfloor, but the best arrangement is to lay it between the subfloor and the finish flooring, or between the subfloor and the underlayment.

To lay a plastic sheet vapor barrier, stretch the sheet out, free of wrinkles, on a clean floor surface. Use sheets as large as you can manage, in order to avoid seams. Where seams must be made, overlap the pieces at least 1 foot at sides and ends. For best results, seal the seams with wide, thin, plastic package tape. Staple the material down here and there to hold it in place. Lap the edges up a few inches onto the wall studs and behind the wallcovering if you can, or up behind base moldings (Fig. 2-12). Make sure there are no slits or punctures after you finish laying the material. If there are, seal them off with tape. Keep traffic off the floor until the finish flooring or underlayment is laid, and repair any damage as you go along.

FLOOR SOUND CONTROL

Much of the transmission of sound down through an upper floor and the ceiling below, into a lower level, is due to impact noise. This is caused by footsteps; objects dropped or scraped along the floor surface; furniture being moved; vibrations from the operation of a refrigerator, dishwasher, and similar appliances; or even water running into a bathtub. The bass thump from a stereo music system is a common annoyance. Some of the sound transmission also originates from airborne sound, which causes objects in the upper rooms to vibrate. This vibration radiates in all directions via the floor/ceiling construction. Although there are two sources of sound in the upper level, the noise that reaches the lower level through the floor and ceiling is structure-borne sound.

Fig. 2-12. A floor vapor barrier is commonly placed between the subfloor and the finish floor as shown here, but may also be placed beneath the subfloor, between subfloor and underlayment, or between underlayment and finish flooring.

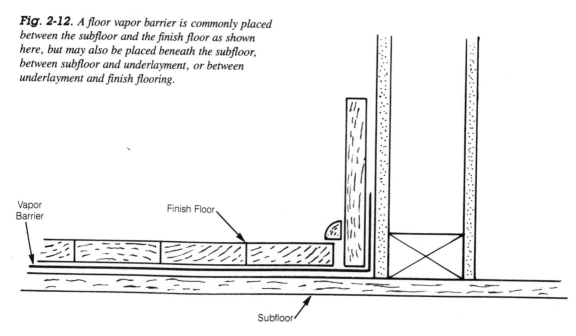

Vapor Barrier

Finish Floor

Subfloor

The floor systems in most houses are not constructed with sound-damping in mind. However, during remodeling and redecorating, there are several steps that can be taken to do just that. Both airborne and structure-borne sound must be considered, and full treatment involves both the upper level floor and the lower level ceiling.

Procedure

The first step in the noise reduction process involves isolating vibrations from equipment resting upon the floor. Refrigerators, freezers, washing machines, clothes dryers, and similar appliances can be set or mounted upon vibration-absorbing pads. Dishwashers and trash compactors should be installed in cabinetry so that they rest on shock mounts and do not directly touch any of the cabinets; surrounding them with fiberglass or other insulation material also helps. Freestanding music system speaker enclosures should be set on stands to elevate them above the floor from 6 to 12 inches (which gives better music response, too). The stands should be isolated from the floor with vibration pads, and the enclosures should not touch the walls.

The second step is to provide plenty of sound-absorbing materials in the upper level living quarters. While this is often automatically accomplished merely by furnishing the rooms, it is still a point to keep in mind. A sparsely furnished, virtually bare room with hard surfaced walls, ceiling, and floors and a lot of closed doors and windows is a big drum that radiates sound endlessly. In rooms filled with soft materials placed in random patterns, such as carpeting or rugs, drapes, upholstered furniture, plants, and fabric wall hangings, sound waves are broken up and absorbed. The impact to the house is minimized as well.

The last and most difficult step is to consider the floor construction itself. Composite floor/ceiling constructions can be tested and rated according to three factors: the sound transmission class (STC), the impact noise rating (INR), and the impact insulation class (IIC). The INR is an older system, the IIC newer—and they are interchangeable. Published rating tables may contain one or the other or both. The higher the rating number under any system, the better the construction in reducing airborne and structure-borne sound transmission.

As an example, the worst-case situation would be $^3/_4$-inch wood flooring and $^1/_2$-inch gypsum board ceiling solidly nailed to 2-×-8 floor joists (Fig. 2-13). This construction has an STC of 30 and an INR of -18—very poor. Adding a $^3/_4$-inch wood finish flooring (Fig. 2-14) improves the STC to 42 and the INR to -12 (IIC about 32). But adding wall-to-wall $^3/_8$-inch nylon carpet and a relatively thin foam padding (Fig. 2-15) brings the ratings to about STC 45 and INR $+5$ (IIC about 56)—a great improvement. Other constructions can be employed to bring the STC as high as 52 or so, and the INR to $+18$ (IIC 69).

The easiest and most effective floor sound-deadening material for an existing house is wall-to-wall carpeting. The least effective carpet is a thin commercial type glued to a $^1/_2$-inch or $^5/_8$-inch plywood subfloor. The most effective, as you might guess, is a heavy carpet laid on a thick pad. Standard 44-ounce (pile weight per square yard) wool carpet, over

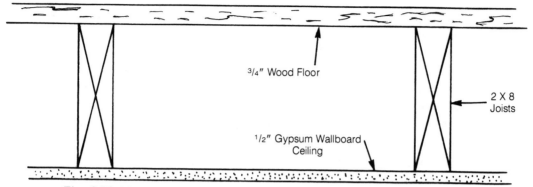

Fig. 2-13. *This basic floor construction has the least soundproofing capability of any.*

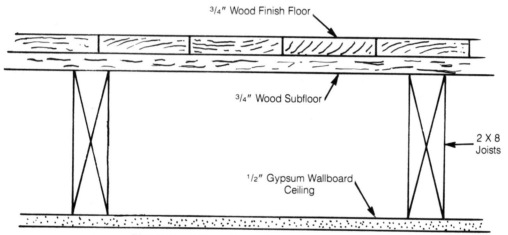

Fig. 2-14. *Adding a layer of ³/₄-inch wood flooring to the basic floor construction improves sound control slightly.*

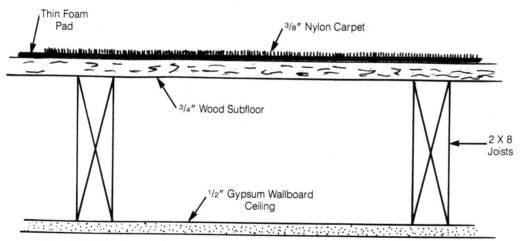

Fig. 2-15. *Further sound control improvement can be added by installing average-thickness wall-to-wall carpet and pad.*

a 40-ounce hair-and-jute pad, laid on $^5/_8$-inch plywood, secured to 2-×-8 floor joists, with a $^1/_2$-inch gypsum board ceiling below results in an IIC of 61. The same carpet over 80-ounce sponge rubber has an IIC of 68.

Another measure is to put down an underlayment of sound-deadening board. This material comes in large sheets and is laid over the subflooring. It must then be covered with a layer of plywood underlayment, upon which a finish floor covering is laid. A variation that affords further improvement involves first laying the sound-deadening underlayment on the subflooring, then placing 1-×-3 or 1-×-4 furring strips down on the board on 12-inch or 16-inch centers. A plywood underlayment—$^1/_2$-inch for a $^3/_4$-inch wood finish flooring; or $^5/_8$-inch or $^3/_4$-inch for a carpeted or resilient flooring finish covering—must be secured to the furring strips. The furring strips should be blocked in at their open ends, to form a solid nailing strip perimeter for the plywood underlayment (Fig. 2-16). Similar blocking beneath the open edges of the plywood sheets will result in a stiffer floor. For installation details, refer to the section entitled "Underlayment."

If the floor framework is open (subflooring taken up) or access can be gained from below (ceiling removed), there is one further step that can be taken at little expense. Install fiberglass or mineral wool blanket thermal insulation between the floor joists. Use either

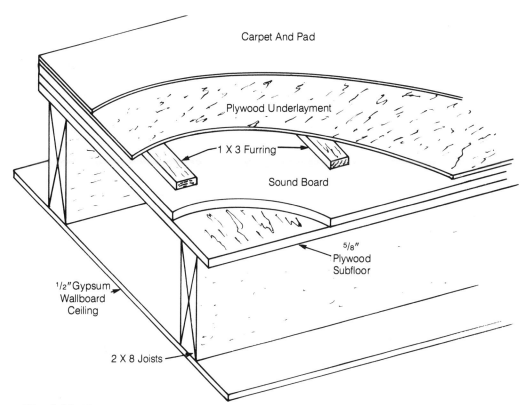

Fig. 2-16. *This construction can be installed in most houses and provides very good noise reduction.*

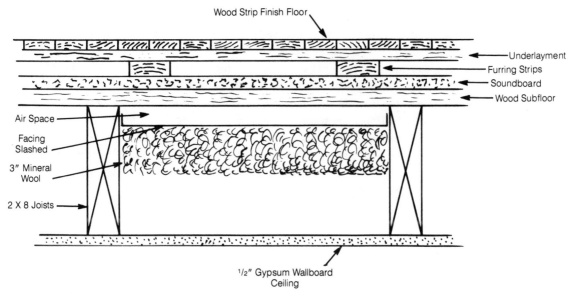

Fig. 2-17. *Where maximum noise reduction between inhabited floors is required, this floor construction is suggested. It is best used in conjunction with ceiling sound control measures.*

batts or blankets, and position the insulation so that its top face lies about 1 inch below the undersurface of the subflooring (Fig. 2-17). The thickness need only be about 3 inches; any more will have no increased value in sound reduction. Before installing the material, slash the vapor barrier facing with a sharp utility knife to allow moisture vapor passage. A vapor barrier is not used between heated upper and lower living quarters. The insulation can be installed either face up or face down.

In conclusion, appropriate combinations of subflooring, underlayments of both sound deadening and standard types, thermal insulation, and finish floorings will serve to markedly reduce impact noise transmission and somewhat improve airborne sound transmission difficulties. Further improvements can be made, either alone or in combination with the above, by making changes in the kind of ceiling construction usually found in houses. This will be covered later.

SUBFLOOR REPLACEMENT

The subfloor is the first layer of flooring, which serves first as a construction platform, then as a base for the underlayment and/or finish floor covering (Fig. 2-18). Replacement of the subflooring in an entire room or area is seldom necessary, because localized damage can be repaired, as described previously, and minor damage, defects, or weakness can be covered over with underlayment. However, in the case of damage by fire, severe termite infestation, or pervasive rot to the floor frame and perhaps the subfloor as well, a new subfloor must be laid after the floor frame has been renovated.

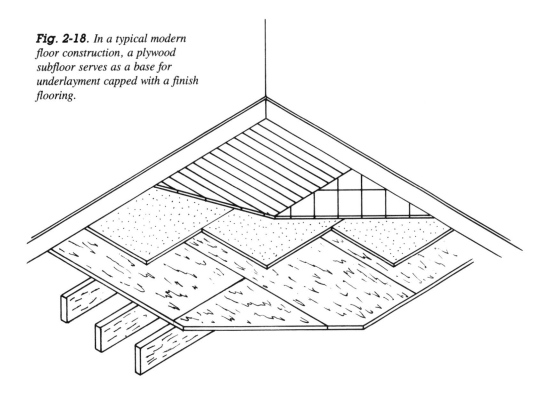

Fig. 2-18. *In a typical modern floor construction, a plywood subfloor serves as a base for underlayment capped with a finish flooring.*

Materials

The material chosen for the new subflooring should match the old in thickness, or a combination of subflooring and underlayment materials should be selected to achieve the same purpose. Plywood is the most commonly used material today. For replacement purposes, Interior Type C-D INT-APA grade plywood is satisfactory. Use minimum $1/2$-inch thickness for 12- or 16-inch joists spacing, $3/4$-inch minimum for 24-inch spacing. Alternatively, you can use nominal 1-×-6 or 1-×-8 boards (actually $3/4$ inch thick), either square edged (S3S or S4S) or tongue and groove. For 24-inch joist spacing, use only the tongue-and-groove type.

Preparation

In some kinds of house framing methods, the subflooring extends only to the inside face of the exterior walls; it may or may not pass beneath partition walls (Fig. 2-19). Where it does not, all of the original nailing surfaces for the subflooring should remain after the old subflooring is taken up. Removal is a matter of prying up the material, starting at any handy point. If only a section need be removed, trim the material off along the centerlines of appropriate joists in one direction, and at any desired stopping point at right angles to the joists.

With other methods, the subflooring passes under both interior and exterior walls (Fig. 2-20). This means that removal entails cutting the subfloor away along the wall lines,

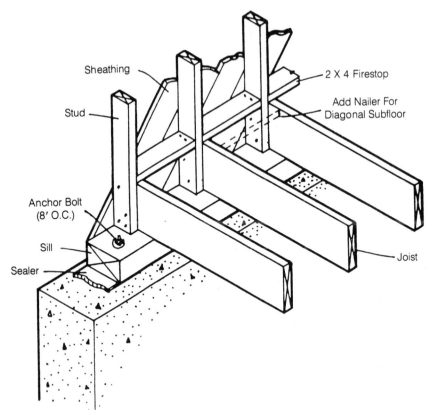

Fig. 2-19. *In a balloon-framed structure, the subflooring abuts the wall studs and firestops. If diagonal board subflooring is used, nailing blocks must be added between the joists for securing the board ends. (Courtesy USDA.)*

as close as possible to the walls. When the old subflooring is removed, this leaves a narrow strip of old material—about 1 1/2 inches wide when cut off with a small power saw—around the perimeter of the room. The new material will butt against the edges of this strip. However, there will be no bearing/nailing surface at this point for the new material.

A new bearing surface must now be provided along the edges parallel with the floor joists in order to maintain the strength of the floor system. One method is to install another joist the same size as the existing ones, set so that its top edge lies halfway under the old material and tight to its undersurface (Fig. 2-21). Another is to block out a length of 2-×-4 from the end joist, set in the same position (Fig. 2-22).

If the new material is to be boards laid diagonally to the floor joists (a strong construction), a new bearing surface will also have to be provided at perimeter edges lying at right angles to the joists. This can be done by nailing 2-×-4 blocking between each joist, halfway under the old subflooring strip and tight to its undersurface (Fig. 2-23). In all cases, the old subflooring should be nailed to the blocking with 6d ring-shank or cement-coated nails.

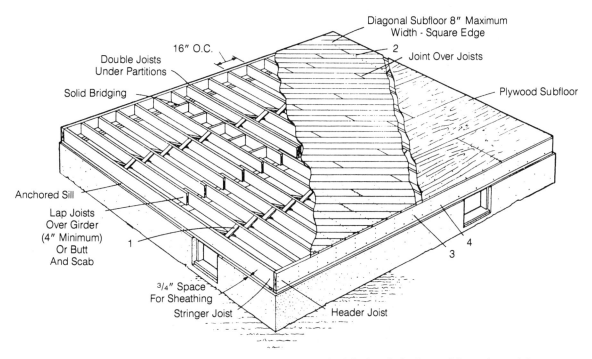

Fig. 2-20. *In a platform framed structure, the subflooring is laid flush with the faces of the perimeter joists before the walls are erected, and thus lies beneath the sole plate. (Courtesy USDA.)*

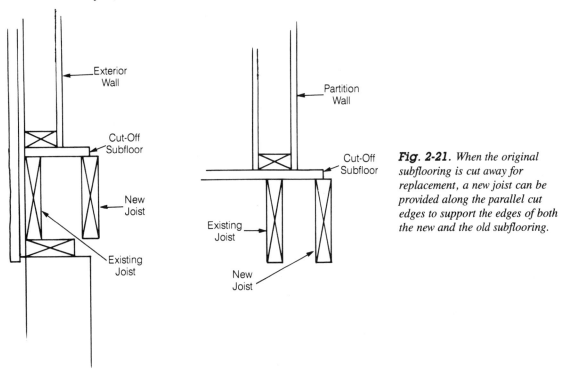

Fig. 2-21. *When the original subflooring is cut away for replacement, a new joist can be provided along the parallel cut edges to support the edges of both the new and the old subflooring.*

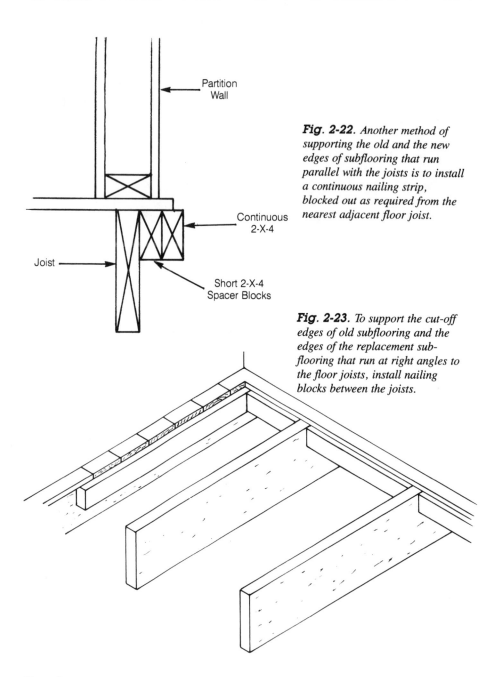

Fig. 2-22. *Another method of supporting the old and the new edges of subflooring that run parallel with the joists is to install a continuous nailing strip, blocked out as required from the nearest adjacent floor joist.*

Fig. 2-23. *To support the cut-off edges of old subflooring and the edges of the replacement subflooring that run at right angles to the floor joists, install nailing blocks between the joists.*

Procedure

To lay a plywood subfloor, start with a full sheet in one corner. Lay the sheet with the C (best) side up and the face grain at right angles to the floor joists. If necessary, trim the outer end to lie on the centerline of the joist. Continue laying the sheets, trimming as necessary. All end joints should be staggered from row to row of the sheets, preferably by two joist spaces. No cut sheets should rest upon fewer than three joists; that is, don't use a small piece from just one joist to another if you can avoid it. Leave a $^1/_8$-inch expansion gap along all edges, and a $^1/_{16}$-inch gap at ends. In consistently humid or damp conditions, double those gaps. To make a tighter, stronger floor that will remain free of squeaks, run a bead of construction adhesive along the top of each joist just before laying each sheet. Figure 2-24 shows a typical arrangement.

Fasten the sheets with 6d nails for $^1/_2$-inch plywood, 8d for thicker material. Common, box, or cement-coated nails are all right, ring-shank nails are better. Space the nails 6 inches apart at all perimeter bearing points, and about 10 inches at interior bearing points. Snap chalklines on the plywood along the joist centerlines as nailing guides.

To lay a board subfloor, start at one end of the room and lay the boards at right angles to the joists. Or start at one corner and lay them at a 45-degree angle. If the boards are tongue and groove, it is generally easier to start with a tongue facing the wall. Once the first few pieces are down, you can work forward from the area just laid. Snug the boards tightly together; stubborn tongues can be driven into place by tapping a hammer against a piece of scrap with its tongue inserted into the board groove. Nail 6-inch boards with two 8d nails, 8-inch with three. All joints should be made over the centerline of a joist. When nailing at joints, angle the nails inward slightly toward the center of the joist. Each piece, except for the short diagonals in opposing corners, should be supported by at least three joists. Figure 2-25 shows typical layouts.

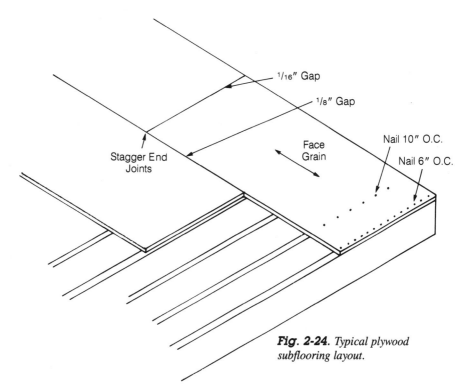

Fig. 2-24. *Typical plywood subflooring layout.*

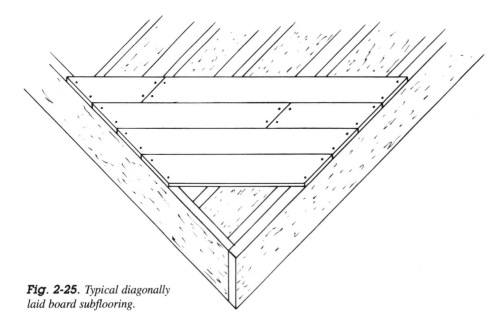

Fig. 2-25. *Typical diagonally laid board subflooring.*

UNDERLAYMENT

In this context, underlayment is a full-coverage, solid, intermediate layer of flooring laid between the subflooring and the finish floor covering. It can be used for several purposes: to bring the strength and stiffness of a relatively thin subflooring, such as $1/2$-inch or $5/8$-inch plywood, to a more satisfactory level; to provide a smooth, clean surface for the installation of a new finish floor covering; to cover an old underlayment or finish flooring in poor condition, and allow installation of a new covering; to equalize or adjust differing floor surface levels from room to room or area to area; and as a noise reduction measure between upper and lower level living quarters.

Types

Several types of underlayment are commonly used. Particleboard is one; it is available in 4-×-8-foot sheets and several thicknesses. For underlayment purposes, a 40-pound density in $3/8$-inch or $1/2$-inch thickness works well. Another choice is interior Underlayment INT-APA plywood, which comes in 4-×-8-foot sheets and $3/8$-, $1/2$-, $5/8$-, and $3/4$-inch thicknesses. A third type is hardboard underlayment, made specially for the purpose. This comes in 4-×-4-foot sheets $1/4$ inch thick.

For sound deadening purposes, special soundboard underlayment is used. This is a fairly soft, resilient, low-density fibrous material that comes in 4-×-8-foot sheets $1/2$ inch thick. Where a new subfloor must be laid, a grade of plywood called Sturd-I-Floor INT-APA might be of value, because it provides both subfloor and underlayment in one $5/8$- or $3/4$-inch thickness.

When selecting an underlayment, it is wise to first determine the manufacturer's recommendation for the type of underlayment to use beneath the contemplated finish floor covering. Not all combinations are compatible. For example, self-stick vinyl tile will not adhere to particleboard. Plywood, however, is effective under any kind of finish flooring.

Procedure

There are no special tricks involved in removing old underlayment. It is invariably nailed down, and usually extends only to the wall lines (except for Sturd-I-Floor). All the underlayments except plywood will break up in chunks as they are pried up. Plywood comes up by the sheet. Just start prying from any handy spot, clearing away the debris as you go. Many nails will remain stuck in the subfloor, and you will have to pull them separately with a claw hammer. In cases where the underlayment goes under a wall sole plate, break or cut the material away, then trim off whatever is left, flush with the wall.

To lay plywood or particleboard underlayment, first remove all mopboards and base or shoe moldings. Or, if there is no shoe molding and the base or mopboard molding is tall and flat-faced, you can lay the underlayment up to that, and later cover the gap between them with a new shoe molding (Fig. 2-26). Lay the sheets at right angles to the lie of the flooring surface or to the joists, if that can be determined. Lay plywood with the best face up. With particleboard there is seldom much difference. Along the wall lines, leave a gap of about $1/8$ inch for plywood, and $3/8$ inch for particleboard. Leave a $1/32$-inch gap between all edge and end joints. Stagger all sheet joints, and cut sheets to fit evenly and cleanly.

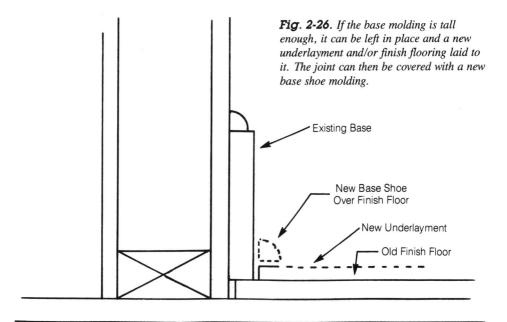

Fig. 2-26. *If the base molding is tall enough, it can be left in place and a new underlayment and/or finish flooring laid to it. The joint can then be covered with a new base shoe molding.*

Existing Base

New Base Shoe Over Finish Floor

New Underlayment

Old Finish Floor

Nail the sheets down with 3d ring-shank nails for $^3/_8$- and $^1/_2$-inch thicknesses, 4d for $^5/_8$- and $^3/_4$-inch thicknesses. Space perimeter nails about $^3/_4$ inch in from the edges. For both materials, the perimeter nail spacing should be 6 inches apart. Intermediate nailing in plywood should be 8 inches in each direction; in particleboard, 10 inches in each direction. If you are applying the underlayment to a subfloor and can determine the lie of the floor joists, you can adjust some of the nailing lines so that you can anchor to the joists as well as the subflooring. Wherever you have an opportunity to nail through, do so with 6d or 8d ring-shank nails.

To install hardboard underlayment, follow the same procedures but place the panels rough side up. Leave a gap along the wall lines of about $^1/_8$ inch, and about $^1/_{16}$ inch between all joints. Fasten the panels with 2d nails. Space them in about $^3/_8$ inch from the edges, and 3 inches apart around the perimeter, and 6 inches apart in both directions in the field.

Combination subfloor/underlayment plywood is laid directly on the floor joists in much the same way as ordinary plywood subflooring (Fig. 2-27). It is available square-edged or with tongue-and-groove edges. Blocking must be provided between the joists under all squared edges, but is not necessary with the tongue-and-groove type. Leave $^1/_{16}$-

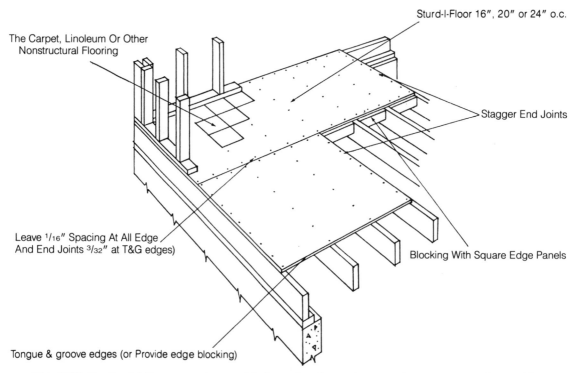

Sturd-I-Floor 16″, 20″ or 24″ o.c.

The Carpet, Linoleum Or Other Nonstructural Flooring

Stagger End Joints

Leave $^1/_{16}$″ Spacing At All Edge And End Joints $^3/_{32}$″ at T&G edges)

Blocking With Square Edge Panels

Tongue & groove edges (or Provide edge blocking)

Fig. 2-27. *The Sturd-I-Floor system is a special plywood product and construction arrangement that provides subflooring and underlayment in one layer. (Courtesy American Plywood Association.)*

inch gaps at all joints ($^3/_{32}$-inch for tongue-and-groove panels). Stagger all panel joints, and fasten with 8d ring-shank or 10d common nails.

Soundboard underlayment comes in several variations and sheet sizes. The generalities of installing this material are similar to other underlayments, but the details vary. Check with your building supply house to see what is locally available, and follow the manufacturer's instructions for installation.

WOOD FLOOR OVER CONCRETE

If an existing concrete floor is actually wet during any part of the year—from ground moisture seeping up through floor cracks or in through foundation walls, or from rain or snow-melt water trickling down the walls—no attempt to install any other flooring should be made unless the problem is corrected. If the floor is slightly damp from the normal hygroscopic action of the concrete, or if it is always dry (protected by external vapor barriers and/or waterproofing and/or a foundation drainage system, or a suspended floor at least 18 inches above grade—a wood floor system can be installed. There are three principal methods to consider.

Installation Methods

Method 1: If the floor is a dry one, cover it completely with a layer of 6-mil polyethylene sheet. Overlap the seams about 1 foot, and seal them with adhesive. Lap the sheeting up the walls about 6 inches. Lay 2-×-4 screeds on top of the sheeting 16 inches on centers and 1 inch away from the walls at ends or edges. Fasten them to the floor with concrete nails, or "shoot" them with a nailing gun. Level the screeds as necessary by shimming or shaving as you go along. If desired, cut and fit sheets of rigid insulation between the screeds. This will help with heating, make the floor warmer, and eliminate the hollow, drum-like sound that would otherwise occur when walking on the finished floor. Plank or strip flooring $^3/_4$ inch thick or more can be laid directly across the screeds (see the section entitled "Wood Plank and Strip Flooring"). For other types of flooring, first lay $^1/_2$-inch or $^5/_8$-inch plywood subflooring and follow up with underlayment if necessary, following procedures previously described. As the next layer of flooring is applied, fold the excess plastic sheeting over and sandwich it between the uppermost two layers of material. See Fig. 2-28 for a typical installation.

Method 2: Clean the floor thoroughly, do any necessary patching, then coat the entire floor with a type of thick mastic that will bond wood to concrete. Then lay short lengths (2 to 4 feet) of 2 × 4s about 12 inches apart down flat in the mastic—either in regular rows or in a random staggered way—with full unbroken rows at parallel walls (Fig. 2-29).

Method 3: If the floor—and basement walls, if present—are not known to be externally protected against moisture, or if there is a small amount of dampness sometimes or always present, use the following method. Clean the floor, then treat it with a waterproof coating. Lay out 1-×-4 screeds on 16-inch centers, leaving a 1-inch expansion gap next to walls, and nail them to the floor. Use either redwood or, preferably, pressure preservative-treated wood. Sheets of rigid insulation can be cut and fitted next, if desired. Spread

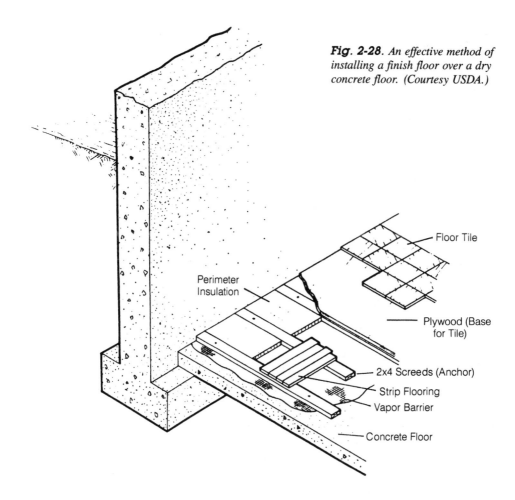

Fig. 2-28. An effective method of installing a finish floor over a dry concrete floor. (Courtesy USDA.)

Floor Tile

Perimeter Insulation

Plywood (Base for Tile)

2x4 Screeds (Anchor)

Strip Flooring

Vapor Barrier

Concrete Floor

sheets of 6-mil polyethylene over the screeds, stretched fairly taut, and staple only enough to hold the sheets in position. Overlap seams 6 to 12 inches and preferably position directly over screeds; seal all seams with adhesive. Nail another row of 1-×-4 screeds (they need not be treated but should be kiln-dried stock) directly on top of the first with 4d cement-coated or ring-shank nails, or secure them with drywall screws. Set the fasteners about 12 inches apart in a staggered fashion. Then proceed with the next layer of flooring (Fig. 2-30).

Provided there is enough headroom, sometimes a more satisfactory arrangement can be made—especially if the concrete floor is broken, irregular, or decidedly off-level—by installing a suspended floor (Fig. 2-31). This requires building an entirely new wood floor frame system, but is not a difficult job.

Start by laying a full-coverage vapor barrier of 6-mil polyethylene sheeting on the old floor. Overlap the seams about 1 foot and seal them. Leave enough extra to curl up the walls about 18 inches, but let it lie loose for the moment. Snap level chalklines on the

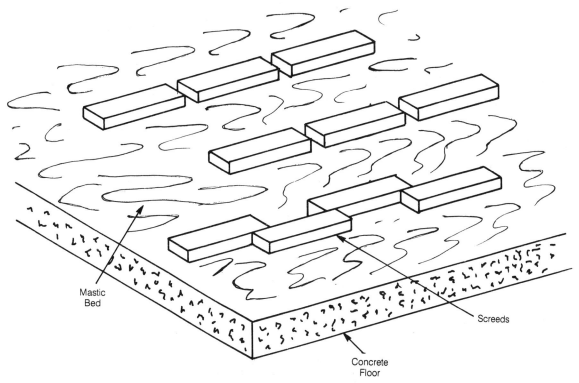

Fig. 2-29. *Aligned or staggered screeds set in mastic spread on a concrete floor can serve as the base for a wood subfloor, which can then be covered with any of several different kinds of finish flooring.*

walls all around the room, 6 inches above the highest point of the old floor. Fasten rim joists of 2 × 6s to the walls, their tops even with the chalklines. If you are working with stud walls, use ordinary construction-grade lumber and 12d nails. If you are using concrete or concrete block, use redwood or treated wood and anchors and lag bolts, or rent a nail gun and set them with concrete nails. As you fasten the joists, position the plastic sheeting behind them.

Next, lay out the floor joist positions on the rim joists; the joists should cross the narrow dimension of the room. Set them on 16-inch centers. Nail a metal joist hanger at each location on the two opposite walls, and secure the joists to them. They can be construction-grade lumber, and should span the entire room, if possible. If not, splice them in overlaps of at least 1 foot, and stagger the splice layout throughout the room. If any of the joists have a crown to them, position the crown upward.

After the joists are set, nail solid blocking of short lengths of 2 × 6 between each joist. Nail down the center of the room if 16 feet or less wide, or in two equidistant rows if wider. Then, using small blocks of redwood and/or cedar shim shingles, place shims between the old floor and the underside of each joist along the blocking row(s). Tap two opposing shim shingles together with just enough upward pressure to make the joists solid, but not to force them out of line. If the shims seem not to be particularly snug,

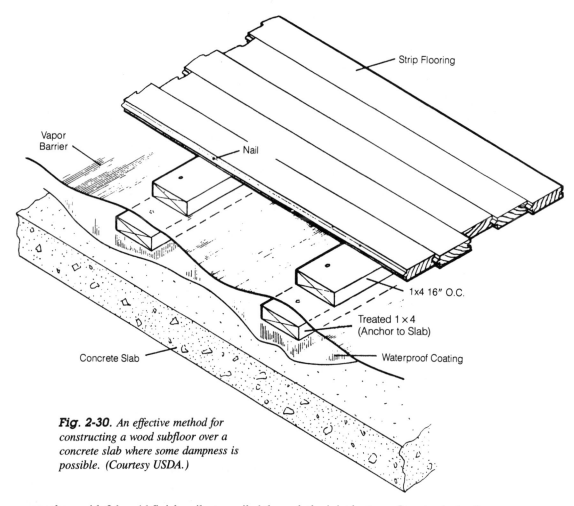

Fig. 2-30. *An effective method for constructing a wood subfloor over a concrete slab where some dampness is possible. (Courtesy USDA.)*

secure them with 3d or 4d finish nails, toenailed through the joist bottom. Or set a daub of construction adhesive in the joint on each side of the joist.

Now check all of the joist tops to see if they are even and level. If any rise above the norm more than about 1/8 inch, plane them off. Install insulation between the joists if you wish, then proceed with the next layer of flooring. Fold the plastic sheeting in between two layers, or beneath the bottom layer.

CONCRETE FLOOR OVER EARTH OR CONCRETE/MASONRY

The most common procedure for covering a dirt floor as part of a remodeling project—in a garage, carport, or basement, for example—is to pour a new concrete floor. Likewise, a floor of concrete, stone, ceramic tile, masonry unit, or brick that is in bad condition or unsuitable for covering with a new finish flooring material because of roughness, dampness or alkali problems, or breakup can be topped with a fresh layer of concrete.

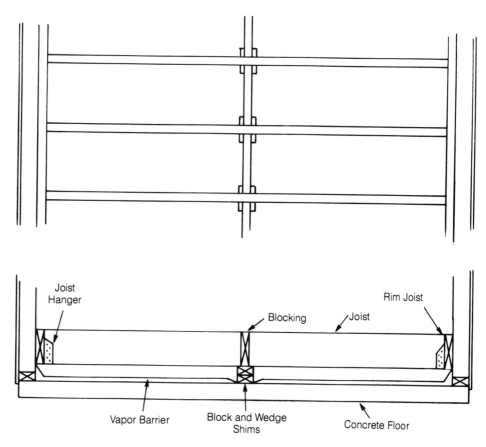

Fig. 2-31. *Using this method, a complete floor system can be suspended over an existing concrete floor.*

Preparations

A new concrete floor over earth (Fig. 2-32) will take up 4 to 8 inches of headroom, depending upon its construction. The concrete itself should be 3¹/₂ to 4 inches thick. It can be poured on the earth surface, but ideally a cushion layer of 2 inches of compacted sand should be spread first. In many areas of the country, it is desirable to add a layer of rigid insulation between the sand and the concrete, usually 1¹/₂ to 2 inches thick. If necessary, remove a suitable amount of earth to compensate for the floor thickness of your design. Also, fill in and complete any foundation gaps as necessary, such as across a garage door opening.

Procedure

Rough-level the surface, disturbing the compact native soil as little as possible. Tamp and compact any low spots that have been filled and leveled with loose dirt. In cold country, set a line of rigid insulation, on edge and about 16 inches wide, around the perimeter of

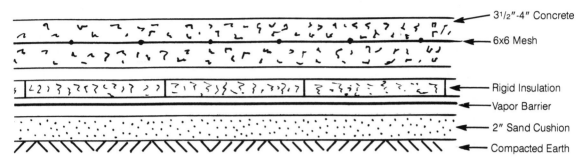

3¹/₂″-4″ Concrete

6x6 Mesh

Rigid Insulation

Vapor Barrier

2″ Sand Cushion

Compacted Earth

Fig. 2-32. *Cross section of a typical poured concrete floor installation.*

the floor, so that the upper edge will be level with the finished concrete. If possible, spread a 2-inch layer of damp sand, then level and compact it. Cover the sand with 4- or 6-mil polyethylene sheeting as a vapor barrier; overlap edges and ends of pieces by about 1 foot. If required, lay sheets of rigid insulation over the vapor barrier. Then lay out lengths of 6-×-6 inch No. 10 wire reinforcing mesh, overlapped at edges and ends by about 6 inches. Snap chalklines around the perimeter walls 4 inches above the surface of the insulation (or vapor barrier if insulation is not installed), to serve as a depth gauge and leveling guide for the concrete.

To determine the amount of concrete you will need for a floor 4 inches thick, measure the floor area to be covered and determine the square footage. Divide the result by 81 to arrive at the required number of cubic yards of concrete mix, then add 10 percent. Order a standard concrete flooring slab mix. Or, if you will mix your own from bulk materials, a 1:2:4 mix consisting of 1 part portland cement, 2 parts sand, and 4 parts stone aggregate (³/₄- to 1-inch size suggested) is about right. This will require approximately 6 bags of cement, 0.44 cubic yards of sand, and 0.89 cubic yards of aggregate for each cubic yard of concrete mix, plus 5 to 5¹/₂ gallons of water per bag of cement for sand of average dampness. If the floor area requires more than 1 cubic yard of concrete, mix with a power mixer.

Start pouring the mix into place at the point farthest from the concrete source. As soon as the mix begins to pile up, screed (level) it with a long, straight length of 2 × 4. Hold the screed flat on edge and work it back and forth across the concrete in a sawing motion. Watch the level as you do so, and keep the fresh concrete at a uniform 4 inches deep as you work back toward the unpoured area. At the same time, hook your fingers into the reinforcing mesh (wear waterproof gloves or you'll have skin problems) and jiggle it upward until it rests at about midlevel of the concrete. After screeding a sizable area, go back over it with a darby. For further information on this and the finishing and curing processes, refer to the next section: "Concrete Floor Over Wood."

A new concrete floor topping can be poured directly over an old concrete, brick, or similar floor that is always dry and has no alkali problem. Clean the old surface thoroughly by sweeping and vacuuming, then washing. Remove all grease, oil, paints, or other contaminants with solvents and trisodium phosphate (TSP, available at hardware

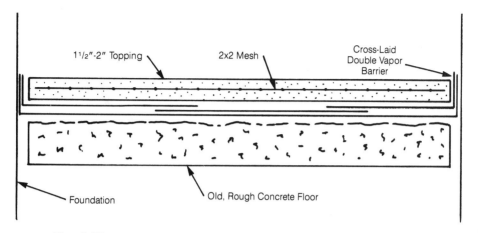

Fig. 2-33. *A damaged or broken concrete floor can be topped with a fresh floor.*

stores and lumberyards) followed by a clear water rinse. For best results, especially if the surface is very smooth, etch the surface with commercial muriatic acid and scrub it around vigorously with a stiff brush or broom. Use several clear water rinses to remove all traces of the acid afterward. This will allow a good bond between the fresh concrete and the old surface.

In many cases, however, a bonded topping is not essential, especially in light-duty residential applications and in small areas. And, if the old flooring has a dampness and/or alkali problem, bonding is undesirable. Instead, pour an isolated topping (Fig. 2-33). Clean the floor of all debris and fill large cracks with patching cement. Then cover the floor with an isolation membrane of 6-mil polyethylene sheeting, overlapped about 1 foot at ends and edges. Spread a second layer either at right angles to the first, half overlapped, or directly on top of the first, depending upon the size of the area and the sheets. Lap the edges up the walls a few inches, and trim them off flush after the concrete has cured.

The new concrete floor should be $1^1/2$ to 2 inches thick, reinforced with 2-×-2-inch No. 16 mesh. The mix is a topping mix, usually 1:2:2, especially if the new flooring will eventually be covered with resilient tile or a similar covering. A 1:1:2 mix is sometimes used when the concrete itself will serve as the finish flooring. The details of pouring this type of concrete floor are the same as for installing one over a wood floor, with one exception: If the old floor is very rough and uneven, add 10 percent or more to the estimated amount of concrete mix needed to pour the floor.

CONCRETE FLOOR OVER WOOD

A concrete floor can be poured over a wood subfloor, either separately or as an integral unit with the frame. It can also be poured on top of an old finish flooring, provided that the frame is strong and stiff enough to withstand the weight. This is done to provide a fresh, clean, smooth surface to which a new floor covering, such as ceramic tile, can be bonded. Or, the concrete itself can serve as the finish flooring. Although there are a lot of variations on the theme as far as details are concerned, two principal methods are covered here.

Preparations

Concrete poured over an old wood finish floor or a subfloor should be a minimum of $1^1/_2$ inches thick: 2 inches is better (Fig. 2-34). There are several different concrete mixes that could be used, but a 1:2:2 topping mix is very workable and will provide a dense surface. This must be made up from scratch—a bagged dry mix is unsatisfactory—by mixing 1 part portland cement, 2 parts damp loose sand, and 2 parts stone aggregate no larger than $3/_8$ inch. Mix the dry materials together thoroughly, then add in water in the ratio of approximately $4^1/_2$ to $4^3/_4$ gallons of water per sack of portland cement—a bit more if added plasticity is needed. Amounts of less than 1 cubic yard can be mixed by hand, but larger amounts should be done in a power mixer. Add the water slowly and mix by hand until you have a completely uniform color throughout the mass, or mix for about 2 to 3 minutes in a power mixer.

To calculate the amount of materials you will need, first determine the floor area to be covered, in square feet. If the topping will be 2 inches thick, divide the square footage by 6 to arrive at cubic footage of concrete; if $1^1/_2$ inches, divide by 8. To make up 1 cubic yard (27 cubic feet) of 1:2:2 mix you will need 8.2 sacks of cement, 0.60 cubic yard of sand, and 0.60 cubic yard of aggregate. For small jobs, it's easier to use $1/_{27}$th of the cubic yard figures and work with cubic feet of materials; 0.304 sack of cement, 0.60 cubic foot of sand, and 0.60 cubic foot of aggregate.

As an example, consider an entry hallway 4 feet wide and 10 feet long—that's 40 square feet. A topping 2 inches thick, then, would have a volume of 6.66 cubic feet. To

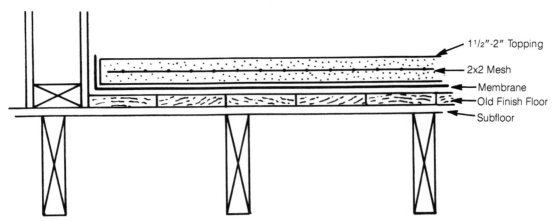

Fig. 2-34. *Provided the underpinnings are sturdy enough, or can be made so, an old wood framed floor system can be covered with concrete to form the base for a new finish floor—typically ceramic tile.*

this, add about 10 percent for waste and good luck, to make about 7¹/₄ cubic feet. Multiplying the material requirement numbers by 7.25 comes out to 2.2 sacks of cement, 4.35 cubic feet of sand, and 4.35 cubic feet of aggregate for this job. The mix proportions should be followed quite closely, so in practice it is easier to round the numbers upward rather than trying to measure out odd quantities. Here, for example, you would use 2.5 sacks and 5 cubic feet each for sand and aggregate, which equals 1:2:2.

Procedure

The first step is to lay an isolation membrane of 15-pound roofing felt over the floor (you can also use plastic sheeting). Overlap the seams about 6 inches and fold the felt up the walls about 1 inch more than the thickness of the concrete. Staple the felt securely along all edges and at a few points in the field as well. Make a single slit at inside corners, fold the excess in, and staple it. At outside corners, make a single relief slit so the felt will lie flat, then wrap another small piece around the corner and staple it, to protect the small bit of uncovered wall.

For small areas, pour the concrete all at once. For larger areas, work continuously from section to section in a preplanned way. A helper or two makes a large job go easier

Fig. 2-35. *Freshly poured concrete must be rough-leveled by screeding or striking off with a straight-edged length of wood that is sawed along the edges of the forms. (Courtesy Portland Cement Association.)*

and faster. Pour a layer of concrete over the membrane to half depth, approximately leveling the surface at the same time. As you finish a suitable area, lay out reinforcing mesh and press it slightly into the wet mix. The mesh should be 2-×-2-inch welded wire, 16 gauge in both directions. Overlap it about 2 inches at ends and edges. Make sure it lays flat, then pour concrete over it to the full depth. Screed (level) the surface off as completely as possible with a length of straight board or 2 × 4 (Fig. 2-35). Dispose of any extra concrete.

As soon as you have finished screeding, darby the surface (Fig. 2-36). A darby is a long length of straight wood with a handle or grip. Place it flat on the concrete and move it back and forth with a sawing motion, continually moving across the surface gently and with no downward pressure. This removes any ripples or rough spots left by the screed, and fully embeds bits of aggregate. The process will also bring some water to the surface. Very little, if any, excess concrete should be worked off. For spots too far away to reach conveniently with a small tool, use a long-handled darby.

There will be a sheen of water on top of the concrete for a while after darbying. As soon as this disappears—usually one to two hours depending upon temperature and relative humidity—start the finishing process. Floating is the first step (Fig. 2-37), and a 12-

Fig. 2-36. *Immediately after fresh concrete is screeded, it should be darbyed or floated to further smooth the surface. (Courtesy Portland Cement Association.)*

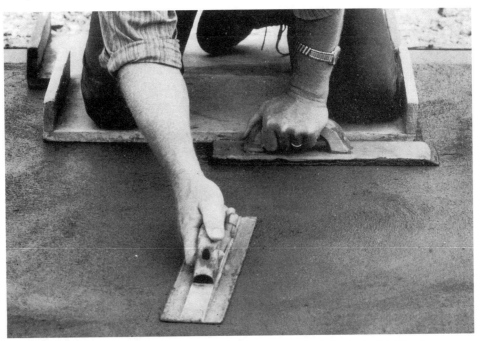

Fig. 2-37. Hand floating is the first, and sometimes the only step in finishing the fresh concrete surface. (Courtesy Portland Cement Association.)

to 16-inch float is the tool to use. A wood float will impart a slightly rough, grainy surface, while a metal float produces a quite smooth surface.

Start at any convenient edge point on the concrete first poured, work backward away from the freshly floated areas. Use walkboards and a kneeling board to work from when out on the fresh concrete. Keep the float almost flat and sweep it back and forth, raising the leading edge just enough to clear the surface. Use no pressure—let the weight of the float do the work. Don't overdo it; as the concrete smooths under the float, keep it moving along into unfloated territory, and try not to rework any areas unnecessarily. However, keep in mind that this is the last opportunity to smooth out any ripples or bumps, and to submerge or pick out any protruding pebbles. Keep a sharp eye on the surface level; some types of finish floorings set in organic mastic require a variance of no more than $1/16$ inch in 3 feet in any direction.

Sometimes the first finishing step is also the last, if you want a sandpapery finish to provide "tooth" for later application of tile, or a nonslip walking surface. But if a very smooth, almost slick finish is desired—as for an organic adhesive-set ceramic tile installation—go over the surface again with a steel trowel. This should be started within a few moments after the floating. Use a 12-inch steel trowel held flat on the surface. With downward pressure just short of forcing the trowel edges into the concrete, stroke the trowel evenly back and forth or side to side. To get an even smoother surface, repeat the procedure with a smaller trowel and as much pressure as you can muster without making the blade dig in. A third pass will create even more density and result in a surface that is almost glassy.

The last step is to cure the concrete. The best method is to cover the whole area with plastic sheeting, tucked tight at all the edges, immediately after completing the finishing operations. Be careful not to mar the finished surface when laying the sheeting out. Leave it in place for at least three days, five to seven if you can. Keep all traffic off the floor for the first two days, and limit it to light foot traffic until you remove the plastic.

Another method of pouring used for suspended concrete floors (Fig. 2-38), starts with the open joist framework itself. The thickness of the concrete can be as little as 2 inches, but may be 3 or even 4 inches. The floor frame and surrounding structure must be very strong to sustain the weight, which is approximately 12 pounds per square foot per

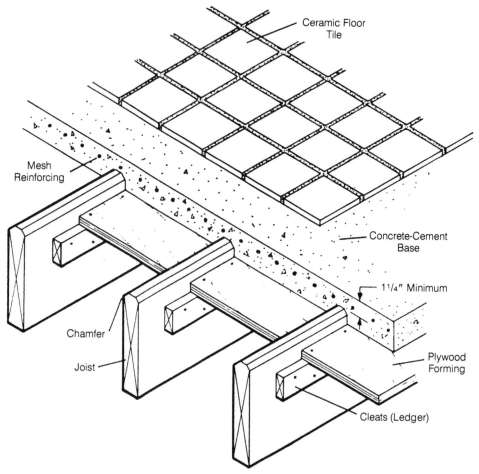

Fig. 2-38. *An effective method for pouring a suspended concrete floor on a wood frame. (Courtesy USDA.)*

inch of thickness when cured, more when freshly poured and still wet. For a thin flooring, the 1:2:2 topping mix can be used, but mixing with 5 to 5^1/$_2$ gallons of water per bag of cement will make a more workable mix. For a thicker flooring, a 1:2:3 mix is fine, or even a 1:2:3^1/$_2$ mix. Using ^{1}/$_2$- to ^{3}/$_4$-inch aggregate.

First, chamfer the top edges of the joists with a power plane or a heavy-duty router fitted with a carbide bit, leaving a chamfer width of roughly ^{3}/$_8$ to ^{1}/$_2$ inch. Secure 1 × 4 ledger strips to the sides of the joists with 6d ring nails about 6 inches apart and staggered up and down. Nail lengths of ^{5}/$_8$-inch or ^{3}/$_4$-inch plywood onto the ledger strips with 8d galvanized nails, spaced about 8 inches apart. The top surface of the plywood should be set so that at least 1^1/$_4$ inches of concrete will cover the joist tops. Pour the concrete between the joists to the level of their tops. Then lay out welded wire mesh at right angles to the joists, and staple it to the joist tops. Overlap ends and edges by about 2 inches. Pour the remaining concrete, and screed, darby, float, and cure it as discussed earlier.

CONCRETE FLOOR FINISHING/REFINISHING PREPARATION

There are two distinctly different sets of procedures that might be involved when one refers to "finishing" a concrete floor. One of these procedures is necessary, the other is optional. The first is a part of the initial concrete laying process, and involves screeding and floating the fresh concrete and then troweling it smooth as it begins to cure. The optional finishing process involves applying a coating to the concrete after it cures, for protection and/or decoration. It is this process that requires preparation of the concrete surface if the coating application is to be successful.

Not all concrete floors are capable of taking an applied coating. If an existing floor surface is chalky, or continually dusty and gritty from wearing off and disintegration, or if spalling or breaking up, a coating application is unlikely to be successful. It will not bond and the concrete could continue to degrade. Likewise, if the floor is damp from moisture migrating upward from the ground and through the concrete; if there is long-term standing water on the surface at any time from foundation leaks or other sources; if there is persistent surface efflorescence or a high degree of alkalinity in the concrete; or if there are widespread deposits of grease, oil, or waxes embedded in the surface, applied coatings will not adhere and will soon lift away. A concrete floor must have a firm, tight, smooth surface that is free of contaminates and loose particulates for an applied coating to provide decent service over a reasonably long time.

Troweling

If it is known ahead of time that a new concrete floor will eventually be painted, do the final troweling of the surface when the concrete is initially finished so that the surface remains just a tiny bit rough and finely grained. Steel troweling to a very smooth glassy finish should not be done; paint will adhere much better to the slightly rough surface. Also, the new floor should be allowed to age for at least 3 weeks in low-humidity conditions, 6 weeks in high-humidity conditions, or as long as possible in any case (a full year is not too much for some kinds of coatings) before a coating is applied. Meanwhile, it should be kept clean and traffic-free as possible. This allows the concrete to cure and dry thoroughly, and permits carbonation of the lime at the surface and renders it noninteractive with the paint.

Examine the Surface

There are several preparatory steps to finish that should be undertaken, if applicable, on any existing concrete floor—whether old or laid within the previous few weeks. First, vacuum the floor thoroughly, then inspect the surface for problems. If there are chips, dings, or gouges in the surface, fill them with a suitable epoxy concrete repair material and smooth the patch surfaces. Cracks that remain stable can be similarly filled, but if they are the kind that open and close slightly with changes in surrounding ground or weather conditions, they should be filled with a paintable elastomeric compound or caulk.

Apply Fillers

If the entire surface is rough, grainy, and appears porous, you might have to first apply a cement mortar surface filler in order to attain a satisfactorily smooth, paintable surface. First scrub and hose the surface thoroughly to clean it and allow all the water to drain away; dampness does not matter. Prepare a mix of one part portland cement to one part fine (passing a No. 50 screen) sand with enough clean water to make a grout the consistency of thick cream. Pour the grout onto the floor in batches and scrub it around with a chunk of burlap, jute, or a stiff-bristled brush. Use a fair amount of pressure and fill all the pores and small depressions. Sweep the excess continually back off the freshly treated surface. When the job is done, none of the grout should be lying free on the surface. If so, use an old squeegee to get rid of it. Finally, damp-cure the entire surface for at least 3 to 5 days, a week if possible. Then you must allow complete curing and drying by letting the surface age for at least 3 or 4 weeks before you apply a coating.

Roughen the Surface

If the floor surface is very smooth, glassy, or slick, it should be roughened up slightly before most kinds of coatings are applied. If the floor area is not large, one effective method is to hand rub the whole surface with a coarse, flat-faced abrasive stone. Wear a dust mask, and vacuum the floor thoroughly afterward. Another common method that is good for large areas requires much less physical effort, but must be carefully and cautiously done. Etch the surface with a 5-percent solution of hydrochloric (muriatic) acid. Follow the manufacturer's safety and application instructions to the letter, and flood-rinse the floor afterward. This process requires drying time for the floor before a coating can be applied.

Remove Efflorescence and Alkali

Efflorescence, a white, floury deposit that leaches out onto the concrete surface, is a common problem. Neither cement filler nor paints should be applied over it. The greater part of such deposits can be removed by scraping, wire-brushing, and vacuuming. To remove the residue, first wet the area, then apply a 10-percent solution of hydrochloric acid. Wait 5 to 10 minutes, scrub the area with a stiff-bristled brush, then immediately flush thoroughly with clean water. Allow plenty of time for drying before applying a coating.

Sulfates of sodium, calcium, and magnesium along with some lesser substances are lumped together under the term *white alkali*. The presence of alkali in concrete is common, especially in the western half of the United States, because of its presence in soils and water. Where alkali is suspected in a concrete floor (can be verified by chemical testing), aging for at least a year before painting is recommended. Also, select a coating that is alkali-resistant: vinyl, latex, and other synthetic paints are suitable. Oil- and varnish-based paints are susceptible to alkali attack, but usually can be successfully applied to floors 12—but preferably 18—months old or more. If these paints must be applied to a floor newer than that, the floor should be pretreated. Mop a 3-percent solution of phosphoric acid liberally onto the entire surface and leave it. Allow the floor to dry completely, then apply the coating.

Clean and Condition Surface

Contaminants on the floor surface, especially oil or grease, will react with the paint and can interfere with proper bonding. The usual cleaning procedure in this situation is a liberal application of strong trisodium phosphate (TSP) solution scrubbed in with a stiff-bristled brush, then flushed away with clean water. Several treatments might be needed, and allowing the TSP to lie on the surface for a short while before scrubbing might help. Waxes and other contaminants can be taken up with strippers or solvents, after which the floor would be thoroughly washed.

Moisture conditioning procedures vary with the kind of coating that will be applied. Paints vary widely in their moisture tolerance, and if the effect is adverse the coating will neither dry nor adhere properly. Regardless of the coating, allow a minimum of 3 weeks for fresh concrete to cure. If oil-based coatings will be applied, the floor surface must be completely dry. Because concrete absorbs and retains moisture easily, the drying time and also the continuous moisture content of a given concrete slab can vary widely with atmospheric conditions and the nature of its surroundings. It is advisable to make a moisture test before attempting to apply an oil-based coating if there is any question about the dryness of the surface. On the other hand, if a latex emulsion paint is to be used, the concrete surface should be dampened slightly (no free water present) before application. Otherwise the concrete, being hygroscopic, will absorb moisture from the paint and will not cure properly. This is particularly necessary if the concrete is porous and/or the working conditions are very dry.

Remove Old Coatings

If the concrete floor has been previously coated, it should be inspected for the conditions just discussed and any problems taken care of. If the old finish is merely worn from ordinary traffic but is still smooth and well bonded, the floor needs only to be vacuumed, washed with a mild detergent, mopped with clean water, and allowed to dry thoroughly before a compatible coating is applied. Stains of grease, oil, or other contaminants should be cleaned up at the same time with solvents or strong detergent.

If the old coating, which may consist of several layers, is peeling, lifting, bubbling, or rough and scaly, it should be stripped off. A flood of strong TSP solution left to stand for an hour or so will often soften an old coating so it can be scrubbed and scraped away. This is particularly effective if grease or oil is causing at least part of the problem.

Another possibility is dry scraping with a putty knife or heavy-bladed paint scraper. You can soften paint as you scrape with a propane torch or a heat gun. For rough, scaly, crumbly paint, a wire brush works well. Sandblasting is yet another possibility and will do a fine job with little physical effort, but presents problems in an interior space because of the tremendous dust buildup.

Regardless of the kind of materials used in previous coatings, if bonding to the floor surface or between coating layers is not complete throughout, the old coating(s) should be completely removed if long, durable service is to be expected from the new coatings. Following the stripping, thoroughly sweep, vacuum, and wash the surface.

APPLIED COATINGS ON CONCRETE FLOORS

In general, decorative painting of concrete floors tends to be less than satisfactory because of rapid wear from traffic. However, residential concrete floors that receive only light traffic can be coated with the expectation of reasonable wearability and durability, provided the job is correctly done. The main object is to apply a thin finish that penetrates into the surface of the concrete. Thick finishes and multiple coatings do not hold up well. They have a tendency to chip and scratch readily and, if they do not bond well, can lift away in patches. The wearability of a thick coating is only marginally better than a thin one, but recoating to a smooth surface is much more difficult once it is worn than is recoating a thin, worn finish. The added new coating simply compounds the potential problems.

Choosing the Coating

Ordinary paints and varnishes are not suitable for concrete floor applications, and in fact only a few coatings are. Note that available products and their formulations and application details change from time to time. This means that it is always necessary to investigate current products and select the locally available one that appears to afford the best results for the particular job conditions. Although the general application procedures for some suitable coatings are given in the following paragraphs, the paint manufacturer's instructions should always be followed explicitly.

One of the most commonly used coatings for either interior or exterior purposes is referred to by the trade terms of "porch and deck paint" or "floor and deck enamel." This is a pigmented varnish vehicle formulation that is relatively tough, durable, and moisture-resistant. It is intended for application on dry, well-aged concrete floors (it can also be used on wood decking). The usual procedure is to apply a first coat consisting of 1 quart of a mix of 2 parts spar varnish and 1 part turpentine, to 1 gallon of the paint. This is followed, no less than 48 hours later, with a full-strength second coat of paint.

There are some rubber-based formulations available for concrete floors, and these are suitable only for interior applications because they will rapidly degrade with exposure to the weather. These are useful for concrete surfaces that are subject to minor dampness problems, and they are alkali-resistant. Usual procedure is to apply two coats. For the first coat, 1 quart of a thinner specifically recommended by the paint manufacturer to 1 gallon of the paint. Work this coat into the pores of the concrete surface, then allow it to dry for at least 12 hours before applying the second, full-strength coat.

Certain polyurethane varnishes and enamels can be applied to concrete floors, interior only. Application details vary, but typically a surface conditioner recommended by the manufacturer must be applied first, to act as a sealer. Then two full-strength topcoats are applied, about 24 hours apart. The polyurethanes are also alkali-resistant and useful where dampness is present in the concrete.

There is another group of paints known in the trade as "masonry paints." These are formulated for either interior or exterior applications on concrete, masonry block, stone, or brick. Details vary according to the particular product; check the manufacturer's specifications and application instructions.

Finally, there are the epoxy paints or enamels, some of which are suitable for concrete floors. If properly applied, they can afford a tough, durable surface. Preparation of the floor is extremely important. The cost of these materials is relatively high, and application is more difficult than with other coatings. They are primarily intended for commercial/industrial applications, are not usable under all job conditions, and are not recommended for do-it-yourself residential projects. However, where application is possible and practicable and the service conditions demanding, the results are very good.

Application

As is true with any applied finish coatings, correct and thorough preparation of the surface is the key to satisfactory results. Follow the manufacturer's recommendations as to coating thickness and the means of application. A certain material, for example, might be best spread at a rate of 400 square feet per gallon. Sometimes brush application is indicated, but often a roller of appropriate nap length and stiffness for the floor surface texture can be used, which saves a lot of time and effort. Atmospheric conditions can also play an important part in how well the coatings bond and cure. Whenever possible, apply coatings when the floor surface temperature is at least 50° Fahrenheit, air temperature is between about 60° and 80° Fahrenheit, and relative humidity is in the 40 to 70 percent range. Avoid direct sun and strong, drying breezes. When working with flammable and/or toxic materials, observe the health and safety precautions. Arrange for plenty of ventilation, wear a proper respirator, use gloves, and avoid skin contact.

WOOD FLOOR FINISHING/REFINISHING PREPARATION

Wood floors in residential applications are typically boards or planks 5 to 6 inches wide or more, strips of about 2 to 4 inches wide or less, or parquet in various styles and patterns. A wide variety of wood species have been used for finish floors, and an amazing array of finishes have been applied to them. But sound old wood flooring of any kind, however previously finished, can be readily refinished with a compatible coating if in good condition. Raw wood, whether it is new or stripped and sanded or unfinished old wood, will accept a wide variety of applied coatings.

Sometimes a perfectly good finish wood flooring has been overlaid with another covering—carpeting or sheet vinyl or linoleum. Often such flooring has previously been varnished or painted. Another common arrangement was to cover the greater part of a floor with one or more large rugs, perhaps in conjunction with smaller throw rugs at doorways, leaving a border of wood finish flooring exposed around the perimeter of the room.

The first step in preparing for a refinish job is to examine the floor to see if any substantial repairs need to be made to any part of the flooring or frame. If so, this must be done first; consult appropriate parts of this chapter. Then check the surface to see if any minor defects need repairing. If the flooring is covered with carpeting or other material laid in place or tacked down around the borders, remove it. Tack, nail, and other small holes, if not too numerous or obtrusive, can be filled. Where adhesives have been used to cement down a finish floor covering, decent refinishing may not be possible. All the glue and residue must be thoroughly scraped away with a paint scraper or shave hook (the latter is likely to do a better job with greater ease), but you must be careful not to gouge and mar the surface. Solvents or heat may help, but then staining or scorching can be a problem. Often sanding is a workable solution. However, this can be a big job and is not always successful, especially if the new finish is to be a clear or semitransparent one. If a good result appears to be questionable, try it in a small area first. If you can bring up clean, fresh, smooth wood with no blotches, stains, or blemishes over the entire surface, then a new finish is practicable.

If the floor system is sound and the existing surface has never had a coating applied to it, and it is smooth and defect-free, it can be treated as a new wood floor. You can go on to the following sections.

If the floor system is sound and the existing surface is in good to excellent condition, is free of major scratches, dents, gouges, and other defects, has no appreciable worn-down spots, does not need any repair work, and is generally presentable and the old finish is sound (even if scuffed, lightly scratched or worn)—not much preparation work needs to be done.

First, if the old finish has or might have had wax on it, go over it thoroughly with a commercial wax stripper. Next, regardless of the kind of existing finish, spot sand any rough places with fine sandpaper. If there are small chips out of the finish, or shallow scratches or scuffs, feather the edges down smooth. Check for any patches of finish that might be loose or peeling away from the wood and scrape them away, then feather in the edges with fine sandpaper. With a knife or other small tool, clean dirt out of all the cracks in or between planks, strips, or tiles. Scrape away any accumulations in corners and along baseboards or shoe moldings. You can remove these moldings by prying them off if that seems necessary—usually it is not.

Vacuum the floor thoroughly, paying particular attention to cracks and corners. Then wash the floor with a damp sponge, warm water, and mild detergent or ammonia. Rinse with cold, clean water and a damp sponge, often and thoroughly. If there are grease spots or heavy dirt accumulations here and there, work on them separately with a strong detergent. Be careful not to soak the flooring, especially where bare wood shows through the old finish.

Examine the floor after it dries, and do any necessary touch sanding; clean up the sanding dust. If the finish is to be paint, fill small holes and defects with any good-quality wood filler and sand the patches smooth. If the finish is to be clear or natural, fill with a wood putty that matches the color of the wood. If a stain is to be used to match bare wood areas, defects in those areas must be filled with a filler that will accept stains (many will not). It's a good idea to experiment on a piece of similar scrap wood first. If the results are unsatisfactory, you can fill small defects by using a matching color wax putty sticks after applying the final coating.

The last step, if the old finish is still glossy, is to either sand away the gloss with very fine sandpaper in a pad sander, or treat the surface with a deglosser—sometimes called "liquid sandpaper." This dulls the finish and creates a "tooth" to which the new coating can bond. Double check to make sure the surface is perfectly clean, and allow the floor to dry completely before applying the new coating.

If the flooring is in bad condition but fundamentally intact, or if it can be repaired to be so, a major sanding job is required.

SANDING WOOD FINISH FLOORING

If an existing finish wood flooring has never been coated or if the flooring is new, raw wood, it should be sanded before any coatings are applied. Sand only enough to fully level and smooth the surface. If the flooring will be covered, as with tile or carpet, it need only be smoothed.

If a finish wood flooring appears basically sound but the surface is in bad shape—worn and scaled, warped, cupped, ridged, uneven, full of defects, or the old finish worn and scarred—the flooring must be sanded down to clean, smooth, bare wood before refinishing. It must be just cleaned off level if it is to be covered with carpet or glued tile. In either case, the first step is to determine whether sanding will be feasible.

Measure the depth of the deepest gouges or other defects and get an idea of the maximum amount of wood that must be sanded away to level the entire surface to that point. Add $1/16$ inch to that figure and check the maximum thickness of the flooring at some exposed spot. Subtract the total from the flooring thickness. If sanding will reduce the overall thickness of the finish flooring to no less than $3/8$ inch (which is a marginal minimum), the flooring can be sanded and finished. If less than $3/8$ inch—especially if the flooring is a softwood on $1/2$-inch thick subflooring and there is no underlayment—you will be treading on thin ice, so to speak. Replacement is advisable.

Preparation

To prepare for a sanding operation, remove all the moldings and trim around the perimeter of the room, including thresholds. Pry them off with whatever tools seem appropriate for the job. If you work carefully you might be able to salvage the pieces and put them back on later, but very often they will break up and have to be replaced. If you do manage to save the pieces, number each one as you go along and put a corresponding number on the wall, to make later replacement easier. If there is a small shoe molding attached to a full baseboard it is sometimes possible to remove just the shoe, which can be easily and inexpensively replaced. This eliminates potential damage to both baseboard and wall.

Remove everything from the room that you can, including heat units or registers, drapery rods, and pictures or wall hangings. Then seal off all openings with masking tape and plastic or heavy paper. Close, tape off, and seal as many doors into other rooms as possible. Bag all light fixtures and seal them off with tape, and cover all electrical outlets with a strip of tape, leaving one free for use. Go over the floor surface carefully for tacks, staples, or nails; remove all that are extraneous and countersink those that are not. Drive flooring nails down into the wood, well below the surface.

Tools and Materials

Most of the sanding is done with a big, heavy drum-type floor sander, which you can find at any tool rental shop. Edges and small areas are done with a smaller, heavy-duty belt sander, also rentable. Whatever can't be done with those two machines must be finished off with a small pad sander or by hand. There are times when a disc sander also comes in handy to take down some hard-to-reach rough spots. You will need a supply of at least two or three (depending upon conditions) grades of sandpaper for the machines, which the rental shop will supply you with, for a fee. Unused pieces can be returned for credit. The rental agent will also show you how to load the machines, and explain how they should be used; there are slight differences in operation for different models. You will also need safety goggles and a good dust mask with a supply of filters—buy your own. Under no circumstances should you sand a floor without a dust mask. Wear old clothes, buttoned up tightly, and a cap. Except for smooth, unfinished (new or old), or single-coated flooring, you will start with 20-grit paper in the floor sander.

Procedure

To begin, open whatever windows and outside doors you can, and begin in one corner of the room. Loop the sander cord over your shoulder to keep it out of the way. Lift the drum off the floor by pushing down on the sander handle, start the machine, and let the drum come up to full speed. Lower the drum gently and slowly, and move straight ahead with the machine, in the direction of the grain of wood. There is one exception to this: if the floor boards are badly cupped and ridged, the first pass should be made diagonally to the grain to cut the ripples off. Move slowly but very steadily. If you stop or change pace, you will leave a chewed-up indentation in the wood that is difficult to erase.

When you have gone as far as you can, lift the drum up quickly, turn the machine around, and go back over the same path. At the end of the run, move the machine over and make the second run, just slightly overlapping the first. Continue until you have done as much of the floor as possible with the big machine. Change the sanding belts often, just as soon as they seem not to be cutting as quickly as they should. Be sure to empty the dust bag frequently, and store the wood flour outdoors in plastic trash bags for later disposal. *Note*: The wood flour is very flammable, and can be extremely irritating to your respiratory system—to the point of incapacitation.

The next step is to load the belt sander with 20-grit paper and tackle all the edges and tight spots, taking the surface down to the level left by the floor sander. Move the machine slowly and evenly, and keep it flat to the surface. It will remove a lot of wood in a hurry and requires close control. Whatever you can't get with this sander must be done with a smaller pad sander, or by hand. In this case it is helpful to first scrape away any old finish with a shave hook or a sharp paint scraper.

Once the whole floor, including the nooks and crannies, have been gone over with the 20-grit, go back to the floor sander and repeat the whole exercise with medium-fine 40-grit paper. This is the starting point for most unfinished or single-coated flooring in good condition and for new wood. Sometimes, however, a 60-grit paper is the first used instead of the 40-grit, especially if the wood is soft. This is also the stopping point if smoothing and cleaning for a new floor covering is all that is needed. Follow up with the belt and pad sanders, and do whatever hand sanding is necessary, again taking the wood down to the new level. Clean up the wood flour, and start all over again, this time with fine, 100-grit paper.

As you go through this final stage, continually check the surface by sighting along it against the light. See if there are any small ripples or rough spots that need to be feathered out. If so, do this by hand or with a pad sander. In some instances, a new wood flooring that is very smooth to start with, or that is a softwood, need only be gone over with the 100-grit paper. Note too that some workers prefer to make one last pass with an even finer grit, such as 180, but entirely with a pad sander rather than the big machine. This results in a very smooth surface that takes semitransparent stains and/or clear hard coatings very well.

At this point the entire room must be cleaned thoroughly. Vacuum all the surfaces, including the ceiling. Wear nonmarking sneakers or couple of pairs of thick athletic socks, and make sure that the vacuum cleaner wheels don't leave marks, or you'll have more sanding to do. Close the doors and windows and let the residual dust settle overnight, then vacuum the whole room out again. Close the room up and don't let anyone in until you have a chance to begin the finishing process: A fresh wood surface scars and marks altogether too easily, particularly if it happens to be softwood.

WOOD PLANK AND STRIP FLOORING

Strip flooring (Fig. 2-39) is comprised of wood strips ranging from $1^1/_8$ inches to $3^1/_4$ inches wide, and from $1^1/_4$ feet to 16 feet or more long. Thicknesses range from $^5/_{16}$ inch to $1^1/_4$ inches. Plank flooring (Fig. 2-40) is comprised of boards $3^1/_4$ inches wide or more and usually 4 feet long or more, but sometimes as short as 1 foot. Thicknesses generally run from $^1/_2$ inch to $2^1/_2$ inches.

Availability

There is a very wide variety of commercial materials available, and the grading and characteristics of all the different kinds is a complex and lengthy subject. Only a few choices might be available from stock at any given building supply house; much strip and plank flooring is obtained by special order, except for the most common kinds and grades. Check with your dealer for specifics. Also, some of these floorings are factory prefinished, but most are not; they must be sanded and coated after installation.

Fig. 2-39. *Hardwood strip flooring is popular as a finish flooring; several different woods and finishes are available. (Courtesy Oak Flooring Institute.)*

Fig. 2-40. *Plank flooring like this bevel-edged oak is commercially available or can be made up from many different species of hardwoods and softwoods. (Courtesy Tarkett, Hardwood Division—North America.)*

Both strip and plank floorings are available in softwoods and hardwoods. The sizing and grading systems differ. White and red oak are the most common hardwoods, although beech, yellow birch, sweet birch, pecan, and sugar maple are also commercially available. The most popular softwoods include yellow pine, Douglas fir, western hemlock, eastern white pine, ponderosa pine, redwood, western larch, baldcypress, and a few others. However, practically any species of reasonable density and workability can be made into flooring by anyone with patience and a well-equipped workshop. Specialty wood dealers can supply walnut, black cherry, elm, mahogany, white ash, butternut, teak, poplar, black willow, and numerous other woods, including exotics. Strip flooring particularly lends itself to the inclusion of feature strips, borders, and designs by insetting woods of different species and contrasting colors.

Estimating Flooring Required

To estimate your needs for strip flooring, determine the area in square feet to be covered, and add 5 percent for a small area, 7 percent for a large one. The material comes in bundles that vary greatly in number of pieces and number of board feet, depending upon species and grade. Order the number of bundles needed for that total, rounded up to the next highest. For plank flooring, figure the area to be covered plus 5 percent for commercially processed flooring material, 10 percent for raw lumber. Then order that total as board feet or square feet.

Procedure

Manufactured strip flooring is typically end- and edge-matched, meaning there are joints at edges and ends (Fig. 2-41). The flooring is blind-nailed in place by toenailing through the tongues. To lay this flooring, first determine the desired lie of the pieces—which should be at right angles to the floor joists, if possible. A diagonal lie is even better structurally, but might not be visually acceptable.

Use a long strip for a first piece, and set it in a room corner parallel with the wall, leaving a $1/2$-inch expansion gap. Place the grooved edge toward the wall and the tongue outward. Position the piece carefully, because the alignment of all the following ones will depend on it. Secure the piece by nailing down through it, close to the grooved edge, with a nail at each end. Space the nails about 10 to 12 inches apart elsewhere. The nail heads will be hidden by the baseboard and shoe molding that will be put on later, or the holes can be filled. Then toenail down through the tongue corner using the same spacing (Fig. 2-42). Use either screw-type flooring nails or square-edged cut flooring nails, 7d for $3/4$-inch thick material, 5d for $1/2$-inch, 4d for $3/8$-inch. Finish out the first row of strips in the same way, keeping them perfectly straight and aligned.

Lay out the next few rows loosely on the floor and shuffle the pieces around so the joints are well staggered. Use longer pieces for the main part of the field and save shorts for row ends, but scatter some of the shorts in the field as well. The correct mix is a matter of judgment, tempered by the nature of the pieces in the bundles. Start the second row by fitting the grooves over the tongue of the first row. Get the joint as tight as possible; stub-

Fig. 2-41. *End-and edge-matched stock is machined to fit together for a clean, trim appearance.*

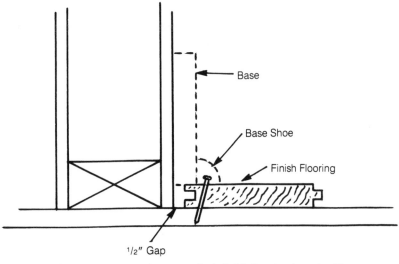

Fig. 2-42. *The first piece of strip or plank finish flooring is set in this manner.*

born pieces might have to be helped along with a hammer and a block of scrap. This time, toenail the pieces down through the tongue corner only, again about every 10 to 12 inches. You can do this by hand, then drive each nail head fully flush with a nail set, or, once far enough away from the wall, you can use a special flooring nailer rented from the flooring supplier. This tool makes the job easier and faster, and drives the strips tightly against one another. Continue the procedure across the floor, rip the last row of strips to the correct width, and set them with a ¹/₂-inch expansion gap. Depending upon width, the last two or three rows might have to be face-nailed and the holes plugged afterward.

Laying tongue-and-groove and/or end-matched plain plank flooring is done in the same way as strip flooring. However, there are a few kinds—sometimes called *ranch plank* or *pegged flooring*—that are fastened down with screws driven through predrilled, countersunk holes. These holes are filled afterward with wood plugs. Another type has smaller predrilled holes, through which wrought-head cut flooring nails are driven. The broad heads are left exposed on the flooring surface. These floorings (Fig. 2-43) may or may not be tongue and groove, and are available both unfinished and prefinished.

Raw boards of many species can be fashioned into plank flooring simply by planing at least one surface, and preferably both edges. The edges can be left square, or jointed in a tongue-and-groove configuration. Random widths and lengths are suggested. Widths from 5 to 8 inches or so are less prone to cupping and warping, but top-fastened widths of even 14 inches or more can be successfully used, and have been for centuries. For best results, the stock should be kiln-dried to 8 percent moisture or less, well aged, and stored loosely for at least a month in the same environment (same room, preferably) where they will be laid.

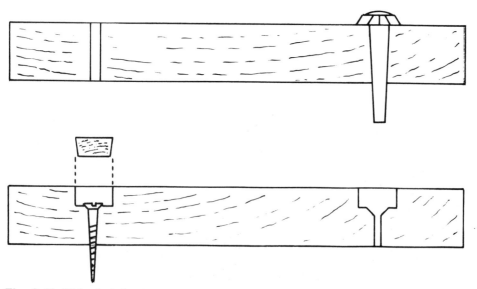

Fig. 2-43. *Wide plank flooring is sometimes fastened with plug-covered screws or with wrought-head flooring nails, which are both highly visible and add to the appearance.*

Lay tongue-and-groove planks just as for strip flooring. Lay square-edged planks tightly together in very high humidity conditions, but just fairly snug or even with a slight crack in very low humidity. Both types should be laid heart side up (Fig. 2-44) to minimize warping and ridging. Plank ends should be squared and abutted tightly. Keep a ¹/₂-inch expansion gap all around the perimeter, and lay the planks at right angles to the floor joists, if possible.

You can secure square-edged planks with the screws-and-plugs method, or by face-nailing with decorative wrought-head cut nails. You can also use standard finish or sinker nails set beneath the surface and fill the holes later. Tongue-and-groove boards can be similarly fastened, or blind-nailed as is strip flooring. Note that it might be necessary to drill pilot holes for nails when nailing near plank ends, to avoid splitting.

APPLIED COATINGS ON WOOD FLOORING

A variety of coatings may be applied to wood finish flooring, and a number of factors have to be taken into account as you make your selection. Chief among those factors are: ease of application, drying time, appearance, overall toughness and durability, relative surface slipperiness, resistance to sunlight, staining, dirt and grime embedment, scuffing, water, and alcohol. For do-it-yourselfers other factors such as toxicity, flammability, and potential for allergic reaction to the solvents or vehicles in the coatings, will be of concern. In refinishing jobs where a new coating will be applied over an old one, the compatibility of the two materials must be considered.

Floor finishes can be placed in two groups. The first is hard coatings, which bond with but form a film of measureable thickness on the surface of the wood. The film is

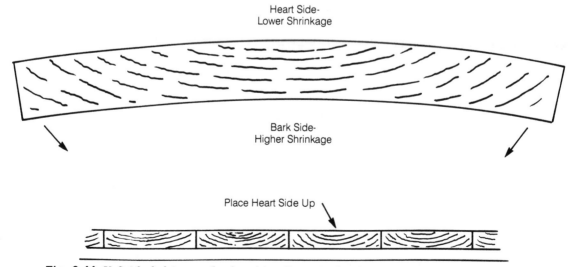

Fig. 2-44. *Unfinished plainsawn (hardwoods) or flat-grained (softwoods) planks should be positioned as shown to minimize cupping difficulties.*

tough and durable to varying degrees, depending upon the specific product. The finish protects the wood from wear and abrasion until it actually wears through, but because it is a discrete layer it is susceptible to scratching and marring. These finishes range in sheen from satin or semi-gloss to very reflective high gloss.

The second group is made up of penetrating coatings that sink into the pores and fibers of the wood. In most cases they harden there, creating a thin top layer of wood of greater density and wearability than the wood alone. These finishes do not show wear and habitual traffic paths as readily as hard coatings. However, they do not protect the wood itself from wear and abrasion because the wood and the finish wear off together. Sheen ranges from flat or dull to a medium satin.

Floor finishes can also be selected according to their opacity. Some clear or natural finishes are as clear as water, while others have a slight brownish or yellowish-brown tinge. This slight coloration does not become apparent until numerous successive coatings have been applied. The clear finishes bring out the grain and figure of the wood without changing its hue, but might obscure some of the more delicate detail. Semitransparent finishes also enhance wood grain and figure, but embed a selected coloration in the wood. This, in turn, changes the natural aspect of the wood and also partially hides its natural appearance. Repeated applications of such finishes will obscure the wood more and more until the coating finally becomes almost opaque and only the most obvious features of the wood show through. Opaque finishes completely hide the wood.

The various options for finish floor coatings are discussed on the following pages. To apply any of them, first do the necessary preparation work as needed for a particular finish. General instructions for applying the finishes are also outlined here so you will know generally what is involved, but always follow the coating manufacturer's current detailed directions and recommendations.

Be sure to also follow appropriate health and safety precautions. Floor finishes, like most others, are compounded from various substances that are to one degree or another toxic, flammable, or irritating to the eyes, nose, skin, and other body tissues. Immediately spread all soaked wiping and cleanup rags outdoors to dry before disposing of them, to avoid spontaneous combustion. Allow ample ventilation; schedule your projects for warm weather so you can open windows and doors. If you must work during cold weather, turn up the heat and ventilate anyway. Wear a suitable respirator, spend as little time as possible in the area, and—if you can—vacate the house for two to three days after a large finishing project to allow outgassed fumes and vapors to dissipate. Pregnant women should avoid exposure to finishes and their vapors altogether. If you suffer allergies, wear protective clothing, including gloves, and take any other precautions necessary for your condition. Don't underestimate potential harmful effects of these finishes.

Linseed Oil

Boiled linseed oil, applied hot and thin, is a traditional floor finish that is most effective on softwoods. The process involves spreading several coats of oil, separated by enough drying time to allow tackiness to disappear. Each coat should be hand rubbed vigorously with rags. Repeat applications until the wood surface is saturated. The finish càn be left alone and re-oiled about once a year, or it can be waxed periodically with a hard paste wax. Over time the oil or wax does build up, and the surface must be stripped of the buildup and freshly oiled or waxed. Also, over time, the floor becomes very dark and in some cases almost black, depending upon the species of wood and the number of oil applications.

This finish is easy to apply, although some effort is needed in the hand rubbing. It is slow drying but eventually becomes quite hard. It is tough and durable, very easy to repair any time at wear points, and easy to keep clean by dry or damp mopping. Once cured, it does not show and is relatively unaffected by dirt or grime. It does not show wear readily, or scuff, mar, or scratch easily. It is not slippery, and it is sun-, stain-, and water-resistant. It is also inexpensive, nontoxic, and has low flammability. The sheen is flat or dull unless waxed, in which case it is a low satin. Linseed oil should be applied only to raw wood.

Wax

Plain wax is sometimes used as a floor finish, but is not particularly effective. A hard paste wax rather than a liquid or a water-emulsion wax is the only kind to consider, preferably one containing carnauba and/or beeswax. The finish is clear unless a specific stain wax is used, in which case it imparts a color tone that varies with the color wax and the wood upon which it is applied. Any wax affords only minimal protection to the wood. After repeated applications an excessive buildup occurs, especially on porous, open-grained woods, and the floor must be stripped and the process started over again. Application and polishing is not difficult, although it does require some physical effort; a power buffer helps in that respect. Paste wax dries almost immediately and hardens with buffing; sheen is a low to medium satin.

Wax has little durability and is suitable only for low-traffic areas. It can be slippery and is not resistant to many stains or to prolonged exposure to water; almost any solvent will attack it. Paste wax is not considered a flammable substance (although it can be ignited), but many waxes release fumes or odors during application that might be mildly toxic and that some people find disagreeable. Waxes can be applied over any other floor finish for protection or to change the sheen.

Stains

Stains can be used to treat any wood floor. Note that this does not in itself constitute a finish; the stain must be topcoated with another floor finish to protect it. Stains should be applied only to clean, raw wood—either new or completely free of the residue from any previous coatings.

Oil-based, color-pigmented, semitransparent (or semiopaque) stains are most often used. Opaque or solid-body stains could be used but seldom are. These stains are advantageous in that they are readily available premixed in a wide range of colors and tones, they are easy to apply, they do not raise the grain of the wood, and they can be easily mixed with certain wood fillers for unobtrusive patching. There are some disadvantages, too.

These stains do not penetrate deeply, even in softwood, and so wear off easily. When this happens they are difficult to match again. They tend to leave a mottled or freckled appearance on some large-pore, open-grained woods, they are relatively expensive, slow to dry, flammable, and emit fumes and vapors of a somewhat toxic nature during application and drying.

Penetrating oil stains are similar, but the oil vehicle carries soluble color dyes instead of pigments. They are easy to apply and are available in a small range of ready-mixed earth tone colors. These stains carry most of the same disadvantages as pigment oil stains. They do penetrate deeper into the wood, however, and they do not have a tendency to streak like the pigment oil stains.

Alcohol or spirit stains are made by dissolving color dyes in alcohol. A modest range of colors is available. They are easy to apply and dry very rapidly, but do not penetrate much into the wood. They are flammable, and release fumes during application. They also have a tendency to bleed through topcoats.

Water stains are acid and natural dyes that come in powder form and are mixed with clear, clean water. Their greatest disadvantage is that the water can swell the wood fibers and raise the grain, in which case the surface must be gently sanded down again. However, if applied carefully and correctly in a damp application—rather than slopping it on liberally—this problem can be overcome. For many users that disadvantage is far outweighed by the advantages: very easy application without any special equipment, great absorption and penetration (which results in better delineation and contrast of figure and grain than any other kind of stain), ease of custom mixing colors and tones, low cost, color-fastness, rapid drying, ease of darkening by applying a second or third coat, non-flammability, nontoxicity, no odor, no clean-up hassle.

To apply pigment oil stains, first vacuum and then wipe the entire floor surface with a tack rag to pick up any loose dust. If there are defects to be filled, you can do this beforehand with a plain filler that will accept the stain and match satisfactorily. Pretest first on a piece of scrap wood. Alternatively, you can mix the stain with the filler so that it matches the completed finish, then fill the defects after the stain has been applied. Use a pad applicator or a pad made up of soft, lint-free cloth. Wipe the stain on in smooth, even strokes, parellel with the wood grain. Apply over just a few square feet and immediately wipe the excess off with clean rags. Blend in the starting and stopping lines between sections before they have a chance to set; overlaps are likely to show as a darker tone, and you must make an effort to avoid this. Also, beware of streaking and runs. If there are cracks between the planks or strips, use these as handy demarcation lines between application areas. Allow several days for drying before applying a topcoat, and two days or more before applying a second coat of stain if the first was not dark enough.

To apply a penetrating oil stain, follow the same procedure as for pigment oil stains. A bit less wiping will be required, lap marks are easier to rub out, and streaks are unlikely to show up.

To apply a spirit stain, follow the same general procedure. However, the stain will dry rapidly, so work in smaller areas. Wiping off the excess is unnecessary because there won't be any. Use smooth, single passes without over-applying the liquid. A second coat, if needed, can be applied almost immediately and a topcoat can be safely applied within a few hours.

Shellac

Shellac is another traditional floor finish that once was the most favored of all, especially in formal rooms. It is made from lac—a resinous substance secreted by the Asian lac insect—mixed with denatured alcohol. Shellac comes in three grades: orange, white (bleached), and wax-free (dewaxed). White shellac is the most commonly available and the one to use for most floor finishing, especially on light-colored woods. Orange is sometimes applied over dark or dark-stained woods to deepen the tone. These two can be purchased either as dry flakes, which are dissolved in alcohol as needed, or as a ready-mixed finish. The first has an indefinite shelf life, the second a relatively short one.

Shellac is referred to in terms of the pound cut. Thus, a 5-pound cut shellac consists of 5 pounds of dry shellac dissolved in 1 gallon of alcohol. A 3-pound cut is suitable for floor finishing. If this is not readily available, it can be made from 5-pound cut by mixing 2 parts shellac with 1 part alcohol, or from 4-pound cut by mixing 4 parts shellac with 1 part alcohol.

A shellac finish is easy to apply and dries rapidly and thus quite dust-free. Usually it can be touched in 15 minutes and recoated in 2 hours. Though the fresh mix appears milky, the coatings are clear and highly lustrous. It is moderately durable and tough—but should be used in light-traffic areas—and has a low ultraviolet resistance. Scratch, scuff, and grime resistance is moderate, and the finish can be slippery—very much so if highly waxed. Water resistance is nil (water leaves white blotches) and alcohol resistance, as you might expect, is zero. Stain resistance is good. The sheen can be anywhere from satin to very high gloss, depending upon the number of coats, whether finish-rubbed, and whether or not it is waxed. It has a powerful odor and emits strong fumes during application, which dissipate fairly quickly. Shellac can be applied over other coatings, and is sometimes used as a seal coat between two different finishes. It is also easy to repair and refinish.

Shellac is moderately easy to apply; it goes on easily but you have to work rapidly. Because it is thin and flows quickly, you have to be careful of pools and trails and strive for an even coating. A 2- to 4-inch brush with fine, soft bristles works well, as does a pad applicator. Patch holes and defects first with a filler that matches the wood, sand smooth, and vacuum the surface. Apply the first coat full strength, allow at least 2 hours for drying, then sand the surface with fine sandpaper—either by hand or with a small pad sander. Vacuum, clean the surface with a tack rag, and apply a second coat. A third coat should be applied on open-grain wood, after a minimum 3-hour drying period. Additional coats can be applied as desired for high gloss. The final coat may be sanded with very fine sandpaper and then waxed for a high-satin finish.

Varnish

Conventional varnish was the replacement finish of choice for shellac years ago, and it is still widely used. These varnishes are oil based and typically compounded of tung, linseed, and soya oils, and alkyd resin. They are clear and nonyellowing. Spar varnish contains phenolic resin; it is tough and more durable, but tends to yellow and/or darken with age. It is primarily intended for exterior and marine applications. Stain varnish, available in several tones, is conventional varnish with color pigments added. Except for the coloration, the characteristics and application details are the same as for clear varnishes.

Varnish is moderately easy to apply and refinish. It bonds well to itself and to most other finishes, and builds up a fairly thick, hard surface skin. Most varnishes are slow drying. Although considered clear, varnish imparts a slight yellowish tinge to the wood, while enhancing the grain and figure. Conventional varnishes have the highest overall toughness and the highest resistance to scuffing of all floor finishes, very high ultraviolet resistance, and very high resistance to dirt and grime embedment. They are virtually impervious to water, alcohol, and stains, but highly acidic liquids left in place for a time can leave slight stains or blemishes in the luster. The sheen ranges from satin to high gloss. Varnishes are flammable and do emit toxic fumes during application and curing.

Despite claims for some of the newer kinds of finishes, conventional varnish remains the most versatile, long-lived, durable, and generally useful floor finish. For best results, however, a medium-oil varnish (manufactured in a ratio of 20 to 30 gallons of oil to 100 pounds of resin) is the best choice. Short-oil varnishes are too brittle and long-oil varnishes are best for exterior purposes.

When applying a varnish finish, first fill holes and defects with a wood filler that matches the wood. Vacuum the prepared surface and clean it with a tack rag. Use a brush 2 to 4 inches wide, set with natural, fine, flexible bristles of medium length. Do not use a brush with synthetic bristles. A professional quality varnish brush is comparatively very expensive but will do the best job. A creditable finish can also be obtained by using the proper type of roller, which makes the job much easier if the area to be covered is large.

For the first coat, cut the varnish by mixing 1 part mineral spirits or turpentine with 8 parts varnish. Stir gently to avoid excessive air bubbles or foaming. Dip the brush into the varnish to only about one-third of its length. Flow the liquid onto the surface from the brush tip. A thin, even coating is most desirable; brush first across the grain of the wood, then level off with long strokes along the grain. Rework the fresh coating as little as possible, and brush from a dry area into a wet one. When using a roller, load it only lightly with varnish and coat small areas at a time with light pressure, so the liquid does not run or pile up in cracks between the planks or strips.

Allow at least 24 hours for the first coat to dry, then sand gently with fine sandpaper by hand or with a small pad sander using light pressure. If the paper clogs or gums rapidly, allow more drying time. After vacuuming and cleaning the surface with a tack rag, apply the second coat full strength in the same way as the first. A third coat is often advisable. Allow 24 hours drying time, sand and use the tack rag again, and apply the coating full strength. The final coat can be waxed as soon as it is thoroughly dry, or left until the surface begins to dull a bit from wear and then cleaned and waxed.

Lacquer

Lacquer is a finish widely used on furniture, and at one time it was also a fairly popular floor finish. After a decline, it is once again being applied as the finish of choice over wood flooring that has received special treatments like fuming, pickling, or bleaching. Lacquer does not change the appearance of the treatment as other finishes tend to. It is formulated from nitrocellulose and solvents such as acetone or ketone, along with various drying oils, resins, and plasticizers. Clear lacquer is generally chosen for floor finishing, but a tremendous array of tones and colors is available. In spray application—difficult but not impossible on a residential floor—many colorful effects can be generated by spraying and misting pigmented lacquers, then overlaying them with multiple clear coats, just as with automobile bodies.

Clear lacquer is the clearest of all floor finishes, and does not change the appearance of the wood at all. Multiple coats can give the appearance of depth, as though looking down through glass. It is moderately easy to apply, but because it dries very rapidly, you have to move quickly and with assurance in order to prevent stroke, bristle, and lap marks. It is easy to repair and refinish. A lacquer finish has moderate overall toughness but only modest resistance to scuffing and grime embedment. It is best applied in low-traffic areas such as bedrooms. Its resistance to ultraviolet is practically nil, but resistance to most stains, water, and alcohol is high—at least equal to varnish. It is very resistant to ordinary acidic substances as well. Lacquer's natural sheen after several coatings is a very high gloss, but this can be downgraded to satin or matte by applying wax or by hand rubbing with various fine abrasives like pumice or rottenstone. Lacquer is highly flammable and emits toxic fumes during application and drying, and must be used with care. It cannot be applied over other coatings.

Prior to applying a lacquer floor finish, plug all holes and defects with a matching color wood putty and sand smooth. Vacuum the surface and clean it with a tack rag. Unless you will spray the finish on, select a brushing grade lacquer. A special primer might be required; this depends upon the kind of wood being coated and the specific lacquer product being used. If so, apply this first according to instructions. At least three coats of lacquer will be needed, and you might want to apply five or six more for a "deeper" appearance and greater durability. You can apply it with a good varnish brush or a short-nap mohair roller.

Apply the first coat full strength, working rapidly and parallel with the grain of wood. Allow about 1 hour for drying, then hand sand lightly, vacuum, and dust with a tack rag. Apply subsequent coats the same way, sanding lightly between each one, but allow 2 to 3 hours each time for drying. The final coat should be cut in the ratio of 1 part lacquer thinner to 3 parts lacquer. This mix has low viscosity and flows like water, but it should be applied rapidly and in as thin and even a coating as you can manage. Allow a full 24 hours of curing time before allowing traffic on the floor or proceeding with waxing or hand rubbing operations.

Penetrating Oil

Penetrating oil finishes, of which there are several, are typically compounded of linseed and tung oils with phenolic or other resins, metallic driers, and possibly other additives. They are thin and penetrate the wood surface to harden there. They are available in clear or color-pigmented types, somewhat like penetrating oil stains.

These finishes are very easy to apply, repair, or recoat, leaving the wood with a natural (or stain-enhanced) appearance and no hard surface coating. They are fast drying and rank moderately high in overall toughness and durability. The coating is not slippery and has a high resistance to water. Resistance to ultraviolet is good. Resistance to scuffing, grime embedment, and ordinary acidic liquids it is low to medium. It does resist alcohol and ordinary household stains quite well. Wear, scrapes, scuffs and scratches do not readily show up. The sheen is flat to low satin. Penetrating oils have a variable degree of flammability, toxicity, and odor during and for a short time after application. They cannot be used over other finishes except for linseed oil, and in that case the penetrating properties would be only partially effective, if at all.

To apply a penetrating oil finish, first plug holes and defects with a type of wood filler that is penetrable, such as a stain-accepting filler. Sand smooth and vacuum the floor. Apply the oil with a pad applicator, wad of cloth, wide paintbrush, or a short-nap roller. Do only small sections at a time that are fully within reach. Allow the oil to soak in for 20 to 30 minutes, then wipe off the excess with rags. The wood will absorb the oil unevenly because of varying density, so a second coat is almost always needed. This is applied in the same way, and can be done as soon as the first coating has been completed or following the time lapse as indicated in the application instructions. Keep traffic off the floor for at least 24 hours. Wax can be applied if desired.

Polyurethane Varnish

Polyurethane varnishes, most of which are of the oil-modified type, have become popular as floor finishes over the past few years, partly supplanting conventional varnishes. They are typically compounded of tung oil, urethane, soya alkyd, and phenolic resin. They are thin, clear, and leave a hard surface coating.

Polyurethane varnish is easy to apply, but following the application instructions for the particular brand is important. Later recoating and repair is difficult because the varnish does not adhere well to either itself or any other finish; thorough sanding is required. It is moderately slow drying and alters the appearance of the wood somewhat in that it tends to obscure finer grain details, even though it is a clear finish. It is durable and second only to conventional varnishes in overall toughness. Resistance to ultraviolet is very good, and it has good resistance to grime embedment and scuffing. These varnishes are impervious to water and alcohol, but somewhat less effective than conventional varnishes in resisting household stains, common acidic liquids, and chemicals. Overall, however, they could be ranked as highly resistant to the whole range of potential minor household disasters. Thus, they are excellent for high-traffic, heavy-use areas. The sheen is either semi-gloss or satin, or high gloss, and the surface can be quite slippery, especially if waxed. These varnishes are flammable and toxic.

When applying a polyurethane varnish, first repair holes and defects with any good wood filler and sand smooth. Vacuum the entire surface, and clean it with a tack rag. You can apply the finish with a varnish brush 2 to 4 inches wide, but a less expensive synthetic-bristle brush will do almost as good a job. You can also use a short-nap roller for fast application. Brush or roll with the grain of the wood, leaving a thin, even coating. The liquid flows and levels well, so overbrushing, lap marks, and streaking are not much of a problem; do watch out for runs or pools, though. Apply all coats at full strength unless otherwise instructed.

Allow the first coat to dry for the specified time—usually about 8 hours—before hand sanding with fine sandpaper. Vacuum the surface and clean it with a tack rag before applying a second coat. A third coat can be similarly applied if you wish. If a satin sheen is desired, use the high-gloss variety first and apply the satin gloss as the final coat. Let the coatings cure for 24 hours or more before allowing traffic on the floor. Waxing is optional, and can be used to change the sheen of the surface and give protection against dulling.

Paint

For many decades paint has been the universal finish for nearly everything. Although it has long since ceased to be a popular interior floor finish, it remains an alternative. Restoration of early period houses, for example, demands this finish, especially if the floors are to be stenciled in original fashion.

From a practical standpoint there is only one paint type available that will give decent service, and that is loosely termed in the trade "porch and deck" or "floor and deck" enamel. There are several formulations and brands. Even this paint does not stand up well under heavy traffic, and should be applied where service conditions are not severe. Only a few muted color tones are available. Epoxy paints are a possibility, but these should be professionally applied. Color lacquer might also be used; it should prove to be at least as durable as paint and affords a limitless color selection. The characteristics and application are the same as for clear lacquer.

For restoration purposes, porch and deck enamels might not do the job because of the limited color selection. However, old-type casein or "milk" paints in authentic period colors can be obtained from restoration specialty supply houses. Also, readily available oil-based enamels can be custom mixed for color. These paints should be used for border or low-traffic areas such as bedrooms.

When applying paint coatings, first patch holes and defects with a hard wood filler, sand smooth, vacuum the floor, and clean with a tack rag. Apply a coat of sealer or primer as recommended by the paint manufacturer. After the specified drying time, sand the surface with fine sandpaper by hand or with a small pad sander. Vacuum and dust with a tack rag again, and apply a full-strength coat of paint. Allow 24 hours for drying, more in humid conditions, and sand again with very fine sandpaper to remove dust nibs and create "tooth" for the final coat. Vacuum and dust with a tack rag once more and apply the last coat. Let the finish cure for at least 48 hours before allowing any traffic on it—if there is any softness the finish will mar immediately. If possible, avoid adding a third coat.

Other Coatings

There are a few other floor finishes of which you should be aware, even though all but one are specialized products not recommended for do-it-yourself application.

There are two relatively new polyurethane finishes. One is water-based and thus avoids the flammability and toxicity problems. The other is a solvent-based compound that hardens with reaction to moisture in the air. Both have the same general attributes of oil-modified polyurethanes, but are difficult to apply.

Swedish finishes—not to be confused with some of the penetrating oil finishes sometimes called Swedish oils—are formulated typically from polyvinyl resin, alkyd urea, and formaldehyde. They combine a natural finish appearance in matte to low satin sheen with excellent durability and easy maintenance. They are very toxic during application and curing, however, and require special preparation and application techniques.

Epoxy seems to have become a modern magic word that implies practically infinite life and durability. And in fact epoxy finishes, compounded of epoxy esters, are very durable and resistant to almost anything. But they are intended for heavy-duty commercial/industrial applications, are expensive, should be applied by professionals, and are not really suitable for residential applications.

Water-based acrylic finishes, made up of acrylic resins, are new to the marketplace. To date they have limited local availability, but this will probably change. This finish is nonflammable and about as nontoxic as a finish can be, and it is easy to apply. Although it has not been in use long, and therefore has no extensive history, the general effectiveness of the material as a floor finish appears to be just a bit less than oil modified polyurethanes.

Fillers

Some woods that serve as finish flooring are open-grained and have numerous large, open pores exposed on the surface; oaks are a good example. Applied coatings only partly fill these pores—even after numerous coats—so the finished surface is not smooth but speckled with tiny pits. Many people consider these a natural aspect of the wood, but others object to the appearance. If this is a problem for you, the solution is to apply a filler (also called a grain filler) to the flooring immediately before applying the first finish coat—or in some cases, the sealer or primer. The process involves wiping the filler on and then wiping it off, leaving the pores filled. There are several different kinds of fillers that might be used, depending upon the kind of wood and the type of finish involved. Application details also vary. Your paint supplies dealer can recommend the right product for your needs.

CARPETING

Carpeting is a complex subject. There is a tremendous array of patterns and colors available, and in addition there is the matter of carpeting type and construction, as well as grade and quality.

Selection

The principal types of carpeting are: woven, which includes the loomed, velvet, Wilton, and Axminster processes; tufted; and knitted. Backings include yarns, sponge or foam rubber, jute fabric, and combinations of jute, cotton, and kraftcord, as well as various latex impregnations. The carpet material itself, known as the pile fibers, may be wool, modacrylic, polypropylene, nylon, or wool. The standards of quality revolve around such factors as pile weight, pile thickness, yarn size and number of plies, number of tufts of yarn per square inch, and the spacing of tufts.

Carpeting selection, then, is not just a matter of finding a carpet that has a nice color and pattern and thickness, but rather a subject for study on your own and something to discuss with knowledgeable dealers. When selecting a carpet, give strong consideration to the following:

- Durability—resistance to abrasion, alkalis and acids, insects and fungus, and burns
- Ease of maintenance—resistance to soils and staining, wet cleanability
- Retention of appearance—resistance to compression and crushing, texture retention

Select a type of carpet that will best serve your specific needs and the conditions under which it will be used, and purchase the highest level of quality and general "wearability" that you can reasonably afford.

Installation

A few kinds of carpeting can be installed simply by trimming the material and laying it in place. A few others can be installed by trimming to the room borders, then sealing the carpeting down with mastic or adhesive. These carpetings have built-in padding—generally of foam or sponge rubber or a heavy latex coating—and are laid directly on the substrate. Most carpeting is designed to be stretched into place under tension, and should always be laid over carpet padding. The padding materials are felted hair, cellular rubber, and rubberized fibers, sometimes joined with jute or burlap. There are numerous weights (in ounces per square yard) and thicknesses. Use your dealer's recommendations as to which particular product might be best for your installation.

To install lay-in carpeting—typically used for small bathrooms—first select a large enough piece to cover the entire floor area plus a few inches up the walls. The floor should be clean. No moldings need be removed. Abut one side or end, or a part thereof, snugly against a wall and weight the carpet so it can't shift around. Many of these carpetings can be trimmed with scissors; otherwise, use a carpet knife fitted with disposable razor-type blades. Make relief cuts at all corners, starting from the top edge and cutting toward the floor. Be careful not to cut out into the field of the carpet.

To trim around objects such as toilet or pedestal lavatory bases, make up paper patterns that you can transfer to the carpeting and cut around. Or, trim along the walls and around objects with repeated cuts, coming closer to the desired final contours gradually with each cut, so that you don't inadvertently cut out into the field. Remember: caution pays off. Meanwhile, make sure that the carpeting remains flat and exactly in place where you have already trimmed. When the whole piece is smooth and snug at all wall or fixture meeting points, you're done.

Installing glued down carpeting is done in much the same way. After preparing the subfloor, roll the carpeting out and abut one edge or end to a wall, or curve it up the wall a bit, if necessary. Make sure the entire piece is properly positioned in the room so that you have plenty of material to trim off all around and won't run short anywhere. With a carpet knife, trim one end or edge to within about a foot of the corners (Fig. 2-45). Fold the carpeting back in half, without shifting the position of the other half. Spread the recommended adhesive over the free half of substrate, to within about a foot of the wall (Fig. 2-46). Loop the carpeting back over the adhesive in a rolling motion, smoothing it as you go. Make sure there are no bulges or ripples. After the carpeting is fully flat, go over it

Fig. 2-45. *The first step in laying the carpeting is to trim most of one edge flush with one side wall line.*

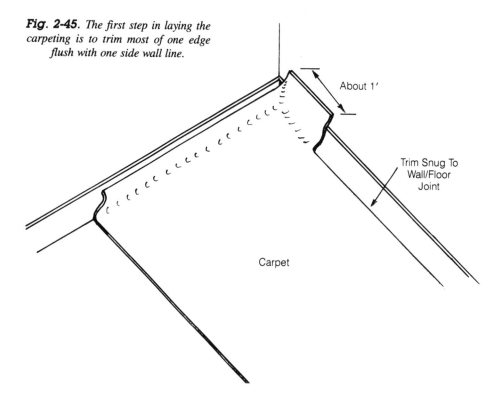

About 1'

Trim Snug To Wall/Floor Joint

Carpet

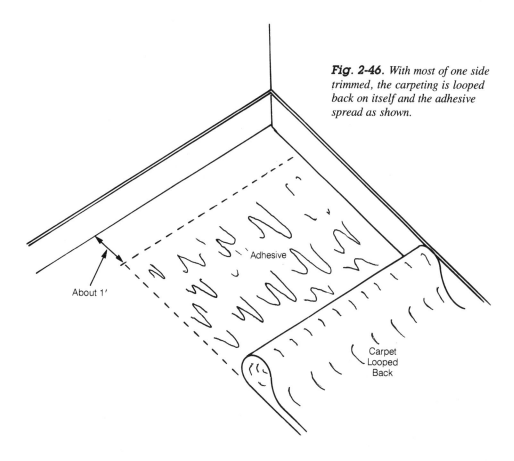

Fig. 2-46. *With most of one side trimmed, the carpeting is looped back on itself and the adhesive spread as shown.*

About 1'

Adhesive

Carpet
Looped
Back

with a pastry rolling pin and apply a moderate amount of pressure partly downward and partly forward, working along the trimmed edge or end and mostly toward the untrimmed portions.

Now you can either trim the untrimmed end or edge you've been working toward, or repeat the glueing process with the other half of the carpeting and trim everything afterward. To trim, make relief cuts at all the corners by slashing downward at the corners from the edge toward the field, or by making successively larger V cuts until the point of the V reaches the floor. Trim around odd-shaped projections by making repeated cuts, working down from the edge, until you achieve a snug contour. Then lay the unglued edge of the carpet back, secure it so it doesn't flop back down, and apply the adhesive to the remaining substrate (Fig. 2-47). Lay the carpeting back down, smooth it, and roll it. If there are two or more pieces of carpeting to be laid, roll the next piece out in place and carefully align the seams, matching the pattern if necessary. Then follow the same procedure, making sure that the seam edges are tightly butted. The whole process is similar to laying sheet vinyl flooring.

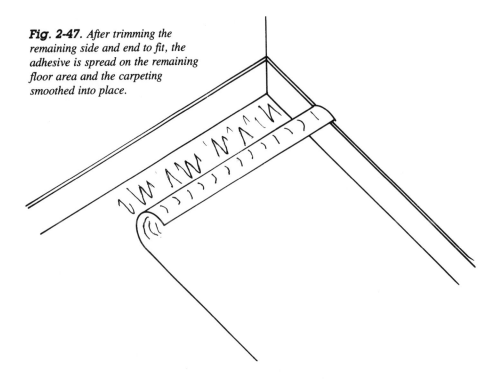

Fig. 2-47. *After trimming the remaining side and end to fit, the adhesive is spread on the remaining floor area and the carpeting smoothed into place.*

Installing Stretched-In Carpeting

When it comes to stretched-in carpeting, most folks prefer to spend the extra few dollars per yard and leave the laying to a pro. Although not technically involved, this is a physically demanding job, and experience does help. The most difficult aspect is the seaming of multiple pieces, and this is indeed best left to a professional, especially if you have selected an expensive, heavy carpeting. However, an inexperienced but physically able do-it-yourselfer can do a creditable job laying a relatively small, one-piece carpet, or laying a seamed piece that has been rough cut to size and pre-seamed by the dealer in his shop.

To begin, make up a detailed sketch of the room or area to be carpeted, with complete dimensions—including all jogs and nooks. Let the dealer calculate the carpeting size and piecing, and provide you with the proper padding, padding fasteners or adhesive, tackless strip, edgings, and laying tools. If you have never used a knee-kicker or a power carpet stretcher, ask the dealer to show you how; it's easy enough once you see it done.

Set the tackless strip first by nailing the strips to the substrate, end to end around the perimeter of the room (Fig. 2-48). Trim the pieces to length as necessary, and keep the teeth pointing toward the wall. The strips should be kept uniformly away from the baseboard a maximum of $1/4$ inch, or the thickness of the carpeting, whichever is less. Nail edging strips across door openings, teeth pointing outward. Nail strips across the outer edge even with the outer face of the door when closed. Nail other edging or transition

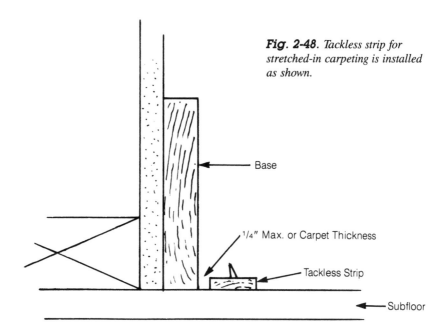

Fig. 2-48. *Tackless strip for stretched-in carpeting is installed as shown.*

Base

¹/₄" Max. or Carpet Thickness

Tackless Strip

Subfloor

strips as required wherever carpeting will abut another kind of flooring at points other than doorways.

Next, lay the padding in strips as long as possible, keeping the edge seams abutted tightly but not bunched up. Keep the padding flat and even without stretching it. Make sure it is right side up. Roll the padding out over adhesive, or staple along the edges and alongside the tackless strip about every 6 inches, plus a few more times at intermediate points. Lap the padding up the walls about 2 inches and trim it off—do not staple it. Trim the padding closely and staple or glue it around other objects such as heating register openings or heating units. After all the padding is in place, trim it carefully all along the inner edges of the tackless strips so that the two just meet.

Working carefully so as not to jerk the padding loose from the staples, bring the roll of carpeting into the room and set it alongside a wall so that the free end laps up the wall about 6 inches. Unroll the carpet fully across the floor; it should lap up the opposite wall, and the end walls, about 6 inches. Leave it in place to acclimatize for at least overnight, longer if it has been stored in a cold area. The floor temperature when the carpeting is actually stretched into place should be a minimum 60° Fahrenheit. Many professionals prefer to leave the carpet lying unstretched with the heat turned up to 80° for at least 12 hours.

To start stretching in the carpet, pick a room corner A, and with the business end of a knee-kicker, hook about 18 inches of carpet over the teeth of the tackless strip (Fig. 2-49). Then go to the opposite corner B, and with a power stretcher or knee-kicker, stretch the carpet along wall AB toward wall BC and hook about 18 inches of carpeting over the teeth from corner B along wall BC (Fig. 2-50). Depending upon the distance between the two

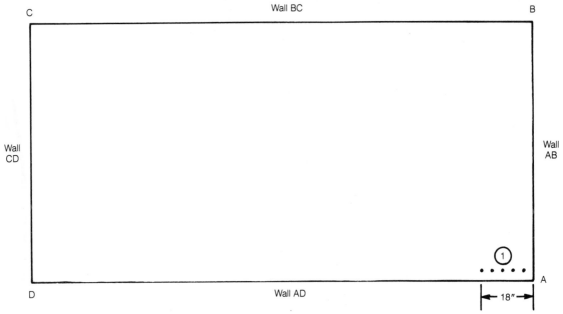

Fig. 2-49. *Step 1: Laying stretched-in carpeting.*

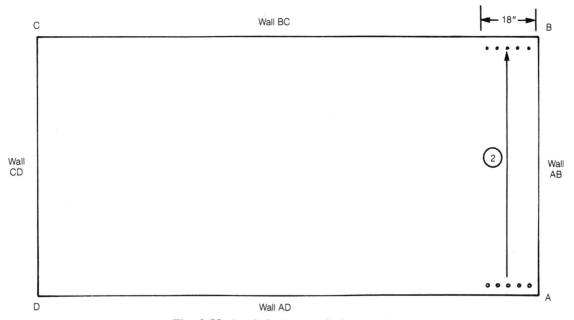

Fig. 2-50. *Step 2: Laying stretched-in carpeting.*

corners, and the stretchability of the carpeting, you might have to do this in several steps. You should have stretched the carpet about 1 inch for every 10 feet of carpeting, and if you can get 1¹/₂ inches, that's even better. You can tell how much you have gained by measuring the amount of carpeting, before stretching and after, that is curled up the wall along that 18-inch section of wall BC.

Now repeat the procedure along wall AD, stretching your way along to corner D. When you reach the corner, hook about 18 inches of carpeting to the teeth along wall CD (Fig. 2-51). You should have stretched the carpeting along this wall just about the same amount as you did along wall AB. If there is a substantial difference, you might have an uneven stretch; stretch again along the wall with the lesser amount of stretch to gain a bit more.

Go back to corner A and stretch the carpeting over the teeth along wall AB (Fig. 2-52) with a knee-kicker, working along a few inches at a time. Keep the angle of the knee-kicker to the wall at about 10 to 15 degrees. The tension on the carpet should be about the same as when you stretched from corner to corner, so that the carpeting hooks smoothly over the teeth. After completing wall AB, repeat the procedure along wall AD, working from corner A to corner D (Fig. 2-53).

Starting at corner D, stretch the carpeting into place along wall CD (Fig. 2-54), keeping the same angle of 10 to 15 degrees. Try to stretch the carpeting proportionately the same as previously, and hook it evenly over the tackless strip. Then repeat the procedure along wall BC (Fig. 2-55). Check the field over for any slight ripples or uneven spots, and take them up as necessary.

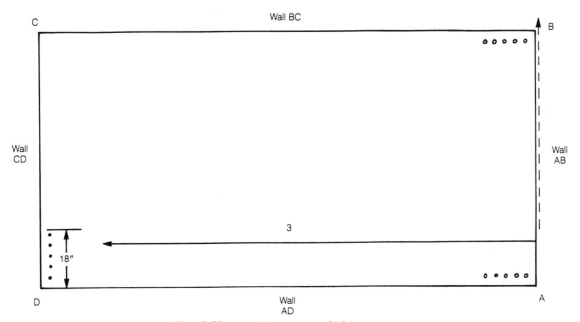

Fig. 2-51. *Step 3: Laying stretched-in carpeting.*

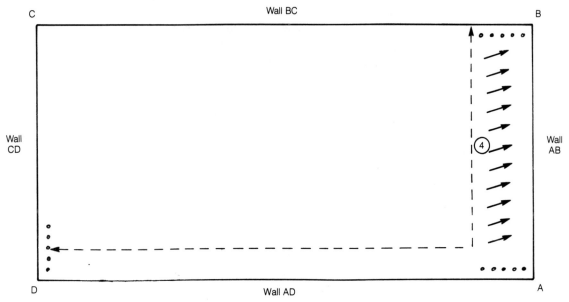

Fig. 2-52. *Step 4: Laying stretched-in carpeting.*

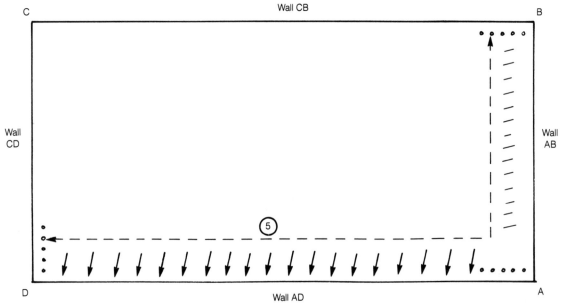

Fig. 2-53. *Step 5: Laying stretched-in carpeting.*

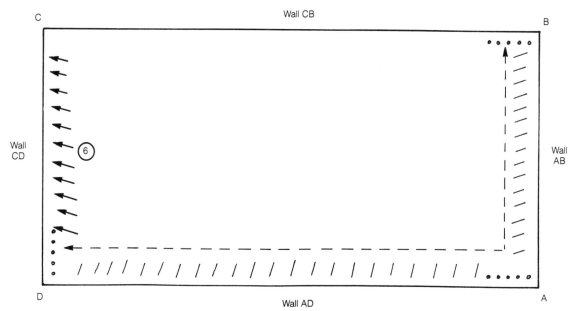

Fig. 2-54. *Step 6: Laying stretched-in carpeting.*

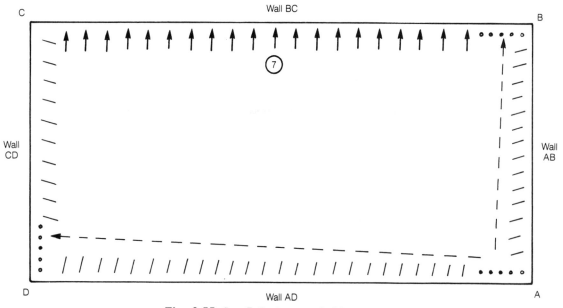

Fig. 2-55. *Step 7: Laying stretched-in carpeting.*

The last step is to trim all the way around the edge of the carpeting with a carpet razor knife. Trim so that about ³/₈ inch of carpeting protrudes beyond the tackless strip (Fig. 2-56), and so that edges fit fully into the metal (or other) edgings. As you trim, snip off ragged tufts of yarn with scissors. With a stiff putty knife, tuck the trimmed edge down between the baseboard and the tackless strip (Fig. 2-57). As you go along, tap the carpeting down firmly onto the teeth of the tackless strip with a mallet. And finally, with a block of scrap wood and a hammer, pound the rims of the metal edging strips down until they close evenly across the carpeting.

If the carpeting must extend under something too low to admit the head of the knee-kicker, such as a radiator, set the tackless strip just outside the object. Place a separate piece of padding under the object, full width and reaching back to the wall line, or at least completely out of sight. Secure it as best you can with adhesive, double-sided tape, or staples. Stretch the carpet onto the tackless strip, then slit and fit the carpet in under the object. Trim it to fit against the wall or fully back out of sight, and let it lie loose.

Stretching carpeting into an irregularly shaped room is done in the same general way. Follow a logical sequence of stretching and kicking-in, working across and around the larger areas first and into smaller ones last, or as they occur as parts of the larger sections.

Installing New Carpeting Over Old

Under certain conditions, new carpeting can also be stretched in over an old wall-to-wall carpet. The old carpet must be relatively thin and in fair or better condition, laid without a pad. Preferably it should be glued to the subfloor or an underlayment, but if laid loose or tacked down around the edges it can be stapled to the subfloor in the same way as

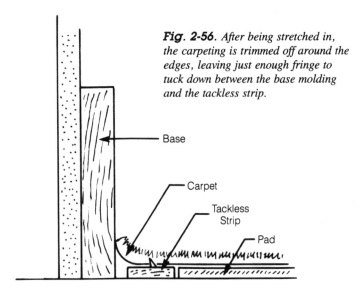

Fig. 2-56. After being stretched in, the carpeting is trimmed off around the edges, leaving just enough fringe to tuck down between the base molding and the tackless strip.

Base

Carpet

Tackless
Strip

Pad

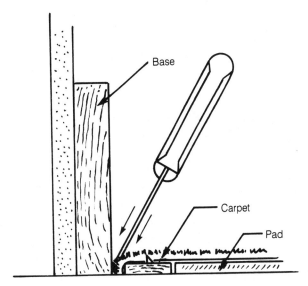

Fig. 2-57. *The trimmed carpeting edge is tucked down with a putty knife or similar tool.*

carpet padding. The carpet should be free of worn-through spots, although some degree of surface wear does not matter, nor do griminess, stains, burns, etc. Rips or separated seams should be pulled together and stapled down securely, and the carpet should be free of loose bumps and ripples.

To prepare for the new carpeting, remove any metal or other carpet edging strips by prying them up with a claw hammer or a small pry bar. Some door stop moldings might have to be removed in order to get the edgings up. Base moldings can be left in place, but base shoe moldings should be removed. Cut away a strip of the old carpet about 2 inches wide all the way around the perimeter, leaving a subfloor or underlayment border exposed next to the base molding and across doorways and transition areas. This is best done with a razor carpet knife. Buy plenty of blades and replace them frequently, because they dull quickly. If there are bumpy accumulations of old adhesive present on the subfloor border, scrape them away with a paint scraper or a shave hook until the surface is fairly flat. Finally, clean up all the debris and vacuum the old carpet with a high-powered machine to get up the worst of the embedded dust, sand, and grit. Shampooing or steam-cleaning are not essential, but are a good idea.

To install the new carpeting, set new edging strips at doorways and transition points, and install tackless strip in the usual way next to the base molding on the carpetless border strip. Then stretch the carpeting in with the same methods outlined earlier. The old carpet will become a carpet pad.

SHEET VINYL FLOORING

Besides being available in an endless array of colors, patterns, and designs, vinyl sheet flooring (Fig. 2-58) is produced in a number of sizes and thicknesses. It can be had

Fig. 2-58. *A typical sheet vinyl flooring installation. (Courtesy Tarkett, Inc., Parsippany, NJ.)*

in rug size or room size sheets that measure 9 × 12 feet, 12 × 12 feet, and so on, or by the running foot from bulk rolls 6 or 12 feet wide. Thicknesses range from 0.065 inch to 0.160 inch. The material may be made in two layers consisting of the wear surface and the backing, or in three, which includes a layer of vinyl foam cushion sandwiched in between.

Types

There are two main types of sheet vinyl flooring: filled vinyl, and clear or unfilled vinyl. Filled vinyl is composed of fillers, resins, pigments, stabilizers, and vinyl chips mixed together and rolled into sheet form, then bonded to a backing. Clear or unfilled sheet is made by imprinting either the face of the backing or the underside of a clear vinyl wear surface with the desired pattern, using colored vinyl inks, and bonding the two layers together. The surface thickness of both types ranges from 0.010 inch to 0.050 inch; the recommended minimum wear surface thickness for residential applications is 0.010 for the clear variety and 0.020 for the filled.

The backing material is typically felt or vinyl, although other materials may be used as well. If the backing is vinyl, the material is often referred to as solid vinyl. The backing determines where the material may be installed: below grade (such as a basement floor), on grade (a slab floor), or on a suspended floor (any floor away from ground contact or high moisture conditions). There is also a wide range of stiffness/flexibility in vinyl sheet, which determines to a considerable extent how easy or difficult installation will be. Be aware that new vinyl sheet flooring products appear regularly, and their installation requirements are variable to some degree. Be sure to get all the details from your supplier that concern the specific products in which you are interested. There are now many choices available that are especially designed for do-it-yourself residential applications, and they are much easier to install than those often laid by professionals.

Preparations

Vinyl sheet should always be laid on a clean, dry, smooth, dust- and grease-free substrate. The thinner materials will eventually imprint any cracks, holes, or other imperfections beneath them unless the imperfections are tiny. Methods of securing the material vary. Some types can be laid flat and stapled or tacked around the perimeter, with the fasteners later covered with shoe molding. Some can be held down around the edges with the molding alone. Others are secured by strips of spray-foam adhesive, or by strips of wide double-faced self-sticking tape. The traditional and most effective method is glueing the entire installation down with adhesive spread on the substrate. The adhesive can be applied with a brush, roller, or notched trowel. Floorings bonded in this way are generally rolled immediately afterward with a 100-pound flooring roller. Use whatever method and particular adhesive the manufacturer recommends, because some of the adhesives have been formulated especially for those particular products.

When you calculate the material needed for the job, try to avoid seams—even if that means using a larger sheet than would otherwise be necessary and ending up with some waste. If seams are unavoidable, try to place them in inconspicuous spots and away from traffic paths. If the material will be laid over an existing wood plank or strip floor, arrange pieces so that as many seams as possible lie at right angles to the planks or strips, and none fall over a crack between them. If the material you select is patterned and you will have seams, be sure to buy enough material so that the pattern can be matched; this always takes a little extra. The easiest way to determine the amount and the piece sizes you will need is to make up a detailed sketch of the area to be covered, with full, accurate dimensions. Give the sketch to your dealer, and let their estimator make the calculations.

Installation

The first step in the installation is to square off the free end of the roll with a framing square and a straightedge. Then cut the piece so that it is 3 inches longer at each end than the distance from endwall to endwall. If the piece is wider than the width of the floor area, cut off one edge of the material so that you will have an extra 3 inches of material along both sides. Otherwise, leave it full width. Loop the piece over from side to side in a loose tube, or fold it back upon itself in a long single lap, whichever is most convenient (a lap is generally easiest to work with), and move it into approximate position in the room.

To fit vinyl sheet, professionals usually employ the scribing method so the sheet abuts exactly against the surrounding vertical surfaces at all points. This, however, takes some practice and experience to do well. For one shot do-it-yourself projects, the knifing method is easier, although it generates some waste material. Lay the sheet out so that it abuts the wall along one edge and the excess curls up the wall elsewhere. If the edge butts well to the wall all the way along, fine. If not, try the other side (unless it will be a seamed edge), then the two ends. If and when one butts well, leave it. If none butt, shift the piece until about 3 inches curl up the wall.

Make safety cuts at the corners or any other places where the material tends to buckle and might crack—down through the 3-inch flashing to relieve the strain. Be careful not to cut out into the field of the material. Make similar relief or safety cuts at other corners or vertical surfaces, until the field of the material can be pressed down flat to the floor without buckling or tearing, and the 3-inch flashing curls freely up the walls. You can make all the cuts with a sharp utility knife, a hooked linoleum knife, or even with scissors. The cuts can be straight downward slashes, but there is a more controllable procedure: Cut a successively wider and deeper V at each spot until the point of the V finally reaches the bottom of the corner at the wall/floor joint (Fig. 2-59). This eases the material more fully and allows you to better see what you are doing.

The next step is to trim away the flashing to make abut fit against all the vertical surfaces. In cases where a molding or cove base flashing will be installed, the butt joint need not be precise—but the cleaner the better. Do the trimming in successive cuts (Fig. 2-60), a little more each time until you have gotten the right line or contour along the wall/floor joint.

Cut around small wall projections in the same way. If the material must be pushed back deep under an object, such as a radiator, make a series of crossover cuts. These are straight slashes from the edge of the material back toward you, about 8 to 12 inches apart,

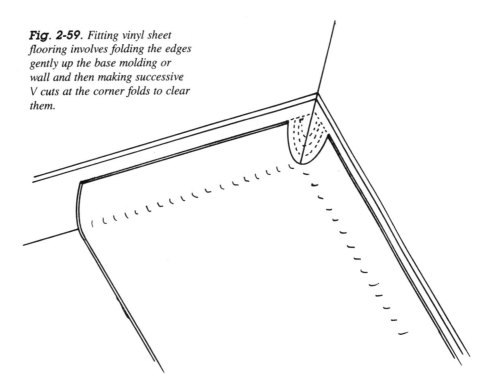

Fig. 2-59. *Fitting vinyl sheet flooring involves folding the edges gently up the base molding or wall and then making successive V cuts at the corner folds to clear them.*

depending upon the flexibility of the material. Make a cut at each outside edge of the object, and more between them to leave a series of strips about 6 to 12 inches wide. Roll the strips back under the object, and trim carefully around feet or pipes or whatever else might be in the way. Sometimes a heavy-duty X-Acto knife is handy for this. The cross-over cuts will be invisible, as long as you end them slightly beneath the object and not out into the field.

In some cases there are very irregular or sizable rounded projections. Knifing around these is difficult, so pattern scribing might be in order. This involves placing a piece of heavy paper next to the irregular contour and transferring it to the paper. You can do this with a drafting compass held vertically. Run the point of the compass along the wall/floor joint contour; the pencil point will mark an identical line on the paper. Cut the paper along the line, tape the paper to the vinyl surface at the proper location, and trace around it. Then trim the vinyl along the line. In lieu of scribing, you can transfer measurements to the paper, then, using scissors, cut and fit on a trial-and-error basis until you get a satisfactory pattern.

Trimming and matching seams can be done in either of two ways: single cutting or double cutting. In either case, if there is a selvage (a factory edge of an obviously different color), this must be cut away. The pieces should be secured to the floor so they cannot slip, but no adhesive or fasteners should be within about 4 inches of the seam.

To single-cut, overlap the two pieces. The top piece should have a pretrimmed, clean,

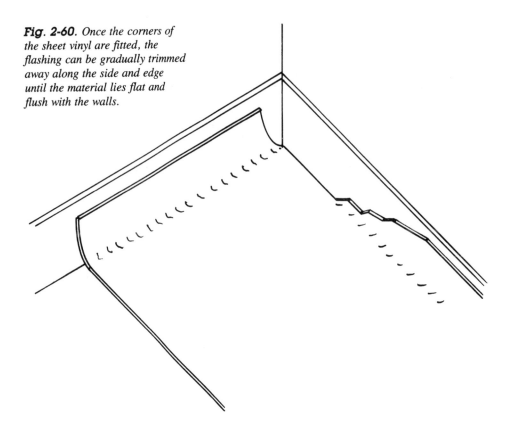

Fig. 2-60. *Once the corners of the sheet vinyl are fitted, the flashing can be gradually trimmed away along the side and edge until the material lies flat and flush with the walls.*

square edge, and cover the selvage of the bottom piece, if present. The overlap should be a minimum of about 1 inch. Make sure the pieces are properly aligned, and use the edge of the top piece as a straightedge (Fig. 2-61). First score along the bottom piece, then cut it with repeated long strokes.

To double-cut, overlap the two pieces by 2 inches or more. Align a straightedge to make a squared line along the top piece, about 1 inch in from the edge of the bottom piece (Fig. 2-62). Score the top piece along the line, then cut down through both pieces and discard the trimmings. Lift the cut edges and apply adhesive or set double-sided tape, then press the edges down. In some cases, a clear vinyl sealer/filler is smoothed into the seam to close it off; follow the manufacturer's instructions.

The exact sequence in which the various trim cuts are made depends upon the room size and shape, the amount of trimming to be done and the number of separate pieces used, and the way the vinyl will be secured to the substrate. Sometimes it is easiest to trim and fit an entire piece, or the whole floor area, then take it up, spread adhesive, and lay the piece in one effort. Fitting a seamless single piece to cover an entire floor area can be tricky and probably will require pattern scribing on one end or side first, plus some gradual trial-and-error knifing. When two or more pieces are needed-and it is sometimes easier to lay the material that way—try this method: Trim one piece, fold it back on itself,

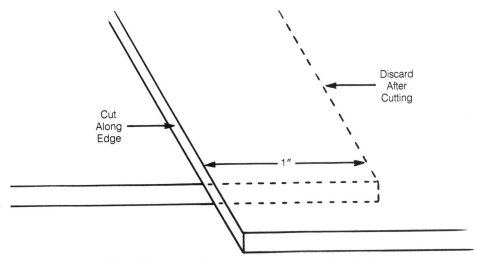

Fig. 2-61. *This method of trimming for a seam is called single-cutting.*

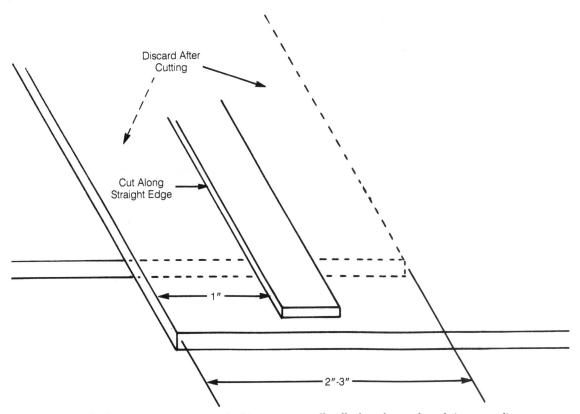

Fig. 2-62. *Double-cutting a seam in this manner usually affords a cleaner, less obvious seam line.*

spread adhesive, lay half the piece, fold the free half back, and repeat the process. Then fit the next piece to the walls and/or the previous piece, and so on. A stapled installation might be done by trimming one end, holding it with a few staples, then trimming the rest. Check for a good fit, then staple all the way around. Some jobs can be done piecemeal: trim a little, glue or staple a little, trim a little more, and so on. Much of the process is a matter of common sense, a bit of ingenuity, and patience. Assess each job according to its own characterisitics.

SEAMLESS FLOORING

Seamless flooring is a type that is used almost exclusively in commercial, industrial, and institutional applications. It is also a relatively expensive flooring. However, the same attributes that make it valuable in such installations apply also to certain residential situations, and it is mentioned here only to make you aware of it as an option.

The material may be of an epoxy composition, or homogeneous polyvinylchloride (PVC). Sheet rubber is another type. The flooring may be laid as a semi-liquid, or in sheets with the seams welded together by a special process. Whatever the procedure, the result is a relatively thin covering that can be laid over almost any substrate—with the proper preparation. It has no seams, it is extremely durable, can be wrapped up the walls for extra protection. It is also hygienic—there is no place for bacteria to hide. It is practically maintenance-free, apart from ordinary cleaning. Various colors are available.

In the house itself, applications would most likely be limited to kitchens and bathrooms. Outside the house, it would be of value in kennels, milk rooms, in any spot where possible contamination might be a worry, or in any area where a tough, long-lived, smooth flooring is desired.

Seamless flooring must be installed by professionals with the required equipment and expertise. Such contractors are likely to be found only in the metropolitan areas. Consult the Yellow pages of the nearest large city, under "Floor Laying, Refinishing, and Resurfacing," or contact the National Association of the Remodeling Industry (1901 North Moore Street, Alexandria, VA 22209) for information.

RESILIENT TILE FLOORING

By far the most commonly used and readily available resilient tile flooring is vinyl, either solid (vinyl-backed) or backed with another material, such as felt. There is a tremendous array of colors and patterns from which to choose. The most widely used size is 12 × 12 inches, though 9 × 9 inches and strips 4 inches wide and 36 inches long are offered on a more limited basis. The most common thicknesses for solid vinyl are $1/16$, 0.080, $3/32$, and $1/8$ inch. For backed tile, the range is from 0.050 to 0.095 inch. As with sheet vinyl products, the wear surface of backed vinyl tile varies; solid vinyl tiles carry the color and pattern all the way through. Long, narrow feature strips are also available for making borders, accent strips, or patterns. Most vinyl tiles have a no-wax finish. There are other kinds of tile available as well, such as rubber, cork, asphalt, and vinyl composition, but their variety and use is more limited.

Regardless of type, resilient floor tiles are all laid in much the same way. Solid colors, or colors and patterns, can be laid to form designs of your own devising, such as checkerboards, border patterns, alternating strips, geometrics, or even mosaics. Many of the products made for residential applications are designed for easy do-it-yourself installation, can be cut and trimmed with scissors, and sometimes come with self-stick adhesive backs. Others are heavier, must be cut with a knife, and are bonded down with special adhesives, but are still easy to lay. For most purposes, especially in relatively heavy traffic areas, bonded tile with a thick wear surface gives better, longer-lasting results. Always follow the manufacturer's recommendations and instructions for adhesives; they vary considerably for different makes of tile and different application conditions.

To calculate the amount of tile you need, determine as closely as possible the square footage of the room or area to be covered. Break the space up into rectangles and triangles as necessary, compute each, and add them up. To this, unless the room shape is very regular, add a 10 percent waste allowance for areas up to 100 square feet, 8 percent up to 200 square feet, 7 percent up to 300 square feet, and 5 percent for over 300 square feet. Then add a few more tiles for good luck, and keep the leftovers stored away for later repair work. Tile is usually sold by the case—though some dealers will break a case and let you buy singles—so you might end up with extras anyway.

Making the Layout

To lay a resilient tile floor, the first step is to locate the center of the room or area. Find the midpoints of two opposite parallel walls and snap a chalkline from point to point, cutting the room in half. Do the same with the two remaining walls, to divide the room into quadrants. Ignore small cutouts, nooks, or closets. If the room is L shaped, break it into two units. The point where the four lines cross will mark the approximate center of the room. Now check the crossing point with a large square to see if the lines are exactly at the right angles to one another; they must be 90 degrees apart (Fig. 2-63). This can be checked with greater accuracy by using the 3-4-5 method described in the "Acoustic Tile Ceilings" section in Chapter 5. If the lines are off angle, snap another line to square up.

The next step is to determine whether this centerpoint will also be your starting point. When the tiles are laid out, the rows in both directions should end at the walls in no less than a half a tile. Measure one line from wall to wall. Assuming standard 12-inch tiles will be laid, if the length is an even number of feet—12, for example—the centerpoint will be the starting point for rows going in that direction: six tiles to each side. If the length is an odd number of feet, such as 11, the same is true: five tiles each way, plus half a tile at each end. If the length is an odd number of feet plus some inches, such as 11 feet $7^{1}/_{2}$ inches, the centerpoint is also the starting point ; five tiles to each side plus a $9^{3}/_{4}$-inch tile at each end. But if the length is an *even* number of feet plus some inches—12 feet and 3 inches, for example—add half a tile width plus half the inches to find the width of the border tile. Move along the line from the centerpoint (either direction) a distance equal to a half tile—

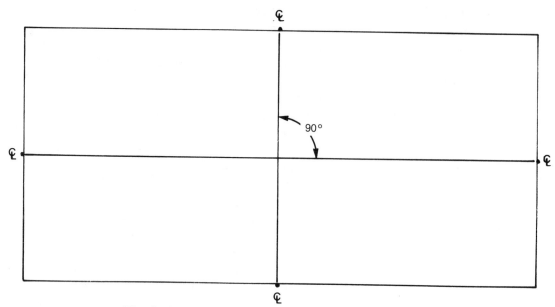

Fig. 2-63. A floor tile layout starts with the initial quadrant lines.

6 inches in this example—and mark a new starting point. This puts five tiles to one side, six to the other, and a $7^1/_2$-inch tile at each end of the row. Mark that point, then go through the same procedure on the other line. Snap two new chalklines parallel with the original ones, setting up the new quadrants (Fig. 2-64).

This is the simplest procedure, used when the tiles will be laid over the entire floor in rows parallel with the walls. If patterns, borders, geometrics, or other designs are part of the plan, more lines will have to be snapped or drawn to correctly locate them within the field, using the starting-point quadrants as a reference. Better yet, you can lay the tiles out temporarily so you can actually see the whole layout as it finally will appear. As you take the tiles up, trace around the patterns or designs and write notes on the floor as to what goes where. Write keys on the backs of the tiles, too, if necessary.

If the tiles are to run diagonally to the walls, find the centerpoint of the room as just discussed. Snap two more reference lines in an X across the original quadrant lines, so that all lines are exactly 45 degrees apart. The easiest and most accurate way to do this is to measure out along each of the original quadrant lines A and B from the centerpoint (a), a distance of 4 feet, then make marks (b), (c), (d), and (e). Drive a small nail through the center of a straight piece of lattice that is about 4 $^1/_2$ feet long, 3 inches from one end. At the other end, drive another small nail through exactly 4 feet from the other one. Drive one nail into the floor at point (b), align the lattice parallel with line A, and scribe an arc on the floor with the point of the second nail. Swing the lattice 180 degrees and repeat. Do the same working from points (c), (d), and (e). This creates points (f), (g), (h), and (i)

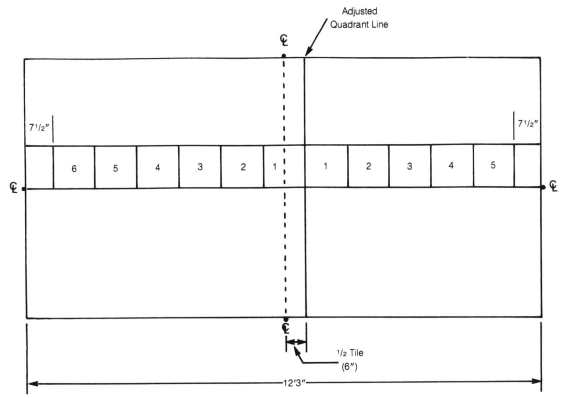

Fig. 2-64. *The quadrants have been shifted to allow proper placement of the tiles and equal border tiles.*

where the arcs cross. Snap chalklines across points (f), (a), and (h) and across points (g), (a), and (i) to form the new diagonally positioned quadrants (Fig. 2-65). Ignoring the original lines A and B, use lines C and D for references and work out the correct starting points as discussed above, so that the end tiles are at least half tiles and the patterns, if any, are correctly positioned.

Although the above procedures are the recommended and most often used, there are circumstances where they do not work out well. A room might be very irregular, or have only two small parallel walls, or no parallel walls. The tiles might be laid parallel or diagonally to only one wall, or perhaps at some random angle unrelated to any of the room borders. In that case, the first step is to lay out a single baseline running across the room or area. This line should be at the room's greatest or most logical dimension lying in the direction the tiles are to be laid. Find the midpoint of that line A and snap another line B at right angles to it (Fig. 2-66). This sets up the quadrants, however irregular they might be, and a centerpoint (a)—which probably isn't the center of anything. Use this layout as your reference, and make more lines or set out tiles to see if the results will be agreeable. If they are not, shift the lines. In such instances a lot of cut tiles at row ends is inevitable and unavoidable, and some of them will doubtless be small and irregular. The object is to

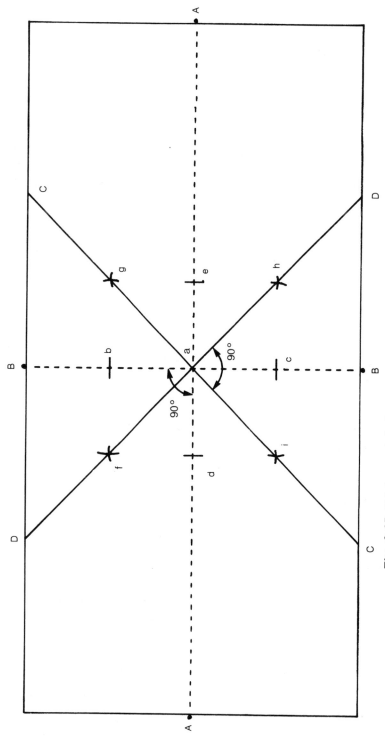

Fig. 2-65. *This layout method is used when a diagonally laid tile field is desired.*

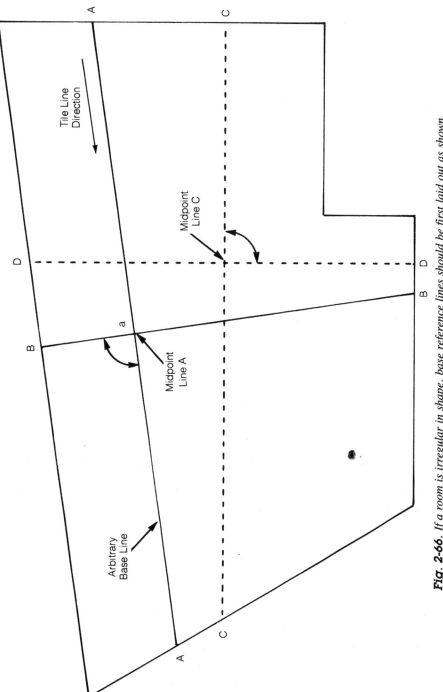

Fig. 2-66. *If a room is irregular in shape, base reference lines should be first laid out as shown. Usually there are at least a few possibilities.*

arrive at a solid starting point that will allow a good tile layout with a minimum of cuts and small pieces.

Laying the Tile

Laying the tile is simpler than making the layout. Tack straight strips of lattice or furring along quadrant lines from the starting point, either straight line or in an L. These strips will straightedge the tiles and keep them from shifting as you start to work. The laying sequences are shown in Fig. 2-67.

If you are using self-stick tiles, peel the backing off the first one, hold it by the edges and slightly curled, and set one corner tip into the starting corner. Position it carefully just above the floor surface and let it drop (or snap down easily) into place. Make sure it is properly aligned, and press it down firmly with your hand. Then set the adjacent tile, then one above, and so on until you fill out the quadrant. As you reach row ends, trim as necessary. This can be done with a utility knife, linoleum or hook knife, or in many cases, scissors. Make your trim cut lines by transferring measurements, by scribing, or by making paper patterns. This kind of tile laying can be stopped and started anywhere, anytime.

If the tiles are adhesive-bonded, the situation is a bit different. Be sure to use the recommended method and tool, so you don't spread too little for the tiles to adhere properly or too much so that the excess bubbles up through the seams.

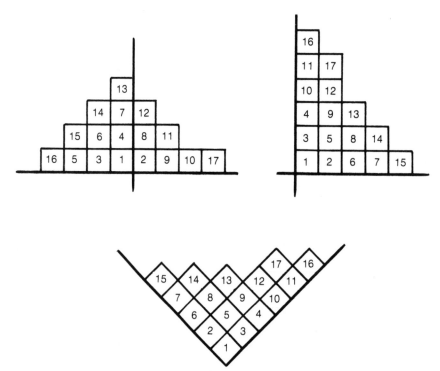

Fig. 2-67. *Tile laying sequences for single, double, and diagonal quadrants or field areas.*

Adhesives vary. Some must be spread and left for an hour to "tack up," then there is a certain time period—the "open time"—during which the tile must be set, before the adhesive begins to cure. Other kinds can be spread and the tile laid immediately. Atmospherics play a great part in how much open time you have to lay the tile over freshly spread adhesive. Minimum working floor temperature varies between 50° to 70° Fahrenheit. High humidity or low temperature, or both, give you the longest open time. Low humidity or high temperature, or both, allows the shortest open time. At 90° Fahrenheit and 15 percent humidity, you might only be able to spread the amount of adhesive you can cover with tile in about 10 minutes, working quickly.

With this in mind, you must organize your tile laying into small sections or subquadrants, depending upon the weather and the adhesive you are using. It also means that, before any adhesive is spread, each area to be trimmed must be actually laid out with the tile. Also, any necessary trimming must be done and all the tiles properly fitted, then taken up and stacked or spread around so you know which tiles go where. Then you can go ahead, spreading a small patch of adhesive and laying the tiles, and repeating the process until the precut and pretrimmed area is covered. Whenever you stop laying tile for more than about half an hour, cut the spread of adhesive off just inside the edge of the perimeter tiles. If adhesive cures on an unlaid area, you will have trouble laying the next batch over it.

As you set the tiles on the adhesive, make sure they are lined up accurately as you drop them in place. Do not slide them into the adhesive as you set them, and move them around as little as possible after they touch down. The usual method is to work from the starting point outward and forward toward the wall, so you will soon be out onto the freshly laid tiles. Don't kneel directly upon the tiles; kneel on a piece of plywood about 2 feet wide and long enough to support your toes as well, and move it along with you. Even this might move the tiles around, so you have to move carefully and keep checking the tile alignment as you go. They must be kept perfectly square to one another or you will find your rows becoming progressively misaligned.

Sometimes it is easier, especially when only small sections can be laid at a time, to work toward yourself instead of away from yourself. Face the quadrant lines and spread adhesive in front of you, two or three tile rows deep and a few tiles wide, on your side of the lines. Set the furthest row of tiles first, then the next nearest, working back toward yourself and kneeling on the bare floor (Fig. 2-68). Then back up two or three rows and repeat the process. When you finally run out of space, resort to the kneeling board, or finish up the next day by working from the set tiles.

Whatever your approach, after a section of tile is laid, go over it with an ordinary pastry rolling pin to set it down firmly and evenly. Keep an even, fairly light downward pressure, and roll toward tiles already laid, not toward the open edges. Or, lay a flat piece of board on the floor and move it along while tapping it smartly with a hammer or mallet—preferably the dead blow type (hollow head partly filled with steel shot) that lands solidly and doesn't bounce back. Clean away immediately any adhesive that oozes out of seams

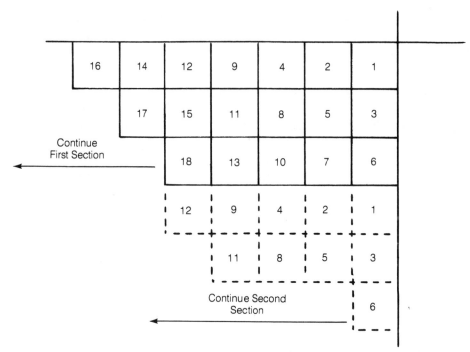

Fig. 2-68. *On some occasions a backward-laying sequence of this sort is appropriate.*

or gets daubed on tile faces. Keep traffic off the floor for 24 hours, and don't wash it for at least a week.

WOOD PARQUET FLOORING

Parquet, or pattern, flooring (Fig. 2-69) comes in two main types. The first is single-slat, which is made up of individual short strips of wood. The other is block, which consists of numerous small pieces of wood that are cut, fitted, and glued or attached to one another in squares or block patterns. Block flooring is further broken down into three types (Fig. 2-70). Slat or finger block is comprised of many short, narrow strips of wood arranged into a pattern. Unit block consists of just a few larger strips joined edge to edge to form a block. Laminated block uses various wood face veneers bonded to two or more layers of solid backing plies.

Both types of parquet are available in a wide range of sizes. Slat flooring typically is about $2^1/4$ inches wide, 6 or more inches long, and $3/4$ inch thick. There are many variations in block flooring, from 4-inch to 39-inch squares to various rectangles, with thicknesses ranging from $5/16$ inch to $3/4$ inch. There are many block flooring patterns. Both types are available unfinished, or with a clear or factory stain finish; some are square edged, while others feature tongue-and-groove joints. Unit and slat block in 12-inch squares, $5/16$ inch thick and prefinished, are probably the most readily available types,

Fig. 2-69. *A typical parquet flooring installation, in this case teak finger tiles in an entryway.*

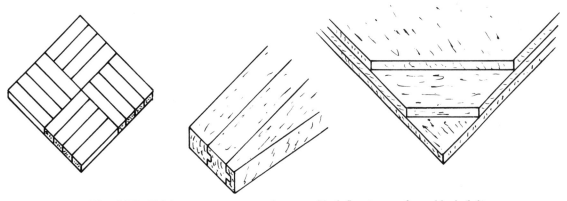

Fig. 2-70. *The three common types of parquet block flooring are finger block (left), unit block (center), and laminated block (right).*

and a good choice for do-it-yourself residential applications. Several different woods are offered; oak is by far the most common, teak is probably second, and other woods such as beech, birch, maple, and pecan can be obtained. Also, slat flooring can be made up in the home workshop from any suitable flooring wood by any craftsperson with a little time and patience.

Installation

Parquet flooring is bonded to the substrate with adhesive. Follow the manufacturer's recommendations for all bonding details. The flooring should be stored for about a week in the area where it will be laid, to allow it to adjust to the atmospheric conditions. Single-slat flooring can be laid in a number of different patterns, much like paving brick. Sometimes the possible patterns of this flooring are dictated by the size and shape of the slats. The patterns can also include designs achieved by insetting slats of contrasting colors or woods as geometrics, feature strips, or borders. Block flooring, particularly the popular 12-inch-square variety, is laid in much the same way as resilient tile flooring. Refer to that section for details. There are a few minor differences, which are noted as follows. In the absence of a laying sequence recommended by the flooring manufacturer, use any of the resilient tile sequences.

You can trim either type with a table saw for straight cuts, or a saber saw for shaped cuts. However, if you have the use of a bandsaw—even the benchtop type—the cutting will go faster and easier. You can make trim lines by scribing or measurement transfer, but be sure to work from upper surface edges when measuring tongue-and-groove blocks, and not from the tongues. End pieces should not abut a wall or other vertical surfaces; the usual procedure is to leave a $1/2$-inch gap all around the perimeter for expansion. With tongue-and-groove blocks, be sure to align the first ones with the edges, not the tongues, along the quadrant lines. When bedding the slats or blocks into the adhesive, plop them squarely in place from a height just above the adjacent pieces, adjust and align them as necessary, and press them down firmly. Tongue-and-groove pieces must be dropped slightly wide of the mark and then gently wiggled into place to fully engage the tongues and grooves. Do this with as little downward pressure as you can, then press down after the piece is fully aligned and locked in. Try not to scoop any adhesive into the joints in the process. In any case, move the pieces as little as possible once they touch the adhesive.

After a section of flooring is done, go over it with a rolling pin to set the pieces firmly, or walk on all the pieces with tiny, flatfooted sidesteps, taking care not to move any of them. In places that are hard to reach, tap the pieces down with a piece of flat scrap wood and a mallet. Keep traffic off the floor for 24 hours. If the flooring is unfinished, it should be sanded, sealed, filled and/or stained as necessary, and given a finish coat before it is subjected to use. If factory prefinished, a coating of the recommended wax should be applied as soon as possible after the curing period.

TERRAZZO FLOORING

Terrazzo flooring is usually considered suitable for bank lobbies and hospitals, but not for residences. In fact, terrazzo has been used for years in residential shower bases and a few other products, but seldom anywhere else. This is regrettable, for although it is an expensive flooring, it is also handsome, extremely durable, utilitarian, and is an excellent choice for thermal mass floorings in solar houses. There is a good variety of colors, all natural tones, and decorative divider strips can be placed where desired. The material can be used either indoors or outdoors. It is excellent for walkways, patios, in swimming pool areas, and in atriums, greenhouses, or sunspaces.

There are two main types of terrazzo: the traditional cement type, and the newer resinous type. Most cement types use portland cement as the binder, very occasionally magnesite. The resinous type may use catalyst-cured resins such as polyester or epoxy, or latex resins like polyvinylchloride, acrylics, or neoprene. To these binders, fillers and/or pigments may be added to make the terrazzo matrix. The characteristic appearance is given by the stone chips—graded bits of marble, onyx, travertine and others—that are mixed with the matrix. Altogether, this is called the topping.

Installation

Beneath the topping other layers might be required, depending upon the nature of the subflooring. A concrete slab may be poured, or an existing one used—which might call for a bonding agent, or perhaps a concrete underbed. There are several possibilities, but the point is that terrazzo flooring of the cement type can be poured on residential basement, sunspace, or any other concrete slab floors. The resinous type can be poured over any kind of flooring that will support the weight, except for slabs subject to moisture or that have a high alkaline content. The cement type of topping need only be $5/8$ inch thick (minimum). If a mortar underbed is needed, that adds at least $1^3/8$ inches. The resinous topping is between $1/4$ and $1/2$ inch thick; in some cases a new underlayment might be needed, adding another $1/2$ inch. At full thickness, the resinous topping weighs only about $5^1/2$ pounds per square foot, nearly the same as quarry tile or paver flooring.

Pouring a terrazzo floor, however, is definitely not a job for a do-it-yourselfer, even one operating at advanced levels of skill and confidence. This is because materials are required that are not available through local building supply houses, equipment is needed that cannot be found at tool rental shops, and both experience and expertise are needed to produce fine results. In addition, the job is a physically demanding one that, once started, cannot be stopped any old place. Several workers are needed even for relatively small jobs, to keep everything moving along. But help is at hand. You can contact the National Terrazzo and Mosaic Association (3166 Des Plaines Avenue, Suite 312, Des Plaines, IL 60018) for details on all kinds of terrazzo flooring and names of terrazzo contractors in your area. Also, check in metropolitan Yellow Pages for names, and with the National Association of the Remodeling Industry (1901 North Moore Street, Suite 808, Arlington, VA 22209) for information.

Terrazzo flooring is also available in tile form. See the section entitled "Stone Flooring."

CERAMIC TILE FLOORING

There are three main kinds of ceramic tile suitable for residential flooring: ceramic mosaic tile in unglazed porcelain and unglazed natural clay, quarry tile, and paver tile.

Ceramic mosaic tile (Fig. 2-71) is $1/4$ inch thick and typically found in 1-inch and 2-inch squares, and 1-×-2-inch rectangles, as well as a few smaller squares, various dot or button sizes, and occasionally some special shapes. There is a wide variety of colors and surface effects available. In the smaller sizes, the tiles are usually mounted into larger units, typically 1 foot square, for easy handling. They may be back-mounted with a paper

Fig. 2-71. *This extensive installation points up the great flexibility of ceramic mosaic tile; note the formed cove at the wall/floor line. (Courtesy Ceramic Tile Institute, Los Angeles.)*

or mesh or other material, which remains in place after laying, or face-mounted with kraft paper that is removed afterward. Unglazed porcelain tiles are very dense, and wear- and stain-resistant. The unglazed natural clay variety is characterized by a dense, rugged, abrasion-resistant body and earth tone colors.

Standard quarry tile (Fig. 2-72) is $1/2$ inch thick, and typically available in 3-inch, 4-inch, 6-inch, 9-inch, and 12-inch squares; 3-×-6-inch, and 4-×-8-inch rectangles; and numerous special shapes such as Moorish, ogee, and hexagonal. Quarry tile is a heavy-duty clay product available in various natural earth tones, mostly unglazed. Some recent types, however, offer colorful matte or texture-glazed surfaces that are sufficiently slip-resistant to make a satisfactory floor covering. A heavier type, $3/4$ inch thick and called *packing house* tile, is also available.

Paver tile (Fig. 2-73) typically comes in 4-inch squares 3/8 inch thick, or 4-inch and 6-inch squares, or 4-×-8-inch rectangles $1/2$ inch thick. This is a dense, tough tile that can be used either indoors or out without danger of damage from freezing. It is made in muted earth tones with a relatively rough and very slip-resistant surface.

Preparation

Ceramic tile is installed in either two ways: thick-bed or thin-set. Thick-bed installation is a complex process seldom used in houses, which involves bedding the tile in a thick layer of mortar. It will not be covered here. Of the several thin-set methods, most professional tilesetters prefer to lay floor tile in either a dry-set mortar or a latex-portland cement mortar. Either of these or the epoxy mortar method (which should be attempted only by a professional) are best for laying tile on an existing concrete slab, but involve special preparations for laying tile over a wood substrate. The organic adhesive or mastic method is suitable for all residential purposes, including small areas of moisture- and alkali-free concrete and all wood floors or underlayments. This is also the easiest and usually the method chosen by do-it-yourselfers.

To lay ceramic tile, you will need at least the following ordinary shop tools: putty knife, can opener, steel rule, folding rule, framing square, try square, chalkline, straight-edge, mallet or hammer, and slipjoint pliers. You will also need some special equipment, which you can either buy or rent: tile cutter, mason's block or brick chisel (for thick quarry and paver tiles), tile nippers, notched trowel of the correct configuration for the mastic being used, and a rubber-faced grout float.

Layout

The first step is layout. You can make your working lines in the same way as discussed previously in the section "Resilient Tile Flooring," adjusting as necessary for the size of the tiles you will be laying. Or, you can start your working lines from the center of one end of the room or area, or even from an entry doorway. This works best if at least two of the adjoining walls are exactly at right angles to one another. A series of working lines is important if different tiles will be laid in decorative patterns. In any case, after setting the final lines, do a dry run on at least a large portion of the area, setting the tiles out without

Fig. 2-72. *A quarry tile floor is as suitable outdoors as indoors, and is a very popular flooring for entryways and kitchens. (Courtesy Ceramic Tile Institute, Los Angeles.)*

Fig. 2-73. *Paver tile is a popular choice for a durable and attractive flooring material, and is available in many forms. (Courtesy Ceramic Tile Institute, Los Angeles.)*

any adhesive and maintaining the required grout spaces between them, to make sure that there are no problems. The recommended grout line widths are as follows (keep in mind, however, that these can be adjusted somewhat, especially in small areas, to gain a better fit or pattern):

- Ceramic mosaic tile: $^1/_{32}$ to $^1/_8$ inch.
- Paver tile (including specials): $4^1/_4$ inches square and smaller—$^1/_{16}$ to $^1/_4$ inch. For 6 inches square and up—$^1/_4$ to $^1/_2$ inch.
- Quarry tile (including specials): 6 inches square or smaller—$^1/_8$ to $^3/_8$ inch. For larger than 6-inch—$^1/_4$ to $^1/_2$ inch.

Cut all tiles that must be trimmed before laying the field, or any particular section of it. Set a few tiles in position adjacent to the ones that must be cut, then mark the cuts by scribing (Fig. 2-74) or measurement transfer. If a pencil doesn't show up well, use a fine felt-tip pen. For straight cuts in thin tile, you can score the finish surface with a tile cutter; pressing down on the handle breaks the tile at the score. If the back of the tile is ridged or ribbed, make your cuts parallel with them. Your tile supplier can show you how to use a cutter.

Another method is to score the surface with a glass cutter. Then place the tile face up with the score directly over a pencil or a dowel, and snap down firmly on each edge (Fig.

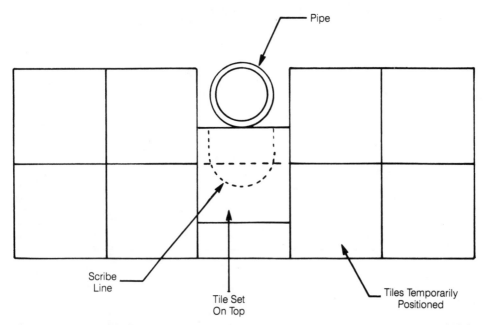

Fig. 2-74. To fit a tile around an irregular object or to an irregular surface, lay the surrounding tiles in their proper positions dry, line up the tile to be trimmed, and scribe a trim line on it.

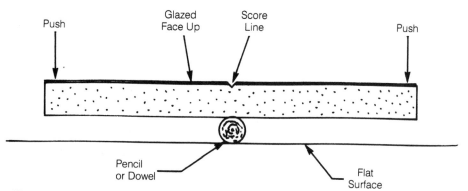

Fig. 2-75. *Do straight line trimming by scoring the face of the tile, then break it by pushing the edges downward over a pencil or dowel.*

2-75). For irregular cuts, use the tile nippers and nibble away at the edges until you get the configuration you need (Fig. 2-76). Scoring first with a glass cutter helps. This takes a bit of practice, so plan on breaking an occasional tile. The thicker tiles can be deeply scored on the surface with the edge of a block chisel or the tang of a file. After scoring, set the tile on a hard, solid surface, and set the edge of the block chisel directly in the score. Hold the chisel vertically, and strike it smartly with a hammer. An alternative, especially if you want smooth, square cuts and no chance of wasted tiles, is to have them cut with a masonry saw. Consult with your nearest friendly mason, or check with your supplier; many tile retailers are set up to custom cut tile.

If your cut tile edges are too rough to suit you, there are several remedies. One is to file the edges flat with a medium-fine metal file. You might wear one or two out, but they are inexpensive. Another method: rub the tile edge back and forth repeatedly on the side of a concrete block. Yet another, useful with inside curves: rub the edge with a carborundum stone of appropriate shape, using water as a lubricant. Let the tile dry before setting it.

Once the dry run has proven out your working lines and layout, and all the required cutting is done, you can start setting the tiles. Take up the tiles, arrange the cut pieces nearby so that you know which ones go where, and set out a quantity of field tiles ready to lay. If there are some special tiles or patterns to be laid in, note their locations on the floor and on the tile backs, if necessary.

Tack wood batten strips to the floor in an L along the first quadrant, to set the starter rows of tile against. Also, have handy some means of keeping the grout spacing between the tiles uniform. If the tiles have spacing lugs, there is no need for this, but most do not. You can lay lengths of string between the tiles, stretched between two nails, or set small bits of wood or other scrap material of appropriate thickness between the tiles as you lay them. You can also buy small plastic spacer crosses that fit at the corners of the tiles. These can be left in place, or pried up after the mastic has cured. If you have a very good eye for linear alignment, you can "eyeball" the tiles in place. Be aware, though, that wandering, crooked grout lines and cockeyed tiles are very obvious.

Fig. 2-76. *After marking the trim line on the tile (a permanent felt-tip pen works best), a nibbling tool makes short work of the cutting.*

Installation

With everything ready, spread mastic over a small section of flooring. Start with a small area to get the feel of the process. As you get used to it, you can cover a larger area. Spread the recommended mastic with a trowel that has the correct notching, then set the tiles, working in either pyramid or L fashion (Fig. 2-77). Work backward, away from the freshly set tiles. If you must work forward from the tiles, use a kneeling board about 2 feet wide and long enough that your toes are supported as well as your knees. Be sure to clean any daubs of adhesive off the tile faces as you go along.

After you have laid a small section of tile within easy reach, take care of the "beating-in" process. Slide a beating block—a flat piece of 2 × 6 or 2 × 8 about 1 foot long, with a scrap of carpet about 3/8 inch thick tacked to it—along the tile surface and thump it repeatedly with a hammer as you move the block. This sets the tiles firmly in the mastic and levels them. If you have to cross freshly laid tiles to get to the next laying area, put out chunks of plywood to step on, and walk gently. After the job is finished, keep all traffic off the floor for at least as long as recommended by the mastic manufacturer—usually 24 hours.

After the mastic has cured, the next step is to apply the grout. Check over the entire surface, clean off any missed spots of mastic, and remove spacers if necessary. Check all the grout lines to make sure that they are clear. If mastic has oozed up between tiles close

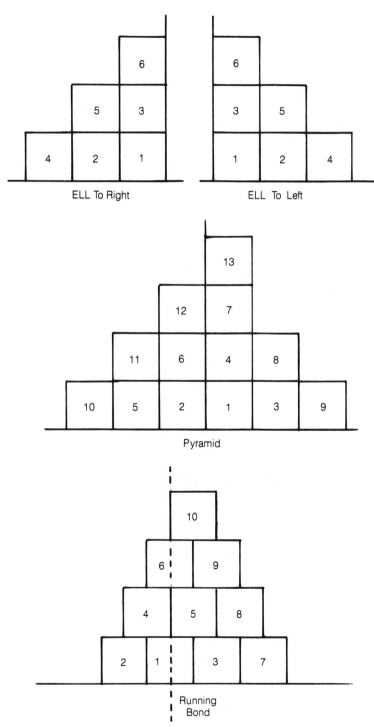

Fig. 2-77. *The various laying sequences for ceramic tile.*

to the surface, dig it out or drive it back down. The grouting is best done in one go, if possible . However, if the area is large you might prefer to break it into two or more blocks—or find someone to help you. If the tile is porous and the grout you will use is a portland cement base, wash the entire surface of the tile with a sponge and fairly liberal quantities of fresh clean water. Allow the water to seep into the grout lines and be absorbed by the tile bodies, which will prevent premature drying of the grout. Your tile supplier can advise you if this is necessary.

There are several kinds of grout that might be used; select the one recommended by your tile supplier. It probably will be either dry-set or latex, both a portland cement base. Regardless of type, you can select natural white or gray, or any one of a number of specially prepared colored grouts.

Mix a suitable quantity of the grout according to the directions. Starting in one corner, dump a good-sized blob of the mix onto the tiles and spread it into the joints with a grout float. Force it well into the joints, packing it as much as possible. Do your best to make sure that each joint is completely filled and there are no air voids, but also move along as rapidly as you can. Keep floating the excess grout toward you and into the joints, leaving behind as little as possible on the tile faces. Swing the float diagonally to the tiles as much as possible.

After you fill all the joints within an easily reachable area, wipe the tile faces with a sponge saturated with clean cold water. Sponge away excess grout, and at the same time, clean the grout lines off about level with the tile surfaces. Rinse the sponge often, and change the water often. Keep the sponge medium wet, and don't let water puddle on the surface or soak the grout excessively. When the surface is as clean as you can get it, continue in the next area with another batch of fresh grout.

After about 30 minutes, go back to the starting point. There will be a thin film of grout dust on the tile surfaces; wipe this off with a dry cloth. Then tool each joint. Using the end of a toothbrush handle, a piece of dowel, a tongue depressor, or some similar implement, go over each grout line firmly. Pack the material and indent it slightly to create a dense, hard surface. A wood or plastic tool is best because it will leave no marks on the tiles (as will a nail head or point), but it will wear fast, so have a supply on hand. As you tool the joints, clean out the corners of the grout lines and brush the excess and scrapings away with a cloth or a clean dustpan brush. Again, refrain from walking on the fresh grout; use a kneeling board and scrap wood walkways.

Cement-based tile grouts fare better if they are damp cured. This can be a chore, but is worth the effort. Check the directions on the grout package. If damp curing is mentioned, do it. After all the joint tooling is done, wash the floor with a clean sponge and clean cold water, leaving a liberal amount of standing moisture on the surface. Immediately lay down sheets of plastic film (available at any hardware store or lumberyard), covering the entire surface. Check at the end of each day to make sure there is plenty of moisture under the plastic. If not, peel the plastic back and wash again, then re-lay the plastic. The covering should remain in place for at least three days, five if you can manage it. Keep traffic off the floor in the meantime. After you take up the plastic, go over the floor with a damp sponge to remove any excess water. After the floor dries, polish it with a clean cloth to remove any residue.

The last step is to seal the tile and the grout. Silicone sealers are most commonly used, but follow the advice of your tile supplier in selecting a suitable product for your kind of tile and grout, and follow the manufacturer's instructions for application. The sealer will penetrate the grout to help seal it against water and grime, and do the same for porous tile faces. It will readily wipe off glazed tile faces and leave no residue. After sealing, you might also wish to wax and buff unglazed tile flooring. This should be done only in dry areas—not entryways, mud rooms, or bathrooms—because the wax will spot or discolor. Apply a recommended wax for your type of tile, according to directions, then machine buff. A good wax layer should last several years under dry conditions, needing only occasional rebuffing.

STONE FLOORING

There are two main types of stone flooring: composite, and solid stone. There are several types of each, available in a variety of sizes and natural colors.

Resilient terrazzo tile is made up of a matrix of thermoset polyester with marble, onyx, and other stones embedded. It looks much the same as a poured terrazzo floor. Tiles typically measure 2 × 3 feet and are $3/16$ inch thick, weighing only 2 pounds per square foot. They can be laid in mastic on any properly prepared, smooth, level substrate, either butted together or separated by narrow divider or feature strips. Also available is a heavier cast terrazzo tile, which can be laid in a flagstone effect. The surface is nonporous and waterproof, with a no-wax finish.

Another type of composite stone flooring consists of a $1/4$-inch-thick veneer of polished marble, attached to a fiberglass and epoxy backing. The appearance is the same as solid stone. These tiles are typically made in 1-×-2-foot and 2-×-4-foot rectangles and 2-foot squares, weighing about $3^3/4$ pounds per square foot. They can be laid on any substrate with proper preparation, usually by setting them in thin-set mortar. Mastic setting can be used under favorable conditions. Several types and colors are available.

The choices in commercially prepared solid stone floor tiles are: marble (imported mostly from Italy and France, some from Mexico), granite, various marble-like limestones, and slate. There is a wide variety of coloring in the marble tiles, including blacks, greens, whites and grays. The limestones are various shades of buff, yellow, brown, and ivory. Granite tiles are in the black and gray range. Slates are usually gray and charcoal, but greens and deep reds are also sometimes available. Random shaped, undressed flagstone-type flooring is also available, depending on the area, in slate and sandstone and sometimes marble and granite. And, of course, you can also lay flooring of random sized and shaped rubblestone or fieldstone that is sufficiently flat-faced and thin enough for your purposes.

Marble tile is typically cut in 6-inch or 12-inch squares, while the limestones range up to 16-inch squares. Thicknesses are usually $3/8$ or $1/2$ inch, and the weight is approximately 8 pounds per $1/2$ inch or thickness per square foot. Granite tiles are typically cut up to 2-foot squares and in $^3/_8$-inch or $^1/_2$-inch thicknesses, weighing about 9 pounds per square foot per $^1/_2$ inch of thickness. Sandstone and slate flags are random in size and thickness, but range from 11 to 13 pounds per square foot per inch of thickness.

Layout

Stone tiles can be laid on any concrete base or any wood subfloor that will bear the weight and is stiff and solid. The first step in laying commercial tiles is to make the layout, which is the same process—adjusted for the size of the units involved—as is used for laying resilient floor tiles. Cutting the tiles can be done by scoring and breaking, but considering the cost and the delicacy of many marble and granite tiles, they are better cut with a mason's power saw. Where the tiles can be successfully laid with mastic as the adhesive, which is often the case in light-duty residential applications, the process is the same as used for laying ceramic tile flooring. However, with many materials the conditions might dictate installation on a thin-set mortar bed, with latex cement or epoxy mortar used for satisfactory results. This is best left to a professional.

Laying random shaped stone flags is a different matter. The easiest approach to the layout is to actually spread the flags out and shuffle them around until you obtain a pleasing pattern. Trim and resize the flags, if necessary, as you go along. For a small area, lay out the whole affair. For a large one, start at one corner and work along the two adjacent walls, filling in the midportions as you go (Fig. 2-78). If you can do this in another room, or outdoors on a same size spot marked off with string or boards, then you can transfer the flags one by one to their proper positions. Otherwise, you will have to spread out a few at a time, pick them up after you have a good arrangement and set them aside, then re-lay them in order.

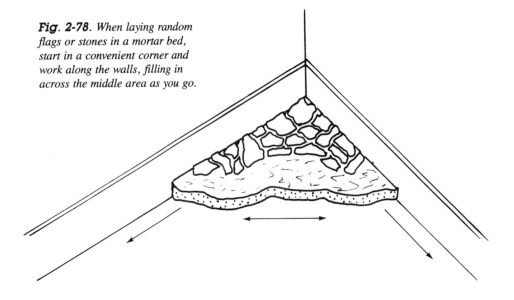

Fig. 2-78. When laying random flags or stones in a mortar bed, start in a convenient corner and work along the walls, filling in across the middle area as you go.

Because of their undressed surfaces and uneven thicknesses, flags or stones have to be laid in a thick mortar bed. Prepackaged dry mortar mix is suitable for this purpose. For greater plasticity, thoroughly mix about $^1/_4$ cubic foot of portland cement with each bag of dry material before adding water. With flags, the mortar bed should be a minimum of $^1/_2$ to 1 inch thick, depending upon how irregular the undersurfaces are. The bed for undressed rubblestone might have to be much deeper, perhaps 2 to 3 inches or more, in order to bed the stones to varying depths and maintain a relatively even floor surface. If you are laying the stone on a concrete floor, the mortar bed can be set directly upon it, and needs no reinforcement. If the surface is wood, first put down an isolation membrane of 15-pound roofing felt, stapled in place with a 6-inch overlap at all seams, and folded up the wall a distance equal to the total thickness of the new flooring. When laying the mortar bed, spread about half the intended thickness of mortar, then embed strips of metal lath in the mortar, and spread the remaining thickness (Fig. 2-79). There are several types of metal lath available; your supplier can provide you with a suitable one.

Installation

To lay either flags or stones, first spread a layer of fresh mortar over a small starting area. Drop the pieces lightly into the bed, then push down gently with a slight back-and-forth twisting motion to set them, while at the same time aligning them. Level each piece as necessary and even up all the pieces with each other. If one sinks too deep, take it completely up, drop more mortar in the hole, and start over. Wherever the mortar does not rise to the surface of the pieces, fill the joints with mortar as you go, and strike off any excess. Use walkways and a kneeling board to move around on the freshly laid stone. As the mortar in the joints around the first laid pieces looses its sheen and begins to cure, go over it with a jointing tool or the tip of a pointing trowel. Pack the surface down dense and tight. After a certain point in the laying process it is possible to lay a few stones, tool a few joints, then lay a few more stones, and so on. After all the stones are set, brush away the mortar crumbs and wash the entire surface several times with a sponge and cold water.

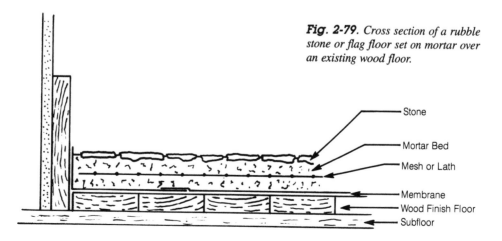

Fig. 2-79. Cross section of a rubble stone or flag floor set on mortar over an existing wood floor.

Stone

Mortar Bed

Mesh or Lath

Membrane

Wood Finish Floor

Subfloor

Change the water often. Cover the entire floor with plastic sheeting for 3 to 5 days, making sure that the surface remains damp during this time. If, after the floor dries completely, some mortar residue remains on the surface, clean it off with a solution of muriatic acid.

BRICK AND MASONRY FLOORING

Brick flooring (Fig. 2-80), which is many instances has a similar appearance to paver tile flooring, is not commonly used for interior residential flooring. It is best planned for and installed during initial construction of the house. It is usually difficult to install later, because of the thickness and weight of the bricks. However, there are instances where a brick floor can be easily laid, and it presents a handsome, durable surface. Bricks can be laid on any concrete slab: the floor of a garage converted to living quarters, a basement floor (if there is sufficient headroom to build the floor up that much), or any slab-on-ground floor in living quarters where adjacent floors can be stepped or brought to satisfactory levels. Brick floors work very well in greenhouses and sunspaces, and can be

Fig. 2-80. Brick flooring can be laid dry, sand-bedded, or mortar-bedded and jointed. Uniform, hard paving brick is best; recycled old handfired brick like this is more attractive but irregular and a bit delicate.

extended into adjoining rooms to convey an air of spaciousness. Brick flooring is also excellent for thermal mass heat retention in solar designs. Brick flooring can be laid on any wood frame floor system that is strong and stiff enough to handle the weight, as well as directly on the ground.

Masonry flooring consists of cast concrete pavers. These are available commercially as bricks, 1-foot squares, and a variety of special shapes—some of which are interlocking. Other than wood plank or single-slat parquet, this is the only kind of flooring that you can readily make yourself. It is done simply by constructing the forms in whatever shape you wish, mixing the concrete (use a bagged premix, to which you only have to add water), and casting the pieces. Commercial units are usually 2 to $2^1/2$ inches thick; home-cast units for interior use can be as little as $1^1/2$ inches thick if they will be laid on a smooth concrete slab. The weight of the units varies, but is about 12 pounds per square foot per inch of thickness. Various earth tones can be obtained by mixing pigments with the concrete. Applications and installation procedures are the same as for brick flooring.

For light-duty residental use, most kinds of bricks will serve for flooring. Many finish floors have been made up of old, recycled handmade bricks that are fairly soft and nonuniform. However, harder and denser bricks are better, and those produced specifically as paving bricks are best. Various types of face bricks with rough glazings in muted earth tones can also be used. Selection is a matter of discovering what specific brick products are available in your area; shipping is very expensive.

Whatever the specific brick type, usually the standard units will measure about $7^3/4$ inches long, $3^3/4$ inches wide, and $2^1/4$ inches thick, and weigh around 5 pounds. These dimensions, however, will vary. Most bricks are made so that when they are laid in the usual fashion with mortar joints between them, the overall dimensions are modular and numerous different bonds, or patterns, can be made up. But if they are laid as a floor covering and butted together, some of the modularity is lost and not all of the patterns will work out. Some paving bricks, however, are sized to be used without mortar joints, and these can be laid in a full range of patterns.

Layout

The first step in laying a brick floor is to determine the pattern you want (Fig. 2-81). If it will be other than a straightforward array, it is often easiest to map out the basic pattern on graph paper first. Then set up a layout and some working guidelines on the floor. Using the layout as a guide, make a dry run and determine what bricks, if any, must be cut and pieced in, and also if the pattern is workable. If you have a repeating pattern, check to make sure that the repeats balance out as desired in the laying area.

In most laying processes there is time enough to make the cuts as you go along, but cutting first is often easier. To cut by hand, score the brick with the corner of a cold chisel or a file tang, then set the brick on a hard, smooth, solid surface. Fit the cutting edge of a masonry block chisel into the score, hold the chisel vertical, and rap it hard with a hand sledge. This results in a rough edge, which can be smoothed fairly well by rubbing it

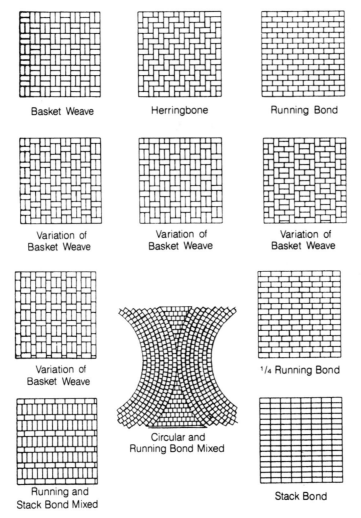

Basket Weave   Herringbone   Running Bond

Variation of
Basket Weave

Variation of
Basket Weave

Variation of
Basket Weave

Variation of
Basket Weave

Circular and
Running Bond Mixed

¹/₄ Running Bond

Running and
Stack Bond Mixed

Stack Bond

Fig. 2-81. *Examples of brick paving patterns; there are other possibilities as well.*

against another brick or a concrete block. For more precise and even cuts, make them (or have them made by a contractor) on a power masonry saw.

Installation

The easiest way to lay a brick floor is to set the bricks dry, and there are a couple of variations. Calculate the layout so that the rows of bricks will fit tightly into place between the walls of the room, in both directions. Put down a layer of sheet plastic over the whole floor first. Then just lay the bricks out in rows, butted tightly together, right on the floor. That's it. This can be done on any smooth and level concrete floor, whether damp or not, on any strong wood floor, or even on a compacted interior dirt floor.

Another method is to lay the bricks with open joints between them, in whatever pattern you choose. After all the bricks are laid, mix up a joint filler composed of 1 part portland cement and 3 parts fine silica sand. Pour this out dry on the bricks and sweep it around, thoroughly filling all of the joints. Walk around on the floor and thump the bricks repeatedly with a broom handle to help settle the filler. Add more mix as necessary. When the joints are full, sweep all the excess up and make sure the surface is absolutely clean. Then spray the whole area with a very light mist of water from a garden hose once or twice a day (depending on how fast the surface appears to dry out), for 3 days. In due course, the filler will harden nicely.

A third arrangement affords better stability, but does not work well on a dirt floor. First put down a layer of plastic sheeting. Then lay the bricks one by one, spreading the butting side and end of each brick with mortar as you lay it. Leave the bottoms dry. Set each brick down flat on the floor and press it against its neighbors until you get the appropriate mortar joint width, then strike the excess mortar off with your trowel. Plain premixed and bagged sand-cement mortar mix works fine for this purpose. You can leave the joints rough, just as they are struck off, or tool them for a denser, stronger joint and a more finished appearance. Start the tooling just as the mortar looses its sheen or wet look, and work section by section. For best results, use a masonry jointing tool of appropriate size and shape. After the jointing is done, spray a light mist of water over the entire surface, then cover it with sheets of plastic. Allow the mortar to cure for at least three days, then let the floor dry out through natural ventilation over a period of time. As an alternative to covering the floor, you can spray it with a light mist of water every few hours for 3 to 5 days.

To lay brick flooring or walkways in an earth floor greenhouse or sunspace, you can bed the bricks in sand. Where the bricks will extend wall to wall, the walls will serve as edge restraints. Otherwise, wherever the brick ends and the earth begins, cut a narrow trench in the earth to a depth equal to the length of a brick—or a bit less if the brick edges will protrude above the surface of the earth. Set border bricks on end and edge to edge in the trench, and tamp around them firmly. Spread a 2-inch layer of sand over the entire area where the bricks will be laid, and rake it out smooth. Then lay the bricks face up in the sand, with or without joints, according to your design (Fig. 2-82). You can fill the joints with plain sand swept into the spaces, or use the sand-cement mixture just discussed.

The two most effective ways to lay an interior brick floor are: provide a reinforced concrete slab $3^1/_2$ to 4 inches thick (or lay on an existing one), or provide a very sturdy wood floor system that has no appreciable deflection—especially under live loads (Fig. 2-83). The concrete should be well cured before the brick is laid. A wood floor must be covered with an isolation membrane of 15-pound roofing felt or 6-mil polyethylene sheeting. Then trowel a setting bed of fresh mortar about $1/_2$ to 1 inch thick into place. Butter each brick with mortar at sides and ends, slip into place, and carefully align and level. Tool the joints, then moist-cure the whole affair for several days.

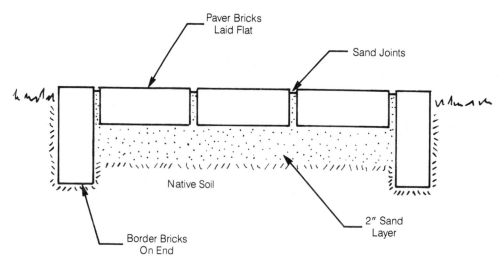

Fig. 2-82. *Cross section of sand-bedded brick paving.*

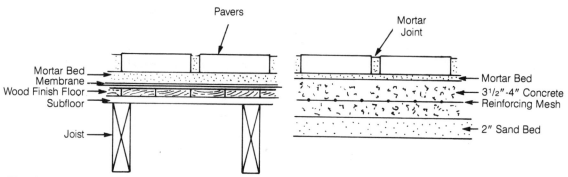

Fig. 2-83. *Cross section of mortar-bedded brick paving on an existing wood floor (left) and a concrete floor (right).*

FLOOR TRIMWORK

The joint between the finish flooring and the finish wall, which shows a construction or an expansion gap, is covered by a molding. This molding also serves as protection for the lower part of the wall, and provides a decorative effect. There are numerous kinds and arrangements of molding that can be used.

Wood is the most common material. The moldings are nailed either to the wall or, in some cases, at a 45-degree angle into the subfloor. They are seldom nailed directly to the finish flooring, because this would interfere with the action of the expansion gap. There are three basic wood moldings: a base molding—also called a baseboard or a mopboard; a base cap molding which is installed atop the base; and a base shoe molding, which is installed on the floor and butted up against the face of the base molding (Fig. 2-84). Sometimes the base is used alone, especially with carpeting that can be abutted tightly against

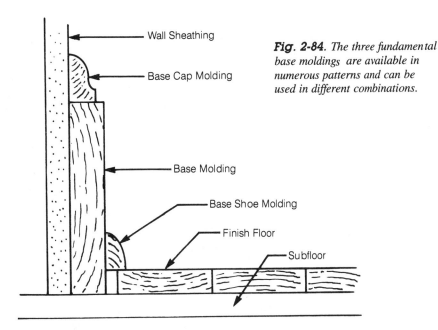

Wall Sheathing

Base Cap Molding

Base Molding

Base Shoe Molding

Finish Floor

Subfloor

Fig. 2-84. The three fundamental base moldings are available in numerous patterns and can be used in different combinations.

it. Often, however, the base is installed at the subfloor level, and a base shoe is added after the finish flooring is put down. The base cap functions only as a decorative afterthought.

The simplest approach is to install a narrow strip of S4S (surfaced four sides) kiln-dried wood stock, or a narrow curve faced molding, often known as *ranch* or *clamshell molding*, as a simple base to hide the floor/wall joint. Such moldings have little decorative impact, especially in large rooms, and are generally used in inexpensive homes. For greater emphasis, the base molding may be nominal 1-inch stock from 6 to 10 inches wide. The face of the base may be plain or custom milled in various patterns, such as veining or beading. The upper edge may be milled to an ogee, a cove, a cove and bead, or numerous other patterns.

The base shoe is generally small and unobtrusive, and may be a curve-faced shoe molding, quarter round, or a small stop or cove molding. Again, there are several possibilities. The base cap is also small: 3/4 inch thick at most, and often less. It may be selected from a variety of molding types, such as cove, screen, batten, bead, or even a small crown mounted upside down. Dozens of base combinations can be made up with moldings available from stock at your local lumberyard or produced in your workshop. (Figure 2-85).

Usually all the base trimwork is matched to and blends with other trimwork in the room, such as door casings, and is painted the same color. However, a natural or stain finish is sometimes used on base shoe moldings that trim out similarly finished wood strip, plank, or parquet flooring, with the baseboard painted. For more installation details on perimeter moldings of this sort, refer to the "Ceiling Trimwork" section in chapter 5.

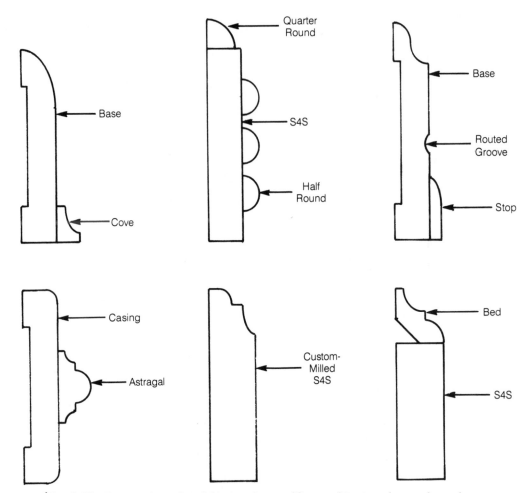

Fig. 2-85. *These are but a few of the many base molding combinations that can be made up from two or more stock patterns or custom molded.*

Another kind of molding that is used primarily with sheet vinyl, resilient tile flooring, and sometimes with ceramic tile, is called *wall base* or *base cove* molding (Fig. 2-86). The most common type is set-on, the bottom flange of which rests upon the finish flooring. It is available in both vinyl and rubber. Vinyl comes in strips and can be molded into or around corners with no trouble. Rubber wall base comes in lengths, with separate inside and outside corner pieces. Both kinds are glued to the wall surfaces with mastic. Several solid colors are available.

The only other trimwork that is associated with finish flooring—except for specific product accessory items, such as metal edging for carpeting—is stencilwork. This is often a necessity in restoration work in old houses, especially Colonial and Early American architecture, where stencilwork was often applied around the borders of painted plank

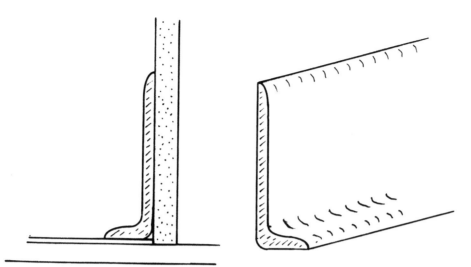

Fig. 2-86. *Molded vinyl or rubber set-on base cove is widely used with vinyl sheet and resilient tile flooring products.*

floors. It is equally effective in some modern decors, and can also be applied on natural finish flooring beneath a clear topcoat.

Various commercial stencil patterns are available, or you can make up your own in any design you wish. The process is simple: merely place the stencil on the floor repeatedly in a border, or whatever other pattern you wish. Paint over it with a stencil brush and stenciling paint. After drying thoroughly, the stencilwork should be covered with at least one coat of a compatible clear finish, such as varnish.

3
CHAPTER

Walls

WHEN IT COMES TO HOME REDECORATING OR REMODELING, THE INSIDE WALLS RECEIVE
the most attention. This makes sense; they are the most obvious part of the rooms, the
vertical boundaries of your living space. They are the most noticeable to yourself as well
as to visitors. Together with the windows and doors, they comprise the greatest interior
surface area of the house. They are the easiest to work on because of their size, position,
and accessibility, and the easiest to work with because there is such a huge range of decorating and remodeling possibilities. With few exceptions, all of the work can be successfully accomplished by any do-it-yourselfer with, at most, modest skills and a little
experience.

REPAIRING WALL SURFACES

Making repairs to a wall finish such as wallpaper, paint, or a fabric covering is almost
never satisfactory. Even a considerable expenditure of time and effort usually results in an
obvious patch. Thus, finish repairs can only be considered a temporary expedient at best.
Structural damage to a wall requires major renovation that often is best done by a professional contractor. This is especially true if the damage has been done to a load-bearing
wall and/or the cause of the damage is uncertain—there might be a major structural problem. However, if not too extensive, surface damage can usually be satisfactorily repaired
in preparation for a new wall covering. The ways in which the repairs are best made
depend not only upon the old wall sheathing or covering, but also upon the nature of the
new covering to be applied. Also consider practicality: repairs can sometimes be more
time-consuming or expensive than resheathing or tearing the old sheathing off and starting
afresh. Sheathing will be covered later in the chapter.

Plaster

If a sound plaster wall is to be covered with a rigid covering such as paneling or planking, no repairs to minor defects need be made. It it is to be papered, you should fill cracks,

holes larger than $1/16$ inch, and all small scrapes and gouges with vinyl paste spackle, then sand smooth. Defects or holes larger than $1/2$ inch or so, as well as large cracks, are best filled with patching plaster. Long and wide cracks at places where a filling is likely to crack again, such as above door or window corners or in wall corners, should be covered with mesh wallboard tape, then coated with joint compound or patching plaster. This might require two or three applications. After the compound cures, sand the patches smooth and feather the edges out onto the original wall surface.

If a large chunk has broken away from wood lath, check to see if the lath, or any of its neighbors, are broken. If the lath is sound, cut the old plaster away from around the edges of the gap, back into sound, firm plaster. Then fill with patching plaster in two or three applications, and sand smooth. If the lath is broken, cut the plaster away along the lath by scoring it repeatedly with a utility knife, back to the nearest wall studs. Cut the lath off at the centerline of each stud with a wood chisel, put in a new length of lath, and fill the gap with patching plaster. If the lath is mesh or expanded metal, it might be dented in, but probably will not be broken. If it should be, wire it together with crisscrosses of soft iron mechanic's wire, twisted tightly. Then fill with patching plaster.

If the wall is to be painted follow the same procedures, but fill all visible holes, scrapes, gouges, or similar defects. Also level off any low spots with spackle and/or patching plaster, and sand off any high spots. The wall surface must be as smooth and defect-free as you can possibly make it.

Gypsum Wallboard

No repairs are necessary if the wallboard will be covered with a rigid material, provided it is sound and well secured to the studding.

If the wall is to be papered, fill all holes, gouges, and other defects larger than $1/16$ inch with vinyl paste spackle. Check for popped nails. Dig the filling away from any lifted nail heads or any filled spots that show cracks around a hidden nail head. Drive the nails in as necessary, and re-cover the heads with joint compound or vinyl paste spackle. Check all of the taped joints. If there is any loose or lifted tape, peel the loose portions away with a knife and retape the area with mesh joint tape. If there are cracks and the tape is torn, remove whatever you can of the old material and retape with mesh tape. Then cover the tape with joint compound in two or three applications. Sand all of the filled spots smooth and feather the edges out.

If a chunk has been broken out of the wallboard, use a keyhole or wallboard saw to cut away the broken edges. Cut back into solid material, in the form of a rectangle. Cut a piece of stiff material such as hardboard, thin plywood, or $3/8$-inch or $1/2$-inch gypsum wallboard to make a backer. This piece should be wide enough to slip through the cut hole on the diagonal and long enough to overlap top and bottom (or side to side) by at least 1 inch, preferably more.

Drill a pair of small holes through the center of the backer and thread a length of soft iron mechanic's wire through the holes in a loop, leaving about 6 inches of free end sticking out of each hole. With your finger, smear a coating of construction adhesive around

Fig. 3-1. *Twisting the string until the stick is tight against the wall surface will hold the wallboard patch backing in place.*

the inside perimeter of the cut hole. Work the backer through the hole, tilt it into place, and pull back on the wire to press the backer into the adhesive (Fig. 3-1). Wrap the wire around a length of dowel or a pencil and twist until the end and the backer is sandwiched against the wallboard firmly. Don't, however, use too much pressure. Then drive several bugle-head wallboard screws through the wallboard and into the backer. Allow plenty of time for the adhesive to cure before proceeding.

If the hole is only 3 or 4 square inches, you can fill it with patching plaster in two or three applications. If it is larger, cut a patch of wallboard to match the hole and bevel the face edges slightly outward. Also, bevel the face edges of the hole (Fig. 3-2) before putting in the backer—a coarse toothed file will make short work of the job. Vacuum away the plaster dust. Apply a few thin smears of construction adhesive to the back of the patch and set it in place against the backer. Secure it with several wallboard screws driven gently into the backer. Then fill the bevel edged seam with patching plaster or vinyl paste spackle and sand smooth, feathering the edges as necessary.

If a very large piece of wallboard has been broken out, a few square feet or more, a patch will be unsuccessful. Cut the wallboard away back to the nearest wall stud on each side of the break. With a chisel, cut down the sides along the centerlines of the studs and remove the damaged material. If the break is near the floor or the ceiling, cut all the way

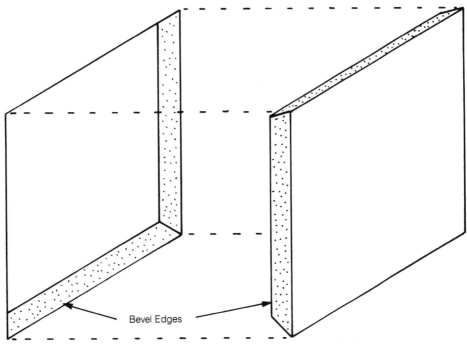

Fig. 3-2. *A snug-fitting wallboard patch with beveled edges.*

down or up until you reach the sole or top plate of the wall frame (Fig. 3-3). If there is horizontal blocking within the wall, cut until you reach that.

If a cut edge has no support behind it and you don't want to remove an entire full-height strip of wallboard between two studs, you can install a nailing support. Construct a bracket of 1- × -4 and 2- × -4 (or all 2- × -4) stock as shown in Fig. 3-4. The bracket should be just wide enough to fit between the studs. Slip the bracket up (or down) under the wallboard so that half of the crosspiece remains exposed and nail or screw it in place (Fig. 3- 5).Nail or screw the cut edge of the wallboard to the crosspiece. Then cut a matching patch from the same kind of wallboard and nail or screw it in place after beveling all edges. Space nails about 4 inches apart, screws about 8 inches apart. Tape the joints and cover them with joint compound, then sand and feather out the surface.

Planking and Paneling

If a planked or paneled wall is to be covered with another rigid material, no surface repairs are necessary. Just make sure that the surface is relatively flat and the old covering is solidly attached.

If the surface is to be painted or repainted, fill and sand smooth all defects. Small ones can be filled with either vinyl paste spackle or wood putty, while larger ones are best filled with a suitable wood filler. Splits or cracks can likewise be filled. However, large cracks,

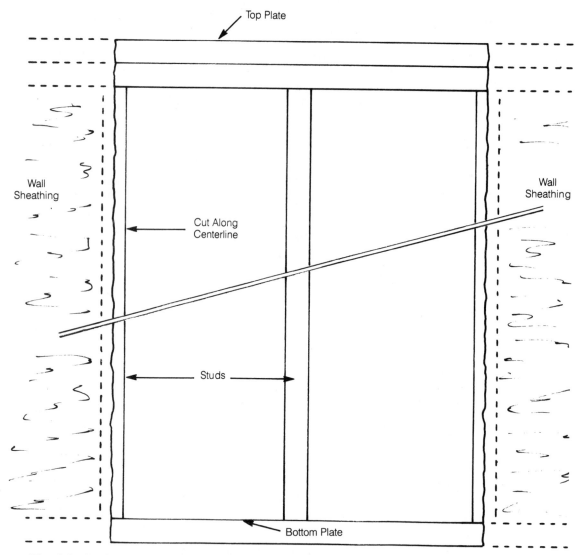

Fig. 3-3. *Cut damaged wallboard away from floor to ceiling and to the centerlines of the next wall studs beyond the damaged area.*

joint separations, and extensive splits often cannot be successfully hidden, and replacement of a panel or plank might be necessary. Broken-out, badly cracked, or shattered hardboard or thin plywood paneling must also be replaced with material of the same thickness fitted snugly in place.

If the surface is to be covered or recoated with a clear finish, the situation is different. Badly damaged hardboard or thin plywood must be replaced with matching material. Serious damage to a wall plank can sometimes be repaired by grafting in a new piece of match-

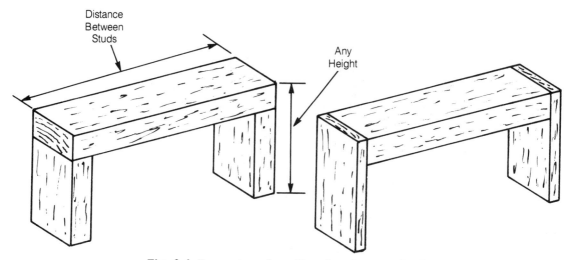

Fig. 3-4. Two versions of a wallboard patch support bracket.

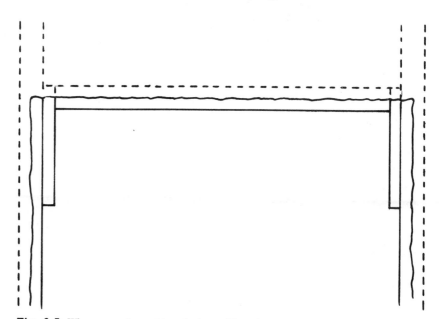

Fig. 3-5. When properly positioned, the wallboard patch bracket affords a nailing strip across the opening for both the old and the new wallboard edges.

ing wood, otherwise a plank must be replaced. If the wall has been stained, the problem arises of matching the stains between the new and the old surfaces, but this can often be done, with care and patience.

Minor defects like picture hook and nail holes, small cracks or splits, dents, and shallow gouges can be disguised in either of two ways. The first involves filling the defects

Fig. 3-6. *Minor defects in stained or natural woodwork can be readily repaired with either colored shellac sticks and a burn-in knife, or a putty stick of matching color.*

before applying the clear coating by melting a special filler into the defects, and at the same time smoothing the surface flush. This requires a selection of appropriately colored burn-in fill sticks (compounded of shellac and resins), used with an electric burn-in knife (Fig. 3-6), or a burn-in knife and an alcohol heat lamp. This will require a little practice. The other method involves first applying the clear coating, then using wax compound putty sticks of matching colors. The former method does a better job of filling shallow scratches and dents, the latter is only suitable for fairly deep but short scratches and holes. Light scratches that have not lifted the fibers of the wood can usually be sanded out well away from the scratch, so as not to create a narrow trench along the scratch line.

Ceramic Tile

The condition of a ceramic tiled wall that is to be covered with a rigid material is largely immaterial. However, if it is to be covered with one of the special wallpaper systems made for the purpose, all loose or broken tiles should be chipped out of their beds. You can do this with a hammer and a small cold chisel, or even an old screwdriver, but proceed gently to avoid breaking up the sheathing. Note: Wear safety goggles while removing tiles, especially with high glaze vitreous tile. Then fill the gaps with patching plaster. The plaster need not be sanded, just smooth it flush with neighboring tile surfaces, using a taping knife or mason's trowel. Two or more applications might be necessary, depending upon the plaster used, its thickness, and the shrinkage factor involved.

REMOVING OLD WALL FINISHES

It is often necessary, or at least advisable, to remove an old wall finish or finish covering before applying a new one. Otherwise, the final result could be unsatisfactory and the effort, not to mention the time and money, wasted. None of this work is difficult, only messy and sometimes tedious.

Paper

Applying new wallpaper over old paper, although occasionally done, is not a good idea. It is probable that hanging the new paper will loosen up the old, so that neither one adheres properly. It is best to remove the old paper.

If the paper is old and/or the thin, uncoated type, you might be able to remove it without any special equipment. First turn off all electrical circuits that feed the room, including those to wall lighting fixtures and wall switches. Tape off the circuit breaker handles or remove the fuses. Take down wall lighting fixtures and remove switch and convenience outlet covers. Cover the floors with plastic film dropcloths or construction plastic; securing the edges to the baseboards with masking tape is also a good idea.

In a bucket, mix up a solution of commercial wallpaper remover according to the manufacturer's instructions. Start at any convenient point on the wall near the ceiling and wash the walls down with the remover and a large spoon. An alternative to sponging is to mist on the solution with a garden insecticide sprayer. Make sure, however, that the container is absolutely clean and free of any pesticide residue. As the solution soaks in, wipe or mist more on until the paper becomes well saturated. Then wait the noted length of time for the solution to loosen up the wallpaper adhesive—usually this takes about 5 minutes.

Start at a seam near the top of the wall and slide the blade of a wide putty or joint knife under the edge of the paper (Fig. 3-7). The paper should peel off easily in large sheets or curls. If if does not, wait a bit longer. If the paper appears to be drying, apply more solution. Eventually the paper should lift off easily, for the most part. There will probably be a few patches that cling tenaciously; just keep sopping them until they break free. As you scrape, be careful not to let the knife blade dig into the surface sheathing. Although the surface of good plaster will remain firm, old hair plaster tends to soften somewhat and gypsum wallboard might become tender in places.

Note: If the sheathing happens to be gypsum wallboard that was not primed before the paper was originally applied, and possibly not glue sized either, you've got a problem. The paper surface of the sheathing will probably become soggy and portions of it will tear away, leaving a rough and ragged surface and perhaps even exposing the gypsum core. Should that occur, you might as well quit, because you won't end up with a surface that can be satisfactorily repapered, and certainly not painted. Your only recourse is to cover the old sheathing—wallpaper and all—with a new sheathing material such as $^3/_8$-inch gyp-

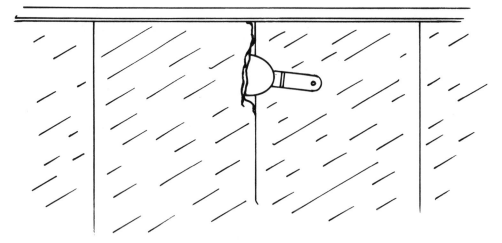

Fig. 3-7. *Start stripping wallpaper with a wide-bladed putty knife close to the wall top at a seam.*

sum wallboard (this is covered later in the chapter). Then apply a new finish covering, or install wall paneling or planking.

If the old wallpaper is heavy paper, foil, vinyl, or vinyl-coated paper, or fabric, dampening it with a sponge for removal will not work. You will have to rent a wallpaper steamer. These machines vary somewhat in operation, so be sure to get instructions for use from the renter, as well as a supply of the proper removing solution to use with it. If the wallpaper is really paper or is fabric, you can go right ahead with the steamer. If it is foil, vinyl, or vinyl-coated paper, go over the entire wall surface first with coarse sandpaper wrapped onto a sanding block. Cut crosshatched grooves through the foil or coating surface into the backing of the material, but not into the sheathing.

Start in a lower-right corner of a wall (lower-left if you're left-handed) with the steamer held in your left hand. Hold if flat against the wall, and move it slowly upward and left in repeated passes. Meanwhile, follow along behind with a wide-blade putty knife in your right hand, peeling the paper off in strips (Fig. 3-8). Keep repeating the procedure until all the walls are stripped clean; it might be necessary to go back over some stubborn patches. With fabrics or vinyls, you might have to stop occasionally and cut away free strips with shears so they don't get in your way.

After all the paper has been stripped off, wash the walls down with a large sponge and clean, warm water at least twice. You can remove thick, stubborn glue patches with warm water and mild detergent or with a commercial wallpaper adhesive remover. Don't, however, scrub too hard. Give the glue a chance to soften in repeated applications, wiping away the loosened portion each time.

Paint

Paint coating—especially those typically used on walls, such as latex—cannot be successfully stripped from plaster or gypsum wallboard walls. However, there is usually no

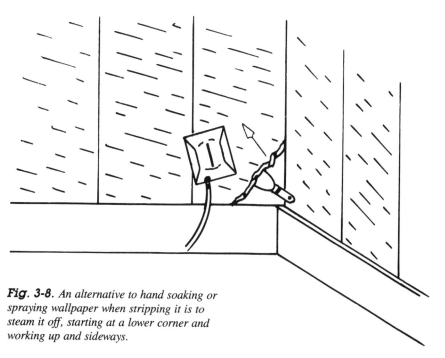

Fig. 3-8. *An alternative to hand soaking or spraying wallpaper when stripping it is to steam it off, starting at a lower corner and working up and sideways.*

need, because the walls can be repainted, papered, or otherwise covered. Paint or varnish, on the other hand, can be stripped from wood, hardboard paneling or planking, or wood trim. Stain can also be removed, though this is a more difficult chore. Some residual stain almost always remains, and must be neutralized by bleaching.

Remove paints or varnishes with commercial strippers, following the manufacturer's instructions exactly. The stripper to use depends upon what is available, the kind of paint or varnish to be removed, and the nature of the undersurface. Be guided by the recommendations of your supplier. If you don't know what kind of finish you'll be stripping, you might have to try two or three different strippers before you get satisfactory results.

Paneling and Planking

Removing old paneling or planking is just a small demolition job. If the material was nailed in place, it should come away easily with the aid of a claw hammer and a prybar. Although you probably won't be able to save much of the old material even if you want to, by proceeding carefully you should be able to avoid any substantial damage to the sheathing behind it (assuming there is any). For the most part this involves prying gently, and setting the prybar on something solid. If you have to pry in an open area between studs, rest the heel of the bar on a fairly long piece of scrap: 1 × 6, for example. This will prevent the bar heel from driving through the sheathing by spreading the pressure over a wide area.

If the paneling or planking was glued to the sheathing, or glued and nailed, your task will be harder. There's not much chance of ending up with a sheathing surface smooth enough for painting or papering, because large chunks will probably tear away with the finish material. However, if you are careful, the sheathing might remain sound enough that it can be covered over with another sheathing or with new paneling or planking. If the sheathing breaks up badly as you remove the paneling or planking, you have little choice but to strip everything away. In some cases, it is easiest to remove all the old finish first, then attack the sheathing. In others, both will come off together.

Regardless, first remove the electrical and plumbing hardware, cover the floors with dropcloths, and remove all the moldings, casings, and other trimwork. If you think there will be a lot of flying dust, seal off doors into the room. Wear gloves against splinters and mashed thumbs, and wear safety goggles and heavy-soled footwear against nails. If a lot of debris will result, you might be able to arrange with your trash collector to supply a dumpster (right outside a handy window is the best setup) on a short-term basis. Clearing the rubbish away almost as quickly as it accumulates makes the job easier, faster, and safer.

Ceramic Tile

Again, this is nothing but a demolition job. There is no hope of salvaging the sheathing or backing in decent enough condition to be re-covered, nor will the tile be reusable, whether it is bedded in organic mastic or one of the thin-set materials. As with glued paneling, it might be easiest to peel the tile off the backing first and then remove the sheathing, or to tear the whole business down at once. This depends upon what materials are involved. By chipping or breaking out a few tiles at some point along a wall stud line where there is good support, you will be able to find out. Once you hack out a starting spot, the rest will follow. One thing to watch out for during the process is pipes and wires hidden within the wall; these can be easily damaged if you get too enthusiastic with the prybar. Leather gloves and safety goggles are in order for this job, because sharp slivers of tile can fly everywhere when they break under pressure or impact.

Wainscoting

Removing false wainscoting is a simple job, because all you have to do is peel the chair rail or belt molding off the wall and discard it. The trick is to do so without excessive damage to the sheathing, which probably can remain to serve at least as a backup for a new sheathing or rigid finish.

To remove a true wainscoting, first remove electrical and plumbing hardware and the base molding. The cap molding along the top of the wainscoting comes off next; it will be nailed to the wall studs and/or the wainscoting material. Thin, glued paneling can be peeled off the sheathing with a claw hammer. The glue sometimes brings pieces of sheathing along with the paneling. By working carefully, however, you should be able to leave it intact to the point that it won't have to be removed, but just resheathed or covered with a rigid finish.

Planking or thick paneling will probably be nailed to blocking within the wall, or to furring strips nailed through the sheathing into the wall studs. Some judicious prying will free the paneling or planking, quite possibly without any important damage. Surface-mounted furring strips can also be gently ripped off, leaving the sheathing reusable after a bit of repair work. The wainscoting might be secured to furring strips attached directly to the wall studs, or to wall blocking, with no sheathing present over that section of wall. In that case you might be able to resheath just the open space, depending upon the exact original construction method and materials, and how you plan to refinish the wall. Otherwise, remove all the sheathing and start afresh.

COVERING OLD WALL SHEATHING

If you wish to repaint or repaper old plaster or plasterboard walls but they are in poor condition, the job will be difficult and the results will be disappointing. A fresh sheathing surface is in order. Likewise, if you wish to paint or paper over old fiberboard walls, hardboard, plank, or plywood that is rough, damaged, or full of defects, a new, smooth surface that is compatible with paint or wallpaper will make the job easier and give better results. If the old sheathing is reasonably sound—not badly buckled or broken out or weakened to the point of practically falling apart—you can probably resheath without much difficulty. The presence of old paint or wallpaper does not matter. However, if the old sheathing is in extremely poor condition, or if you wish to install insulation within the wall cavities or reinsulate, or add a layer of rigid insulation to the inside of exterior walls, you will have to remove the old sheathing completely and start over.

Materials

The sheathing most commonly used for this purpose is gypsum wallboard (gypboard, drywall, Sheetrock) which has tapered edges. If the walls are sound and very smooth, and the old sheathing is tight and free of holes or breaks, the $1/4$-inch thickness can be used. In most instances, however, use $3/8$-inch minimum; $1/2$-inch is even better and is universally available. If you are working alone, the 4- $\times$ -8 foot sheet size is the easiest to handle; set them up vertically. If you have a helper, you can use lengths up to 14 feet (same width) and run them horizontally, which leaves less joint work. You can secure the sheets with plasterboard nails, but the special Phillips bugle-head screws made for the purpose are much better, and can be quickly and accurately set with a power screwdriver.

Preparation

To begin, remove the moldings around the wall/ceiling joint and the wall/floor joint. Pry them off carefully. If you can save them for later replacement, number them in sequence on their backs so you will know where they go. Also, remove all the window and door casings (see chapter 4); these will probably have to be replaced. Turn off all the electrical circuits that supply the outlets. Check each one with a lamp to make sure it is indeed off, then remove the relative fuses or put a piece of tape over the circuit breaker handles. Make sure all wall switches are deactivated, too. Investigate to see if you will be violating a local building code by doing your own electrical work. If so, or if you are leery of it, hire an electrician to do that part of the job. If not, remove the outlet covers and set them aside, then take the mounting screws out of the outlets. Pull the outlets out of their boxes, twist them gently around sideways, and stick them endwise back into the boxes (they won't go all the way in). Do the same with the wall switches.

The next step is to build out the door and window jambs. Purchase molding stock or rip out flat strips of wood about $3/4$ inch wide and the same thickness as the plasterboard you will be putting up. Fasten these strips to all the window and door jamb edges with brads or 2d finish nails, to extend the jamb edges out flush with the surface of the new sheathing. You can set the strip edges flush with the jamb faces, but for easier and more presentable finishing, set them back about $3/16$ to $5/16$ inch from the face, depending upon your preference (Fig. 3-9).

Unless the old sheathing nails are visible, you need to locate the wall studs. Exception: If the old sheathing is $3/8$-inch or thicker plywood, or $3/4$-inch planking well secured to the wall frame, you can go ahead and fasten the new sheathing to the old. If not, you must fasten the new sheathing and at the same time refasten the old directly to the wall studs. An inexpensive magnetic stud locator will usually locate the studs, but a more expensive solid-state or electronic one will work better and faster, and will also pick up pipes and wires. Locate each stud, and make a small mark on the floor and ceiling to denote the centerline of each one. They will probably be on regular 16- or 24-inch centers in houses 30 to 40 years old or less, and probably won't be in very old houses. In any case, at least a few of the studs will be at odd spacings; look for them. Window and door openings will be fully surrounded by studs and headers.

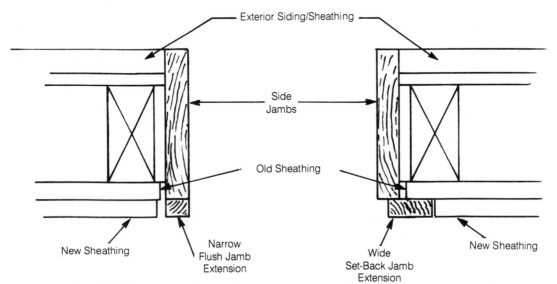

Fig. 3-9. *This top view shows two methods of extending window or door jambs flush with new wall sheathing laid over old.*

Layout

Now pick one corner of the room and make a mark on one wall or the other, 1 inch out from the corner at the ceiling level. Drop a plumb bob down to the floor with the line held on that mark, and make another mark at floor level. If it is also 1 inch from the corner, the wall adjacent to the one you marked is plumb (vertical) and that's the wall to start on. If the measurements differ, try another corner. If none of the walls are plumb, you will have to trim the corner edge of the first sheet of plasterboard to compensate, so that the sheets will stand plumb as you set them.

If you find a corner where one wall is plumb to the adjacent wall, measure out along the floor 4 feet and see if there is a stud centerline there. If so, you're ready to start. If not, measure out from the corner until you get to the stud centerline closest to 4 feet away. You will have to trim from the sheet an equal amount to the difference between that measurement and 4 feet. For example, if the stud centerline is 43 inches from the wall, you will need to trim 5 inches off the sheet. This should be done to the edge that will fit into the corner, not the one along the wall (Fig. 3-10).

If no walls are plumb (not unusual), lay a sheet of plasterboard on the floor where you can work on it. Measure the floor-to-ceiling height, subtract about $1/2$ inch for working clearance, and transfer the measurement to the sheet. With a straightedge, draw a line across the top of the sheet to indicate the trim cut. Determine how far out of plumb the wall is that is adjacent to the one on which you want to start. For instance, the wall might lean so that the top is 1 inch out of line with the bottom. If the lean is outward at the top, measure off 1 inch from the bottom corner of the sheet—from the side edge that will fit into the wall corner and along the bottom edge—and make a mark. If the lean is inward at the top, mark off 1 inch from the edge along the height trim-cut line. With a straightedge, draw a line connecting the mark at the bottom edge with the intersection of the height trim-cut line and the sheet edge, or the bottom corner with the mark on the height trim-cut line. This is the initial width trim-cut line. Then, determine the distance from the wall corner to the wall stud closest to 4 feet (or less) away. If it is 42 inches, for example, subtract that from 48 inches and make a new width trim-cut line 5 inches from and parallel with the initial one. Figure 3-11 shows a typical layout.

Installation

Cut the sheet to fit. With a straightedge and a sharp utility knife, score through the paper facing of the sheet and into the gypsum along the height trim-cut line. Stand the sheet on edge with the finish surface facing away from you. Hold the sheet upright with one hand, grasp the top edge with the other, place your knee against the back of the sheet right at the score, and snap the waste piece back sharply against the pressure of your knee. The gypsum will break and the paper backing will fold over. Cut the backing along the fold with the utility knife and discard the scrap piece. The cut edge will be rough; smooth it off with the knife, keeping the blade almost flat, and pull it toward you as though peeling bark off a stick.

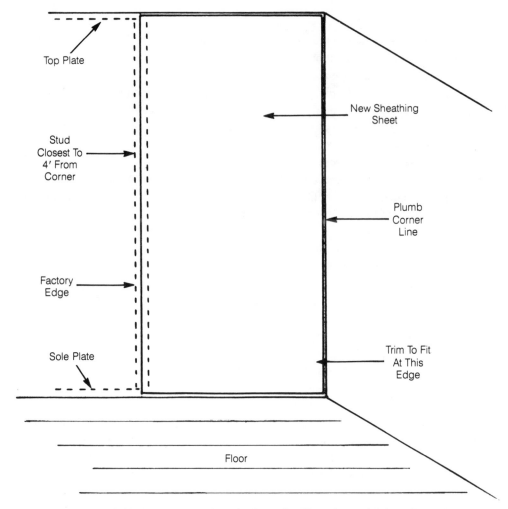

Fig. 3-10. *Trimming and fitting a sheet of wallboard to a plumb corner.*

Lay the sheet flat again, and score along the width trim-cut line. Stand the sheet on edge again, and snap the waste piece back. Smooth the rough edge. Then stand the piece in place and tuck it into the corner to see how it fits. A few gaps or irregularities are to be expected, but their width should be no more than about half the thickness of the plaster-board. If the adjacent wall surface is bowed or wavy, you might want to shave the edge somewhat for a better fit. Be careful not to take too much off, however, or you will narrow the sheet to the point where the other side edge will not lie over a stud and you'll have nothing to fasten the sheet to. If you have measured correctly, you should have a decent fit at the wall corner, and the opposite side edge of the sheet will be plumb and aligned with the centerline of a wall stud.

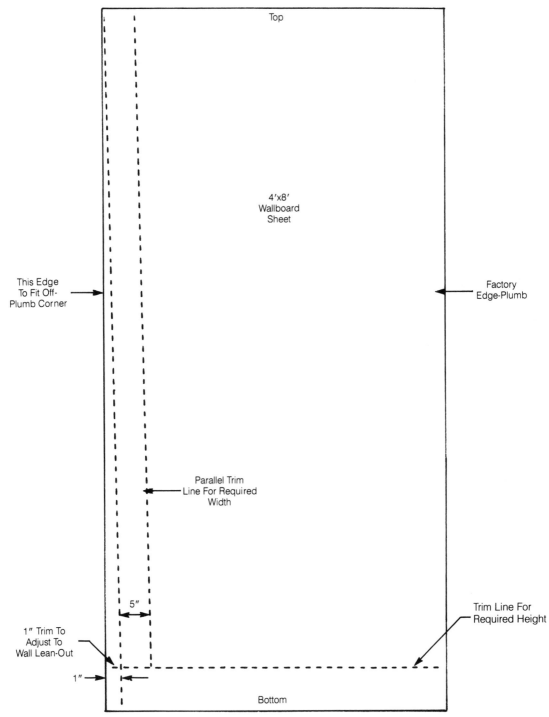

Fig. 3-11. *A layout example for trimming a wallboard sheet to an off-plumb corner and a ceiling of less than full-sheet height.*

With the sheet properly fitted, tack it into place with a few nails or screws at each side. With a straightedge (a straight piece of furring strip also works well), line up each top and bottom set of stud locating marks that you made earlier, and draw a light pencil line on the plasterboard as a guide for driving the fasteners. Start about mid-height at one of the intermediate stud guidelines and fasten upward to the top, then downward to the bottom. Repeat the process at other intermediate studs, then go to the side edges. Finally, fill in the gaps along the top and bottom edges.

Space nails approximately 8 inches apart on centers, screws about 16 inches on centers. Keep these at least $3/8$ inch from the edges. Drive nails with a hammer that has a slightly crowned or rounded face, which will leave a dimple in the facing of the plasterboard. The nail head should be slightly recessed and at the center of the dimple, but not so deep as to break the paper (Fig. 3-12). Drive screws with a power screwdriver, preferably clutch-equipped, so that the heads are recessed just below the surface.

Once the first panel is fitted, aligned, and fastened, all the others follow in progression around the room (Fig. 3-13). Electrical outlets and wall switches must be cut in as you go along, by transferring the location and measurements to each sheet as necessary. If there are several to do, but an outlet box of the same size at your hardware store and use it to trace around. Once outlined, you can make the cutouts with an ordinary keyhole saw, but a short, tough, plasterboard saw is easier to handle and will stay sharp much longer.

When fitting to windows and doors, or around heat registers, built-in shelving, or similar items, transfer the measurements to the sheets as necessary. Leave about $1/4$ inch

Fig. 3-12. Wallboard nails should be driven in so that a dimple is formed by the hammer head, but not so deep that the paper covering is broken and the gypsum crushed.

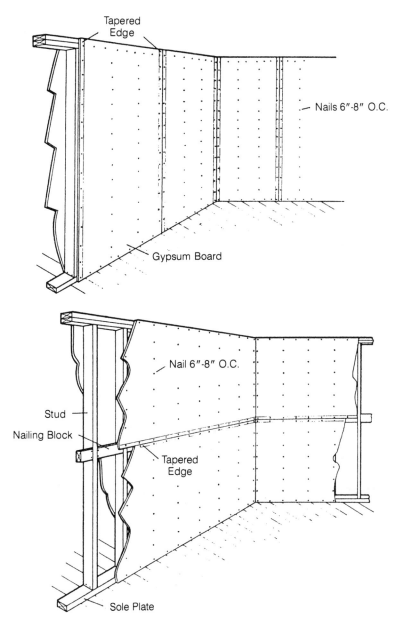

Fig. 3-13. Properly applied gypsum wallboard. (Courtesy USDA.)

or so of leeway; moldings will cover the gaps. The taper-edged joints between sheets should be snug. Do adjustment trimming along the top or bottom edges or the cut edges of the sheets, as necessary. Try to avoid butting two cut edges, or a cut and a tapered edge, even though this will be necessary at many corners and odd-piece joints. At outside corners, overlap the sheets so that the edge of one is flush with the face of the other, top to

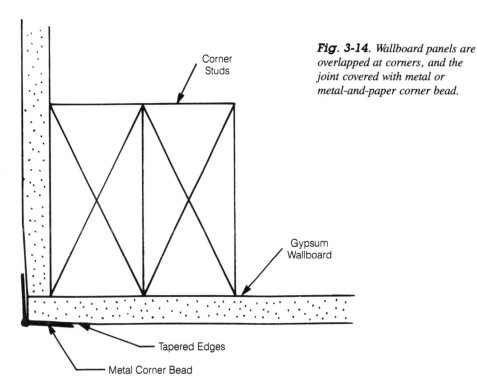

Corner
Studs

Gypsum
Wallboard

Tapered Edges

Metal Corner Bead

Fig. 3-14. Wallboard panels are overlapped at corners, and the joint covered with metal or metal-and-paper corner bead.

bottom (Fig. 3-14). Cover each outside corner with metal cornerbead, nailed into place. Inside corners need no added treatment.

Finishing the Joints

After all the plasterboard has been put up, the joints between sheets must be taped and finished. For this you will need some special tools and supplies (Fig.3-15). Three joint knives are required, which you can purchase separately or as a set. The blade width selection varies, but a 3-inch, 6-inch, and an 8- to 9-inch work well. You will need a supply of joint compound; use the ready-mixed type that comes in 1- or 5-gallon pails. You will also need a roll or two of joint tape. Perforated paper tape is the old standby, but if you have never done any taping before you might find the newer self-stick mesh tape easier to apply.

To tape the flat joints (those between two edge-butted pieces) with paper tape, hold your narrowest joint knife at a 15- to 20-degree angle and butter the joint from top to bottom with compound (Fig. 3-16). Keep the thickness even and the surface smooth, and stay between the edges of the tapers, where present. Make sure the joint itself is well filled, but don't make the layer of compound too thick; a little practice will show you how much is needed. Work from top to bottom in as few long, steady strokes as you can manage, and rework as little as possible. Then lightly press the strip of paper tape against the wet compound, starting at the top and smoothing it gently into place down the centerline of the joint (Fig. 3-17).

Fig. 3-15. *This selection of tools and materials is all that is needed for a joint taping job.*
The black-handled tool, not essential but handy, is for smoothing right-angle corners.
Either paper or mesh tape can be used.

Immediately start at the top. Hold your mid-size knife at a 45-degree angle and press the tape firmly into the compound, smoothing it as you go along (Fig. 3-18). Work out any air bubbles and make sure the edges are down tight, but not too tight. There must be at least $1/32$ inch of compound under the tape at the edges to make the proper bond. Then, before any drying can take place, use the same knife held at the same angle and cover the tape with a thin skim coat of fresh compound. At the same time, remove any excess compound that oozes out, but do not apply so much pressure that the tape is disturbed in its bed. The tape should be just visible under the skim coat, and the edges of the compound should extend beyond the edges of the tape but still be within the edges of the taper in a band roughly 3 inches wide (Fig. 3-19).

Allow the joint to dry thoroughly, then apply a second layer of compound with the mid-sized joint knife. Feather this coat out 2 inches beyond the edges of the first one. Use long, smooth strokes, and rework only if unavoidable. The tape should now be invisible. Again allow the joint to dry, and apply a third coat of compound with your widest knife, feathering it out 2 inches beyond the edges of the second coat. This coat fills the taper

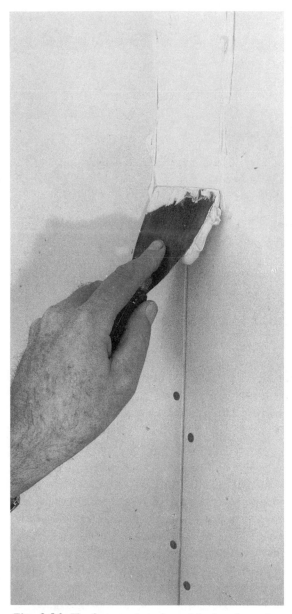

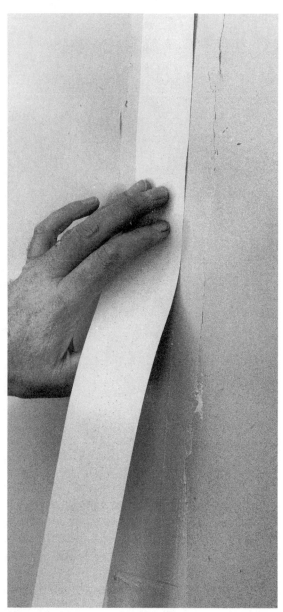

Fig. 3-16. *The first step in taping a joint is to butter it with joint compound.*

Fig. 3-17. *Step 2: Gently press the paper tape onto the fresh joint compound.*

depression and extends somewhat beyond the taper borders in a very thin skim coat (Fig. 3-20).

Use mesh tape in much the same way, except do not butter the joint with compound first. Instead, unroll a few inches of tape from the roll and stick it to the wall at the top, centered on the joint. Then unroll the tape along the joint, sticking it down as you do so,

Fig. 3-18. *Step 3: Press the tape into the joint compound with a joint knife and smooth it, meanwhile squeezing out excess compound and any air bubbles.*

Fig. 3-19. *Step 4: Immediately spread a thin layer of joint compound over the freshly bedded tape, smoothing it as you do so, leaving a band about 3 inches wide.*

Fig. 3-20. *After a second application of compound has cured, apply a final, thin smoothing coat with your widest knife.*

Fig. 3-21. *Fill nail dimples with a swipe of a joint knife; two applications are usually sufficient.*

and cut it off at the floor line. Apply the first layer of compound so that it fills the mesh openings and any cracks or gaps beneath, and just barely covers the mesh itself. Apply the second and third coats the same way as with paper tape.

To tape inside corners, apply a narrow butter coat of compound down each side of the corner. Fill any gaps in the process, and keep the corner angle as sharp and clean as you can. Then fold a strip of tape in half lengthwise and set it in the compound. You can bed the tape by smoothing it on each side with your small joint knife, or you can use a special corner joint knife that works both sides at once. In either case, you will have to take care to keep the tape smooth and even on each side and to keep the center from wrinkling or driving into the joint. When applying the second and third coats, do one side of the tape, let it dry, and do the other side. Or do it all at once, whichever seems to work best for you. To do the outside corners, just fill across the cornerbead in three successive coats and feather out about 5 to 6 inches in each direction.

The dimples and fastener heads between the joints must also be covered with compound. Use your small joint knife at about a 45-degree angle and skim across the dimples, filling them (Fig. 3-21). After they dry, apply the second and third coats, just as with the joints.

The minimum temperature for taping joints is 55° Fahrenheit, and should preferably be at least a bit higher. Try to avoid temperatures above 85° Fahrenheit, especially if the humidity is low, or you will have trouble with the compound curing too fast. In dry conditions (below about 40 percent humidity), keep all windows and exterior doors closed, and humidify the room if possible. In damp and/or cool conditions, allow plenty of time for drying before applying fresh compound coatings; sometimes two or three days are needed.

The last step involves sanding the joints and the fastener spots smooth. Use fine sandpaper: 150-grit or 180-grit is all right if the wall will be papered, but 220-grit or even finer is best for walls that will be painted. Wrap the paper around a rigid sanding block about 6 inches long and 3 inches or so wide. Sand up and down along the joint and smooth out any ripples. At the same time work the compound flat along the joint line. Be careful not to scuff up the paper surface of the plasterboard, especially if the wall will be painted later. Inside corners also require careful sanding so that one side does not get gouged as the other is smoothed. In tight corners it is often easier to use a folded piece of sandpaper held flat in your fingers. Proceed with caution: joint compound cuts away quickly.

The final step is to replace all the electrical and other hardware and the moldings and install new casings on the windows and doors.

REMOVING OLD WALL SHEATHING

In certain circumstances removing the old wall sheathing is easier or more effective than trying to repair or re-cover it. This is the case, for example, when old plaster or hair plaster has separated from the lath and perhaps fallen away in great chunks, when plasterboard walls have been damaged or broken out, or when removal of wall coverings has caused widespread changes to the old sheathing. Removal and replacement might also be necessary during an extensive remodeling project involving the shifting of doors or windows, replacement of or addition to exterior wall insulation, or inclusion of soundproofing. Whatever the case, the procedure is as follows.

Preparations

Regardless of the kind of sheathing involved, the first step is to turn off all electrical circuits into the room. Check each outlet and switch to make sure they are indeed off, then remove the pertinent fuses or tape off the circuit breaker handles. Remove outlet and switch cover plates, take out the mounting screws, pull the devices out of their boxes, and disconnect the wires from the terminals. Then pull the boxes out of the walls and free the wires from them. Set the boxes aside for later reinstallation. Similarly, take down any wall-mounted lighting fixtures and remove their outlet boxes.

Disconnect and remove any objects that would interfere with the job, such as radiators, wall heaters, or heat duct grills. Cabinets must be dismounted. Sometimes built-in cabinets or bookshelves can be worked around, but often they must be removed, too. Disconnect and remove plumbing fixtures, leaving just the pipe stubs. Water lines might have to be temporarily capped off, and drain pipes should be sealed off with duct tape to prevent sewer gasses from coming into the house.

Once all the walls are free and clear of any impediments, remove all the moldings around the floor-wall joints and the wall-ceiling joints. If they are salvageable, number each one in sequence so you can more easily replace them later. Also, remove all of the door casing trim and window trim (see Chapter 4).

Procedure

If the wall sheathing is plywood, you will have to remove it in full-sized sheets or trimmed pieces (although 1/4-inch or 5-millimeter plywood can sometimes be broken into smaller chunks as it is removed. This takes some care so that the floor ceiling doesn't get scraped or dented. Try to start with smaller cut sections next to a window or door. Pry a section loose, starting at any convenient spot—preferably along a lower side edge. Often you will find a good starting spot where you have removed a piece of trim. Pry the piece loose gradually, working upward toward the ceiling and pulling all the nails you can as you go along. Then just work from piece to piece.

Plank sheathing is removed in much the same way: finding a convenient starting point and prying the first board out, then moving along from there. If possible, pry the bottoms loose first, and work upward toward the ceiling while pulling the planks away from the wall from the bottom. This will give you the best leverage against the fasteners.

If the sheathing is hardboard, start prying at any handy point. Then you can break large chunks off by snapping the loose edges outward away from the wall. If the sheathing is fiberboard, often the easiest way to start is to poke hole in a sheet, then break the material away from the nails. No matter what your approach, the nails will pull through the fiberboard and remain in the studs, so you will have to go back and pull all the nails and clean away the bits and pieces.

The procedure for plasterboard walls is a bit different. It's a good idea to seal off interior doors with tape, because you will probably raise a lot of dust. Wear old, buttoned-up clothing, a cap, heavy-soled shoes (plasterboard nails are nasty to step on), and a dust mask. Gloves are also suggested, especially if you have sensitive skin. With prybar or wrecking bar, smash a hole through the plasterboard, then hook the curved end of the bar onto the back of the plasterboard. Yank the stuff apart in chunks, opening the walls up in bay after bay between the studs. This will leave a lot of bits and pieces and most of the nails still attached to the studs; strip each one clean with the claws of a carpenter's hammer. If the plasterboard was attached with screws—a recent method not likely to be found in walls more than a few years old—the approach is the same. You will not be able to unscrew the screws because the slots will have been filed with joint compound (and that would take forever anyway). You will have to tear them free with a wrecking bar.

Removing plaster or old hair plaster (made by including horse or other hair in the wet plaster mix, common in old houses) on wood lath is a miserable job. A dust mask, safety goggles, leather gloves, and heavy-soled boots are absolutely necessary, along with old, well-buttoned clothing and a cap. Seal off all doors and open whatever windows you can. Unless the floor is to be redone too, protect it with sheets of plywood, preferably fitted over the entire floor area, with the joints sealed off with duct tape.

Wood lath, especially if very old, is likely to be surprisingly tough and springy. It might be solidly attached to the studs with square cut nails that are rusted in place. Start about waist high alongside a door or window where the trim has been removed and some lath ends exposed. Drive the end of a small, wide-bladed pinch bar under a series of four or five lath ends and loosen them up. Then insert the crook of a long wrecking bar under one of the loosened laths, stand as far away from the wall as you can, plant your feet, and pull hard. With luck, you'll be rewarded with a pistol shot crack as the lath snaps off along with a shower of plaster chunks. That's the process.

Work your way upward and downward to clear out one stud space, then move along to the next. Most of the laths will probably come hard, sometimes snapping in half, sometimes pulling free, sometimes coming away two or three at a time. Before long you'll hardly be able to see across the room for the dust. Shovel up the debris as you go along and put it in trash cans (the stuff will poke right through bags). Only fill the cans halfway or you won't be able to lift them.

Removing plaster over metal lath is an equally difficult proposition. The metal lath is nailed to the studs in sheets, floor to ceiling and across at least three studs. It is very sharp, tough, springy, and sometimes has a mind of its own. Usually the easiest place to start is alongside a door or a high window, as close to the ceiling as possible. There will be a plaster ground present, a strip of wood that acts as an edging for the plaster. Pry this strip off to expose the edge of the raw plaster and the underlying metal lath. Then pry under the lath along this edge, upward toward the ceiling, and across the top. As the corner comes free, pull the lath diagonally across and downward, and at the same time keep prying it loose along the top edge. When the top of one full-width sheet of lath is free, keep pulling it straight downward and prying the nails loose—like pulling a blanket off a bed. Knock plaster loose from the lath as you peel it from the studs, and crimp it into a manageable shape. Some of the nails will pull free from the studs, others will remain and must be pulled later. You might have to cut the lath apart in places for easier handling and removal; use electrician's diagonal wirecutters or heavy-duty angle jawed tin snips.

Dispose of all old wall sheathing materials with some care, because nails and dust will scatter in all directions. Trash barrels will work well in some instances, but a pickup truck backed up to a convenient door or window is the handiest method. For large quantities of debris, you might have to make prior arrangements with a trash hauling service for quick and easy (and not terribly expensive) disposal.

REMOVING WALLS

There are two classes of walls in a house: load-bearing, and nonload-bearing. Load-bearing walls form a part of the structural skeleton and support a certain amount of the total weight and stresses of the building. Nonload-bearing walls support only their own weight, and perhaps the incidental weight of a few joists or other members which, if the wall were not present, could be installed to be self-supporting.

Precautions

Removing a nonload-bearing wall presents no problems beyond the removal process itself. However, taking out any portion of a load-bearing wall larger than a door or small window opening has to be properly planned and executed. In some instances, it is not advisable at all. Disruption of the structural integrity of the building can lead to great problems and can sometimes be downright dangerous, especially if not correctly done.

There is no general rule as to which particular walls of a house are load-bearing and which are not; this depends entirely upon the structural design and the construction method used. Short, discontinuous interior partition walls are usually nonload-bearing. Partition walls that run parallel with the ceiling joists, whether beneath a joist or between a pair of them, are also nonload-bearing. But in either case, a partition top might be used to support spliced ceiling joists. Most exterior walls are load-bearing, although some— such as the endwalls of a gable-roofed structure—are not. In post-and-beam and timber framed construction, the exterior walls are curtain walls, merely set in place and secured to the post frame. In any instance where a portion of the weight of the structure bearing down from the roof or an upper story rests upon or exerts lateral force upon a wall, it is load-bearing and must be handled with care.

Always, some structural means must be introduced to properly bear and disperse the load imposed upon the original wall, and that load must be adequately supported with temporary props while any changeover is being made. If you are at all unsure as to whether a wall is load-bearing or not, or exactly how to support and rebuild such a wall, get advice from a professional architect, engineer, or builder. Otherwise, you could have some big problems.

Procedure

Whether a wall is load-bearing or not, the first step in removing it is to take out doors and/or windows. Then take down any finish wall covering, such as paneling, wainscoting, or stone/brick veneer and strip off the sheathing. (All these procedures are covered in this chapter.) In some cases the sheathing and finish might be one and the same, which simplifies matters. Remove all the covering materials on both sides of the wall. All pipes and electrical wiring must also be taken out. Pipes can be either removed back to their point of origin, or capped off at some convenient point. Wiring should be either removed completely or rerouted—if in good condition—for use at other locations. Professional assistance might be advisable here.

If only a portion of the wall is to be taken out, remove the sheathing past the borders of the opening to the next convenient joint on one or both sides—if the old sheathing/covering can be matched in satisfactorily during the rebuilding. If this does not seem feasible, strip the entire wall frame to the corners and plan for new sheathing and/or finish. Note that this does not have to be the same construction as the rest of the room: Considerable decorative impact can be gained by finishing one wall, or parts thereof, in a completely different fashion than the others.

Nonload-Bearing Walls. If the wall is nonload-bearing, removing the wall frame is simple enough. In conventional interior stud frame construction (Fig. 3-22) there will be a series of upright studs, most likely 2 × 4s, nailed to a 2-×-4 sole plate at the bottom and a single or double 2-×-4 top plate. Extra members might be included to box out door or window openings, and there might also be blocking between the studs. These are short horizontal pieces nailed in about midlevel. Start by knocking the blocking out, if neces-

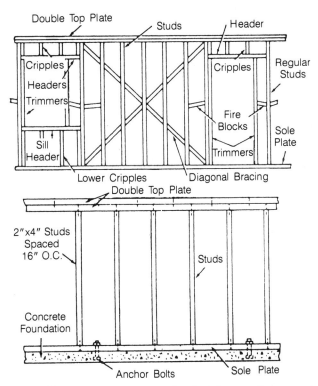

Fig. 3-22. *The component parts of typical frame wall construction.*
(Courtesy U.S. Department of the Army.)

sary, with a heavy framing hammer or a 2-pound hand sledge. A few sharp downward blows at each end of the blocks will do the job. Then select any one of the individual studs and drive the bottom sideways off the nails driven through or into the sole plate. When it is loose, pull it out and pry the top away from the upper nails. Clean any nails out of the studs and save the studs for later use elsewhere.

If there are window or door openings framed by doubled studs, use a prybar to separate the outer studs on each side from the inner ones. Then knock the bottoms of the inner studs outward and, in the case of windows, drive the short cripple studs out and the rough sill down until the various members free up from each other. Pry the inner studs apart from the header and pry the header downward from the top plate, or away from upper cripple studs. The studs at the ends of the wall will be nailed to other walls, and must be pried loose. The sole plate can then be pried up from the floor. The top plate might or might not be secured to anything; it could come away as the last studs are removed, or it might have to be pried loose from a joist or two. If only a section of wall is being removed, cut both top and sole plates off at the proper points and install new end studs. Nail them to the cut plate ends. It is advisable to double these studs for good rigidity, with the inner stud placed between and toenailed to the plates and the pair nailed together (Fig. 3-23).

Load-Bearing Walls. The procedure for removing load-bearing partition walls is somewhat different. You can safely resupport wall sections up to about 8 feet long in a

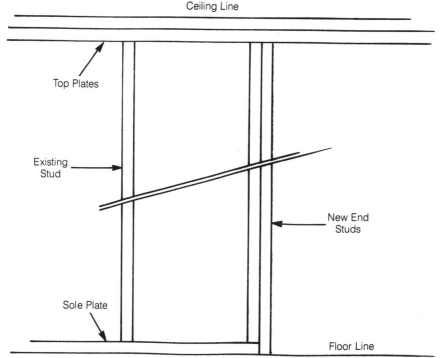

Fig. 3-23. An open wall end should be double-studded, the studs placed as shown.

conventionally framed wall, but for larger sections or other constructions get professional advice and help. Even a short wall section cannot be completely removed; a header must be installed to replace the support lost when the wall studs are taken out.

To begin, remove the wall covering and/or sheathing on both sides of the wall, to the point of the desired opening or back to the next nearest stud on each side. If the opening extends to an existing stud on each side, this is the easiest course. If not, fit a new stud and toenail it in place top and bottom, $1^1/2$ inches to the outside of the rough opening dimension at one or both sides (Fig. 3-24). Then place a new temporary stud about every 3 feet between the existing studs. They need not be nailed, but should fit tightly enough between

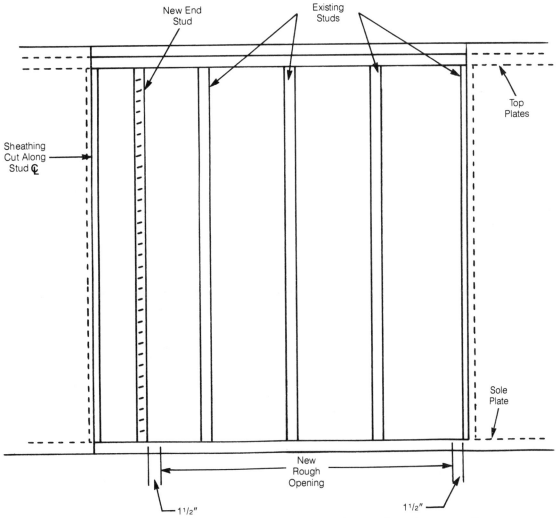

Fig. 3-24. *After cutting away the wall sheathing, the first step in cutting out a wall section is to install a new wall end stud.*

the sole and top plates that they have to be driven into place with a hammer. Position them as shown in Fig. 3-25, and check them with a spirit level to make sure they are plumb.

Now you can remove the old studs. Rather than bashing them out, it is advisable to cut them off about 3 inches above the sole plate, then pull the long pieces free and pry the short ones up (Fig. 3-26). Bend over or cut off with a hacksaw any nails sticking out of the sole plate so you don't step on one. Cut off flush any unpullable nails protruding down through the top plate. Then cut through the sole plate at the rough opening marks on each side, which should be 1¹/₂ inches to the inside of the bordering studs. Make two more cuts further along the sole plate so that you can remove a short section at each side without disturbing the temporary support studs, and pry those pieces out (Fig. 3-27).

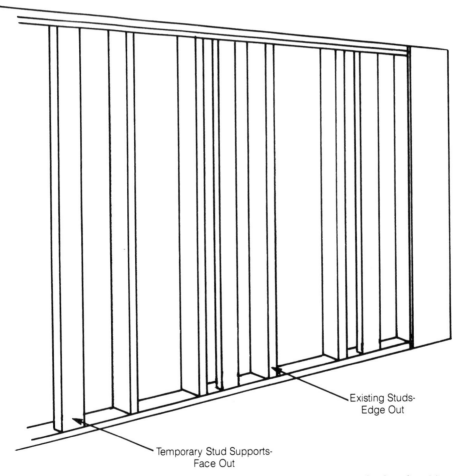

Existing Studs-
Edge Out

Temporary Stud Supports-
Face Out

Fig. 3-25. Before dismantling any of the wall structure; jam temporary props plumb and positioned face out between the top and bottom plates, in between the existing studs, to take up the load.

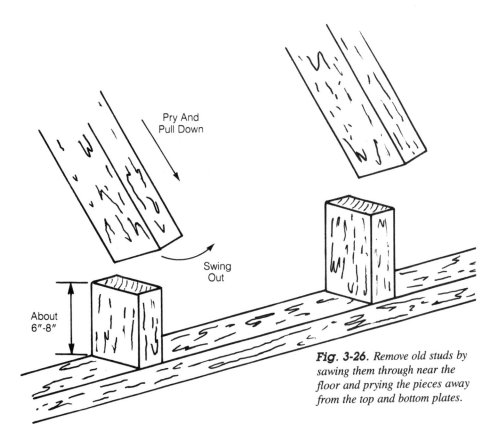

Pry And
Pull Down

Swing
Out

About
6"-8"

Fig. 3-26. *Remove old studs by sawing them through near the floor and prying the pieces away from the top and bottom plates.*

Cut the two header sections to length so they will fit snugly (not a driven fit) between the studs that border the opening. The stock should be kiln-dried construction grade. If the span between the bordering studs is 4 feet or less, use nominal 2-×-6 stock. If the span is from 4 to 6 feet, use 2-×-8 stock, from 6 to 8 feet use 2-×-10 stock, and from 8 to 10 feet use 2-×-12 stock. To the inside face of one of the members, nail a full-sized spacer (it can be made up of several small pieces) so that when the two header members are nailed together the header will be the same thickness as the width of the studs.

Boost the header half, spacer facing inward, into place across the top of the opening, taking care not to disturb the temporary supports. Wedge a 2 × 4 under each end to hold it firmly, and secure it by nailing through the face of each border stud into the header end with 16d nails (Fig. 3-28). Then toenail it to the top plate with 10d or 12d nails about every 18 to 24 inches. Keep the outside face aligned with the outer edge of the sole plate. Remove the props at the header ends. Fit a jack or "trimmer" stud into place beneath each header end and against the border stud faces. It should be a snug fit (Fig. 3-29). Nail each pair of studs together with 10d nails staggered on 16-inch centers.

Now you can remove the temporary support studs and take up the remainder of the sole plate. Set the second header member in place atop the jack studs. Nail it in place as

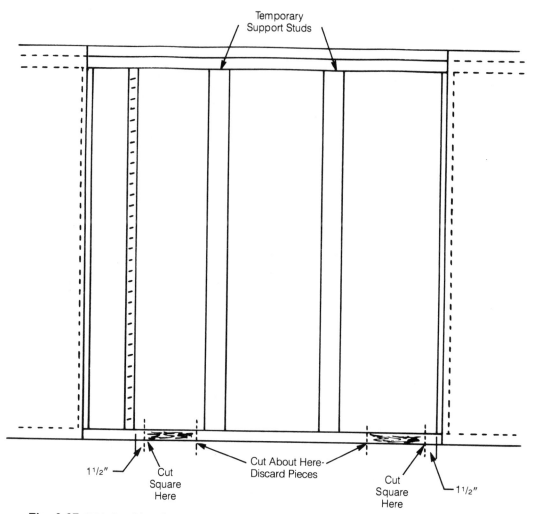

Temporary
Support Studs

1¹/₂″ —

Cut
Square
Here

Cut About Here
Discard Pieces

Cut
Square
Here

1¹/₂″

Fig. 3-27. *With the old studs removed and the props still in place, the sections of sole plate at the sides of the new opening can be cut out as shown.*

you did the first one, and then face-nail the two header members together with 12d nails staggered about 16 inches apart, from both sides (Fig. 3-30).

Much the same procedures are used when removing portions of exterior walls. Small openings, as for doors or windows, pose few problems. However, larger openings might. Much depends upon the construction methods used and the structural design, and there are considerable differences between timber-frame, post-and-beam, balloon, platform, and the very old types of framing. Log, masonry, or stone walls present their own peculiar difficulties. Therefore, if you are contemplating tearing an exterior wall apart, it is best to get some expert advice first and some professional assistance during the job.

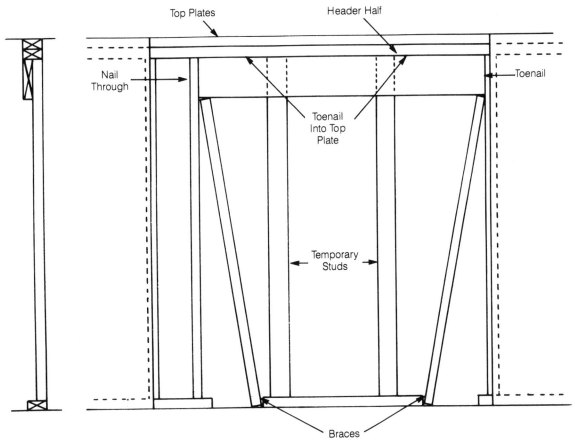

Fig. 3-28. *Here the first header half is in position against the top plate and side studs, supported by temporary braces jammed (and nailed if necessary) against the floor.*

BUILDING NEW WALLS

Constructing a new partition wall is not a difficult job. Often the hardest part is to situate the wall so that it can be securely anchored. Attaching both ends to abutting walls, as well as securing the top and bottom, provides the most rigid installation. If the wall can be well anchored to the floor and to joists at the top, one or both ends can be left free and open. However, a stem partition wall, which extends only partway to the ceiling (most stop approximately counter height) should be stabilized by solid attachment to abutting walls or by full-height support posts at free ends.

Preparations

The first step in building a full-height partition wall is to select the best possible position for it, so that you will have something solid to which it can be attached. If the wall will join another framed wall at one or both ends, position it directly against a wall stud. If

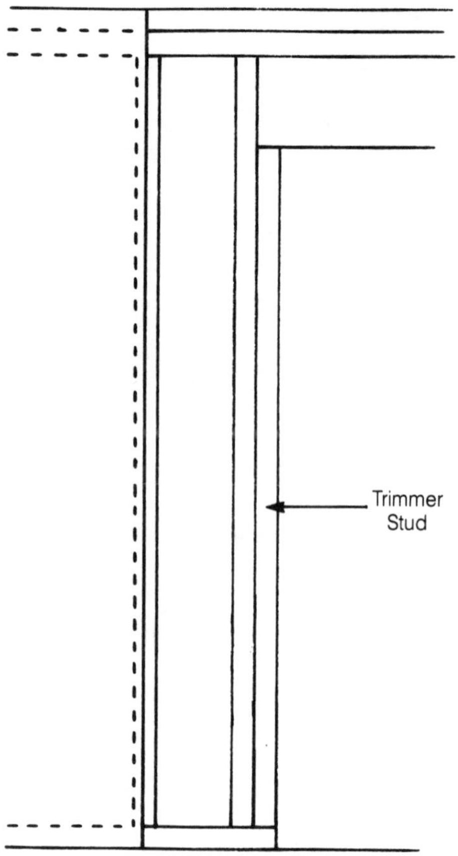
Trimmer
Stud

Fig. 3-29. Install a trimmer stud to support each end of the header and secure it to the side stud.

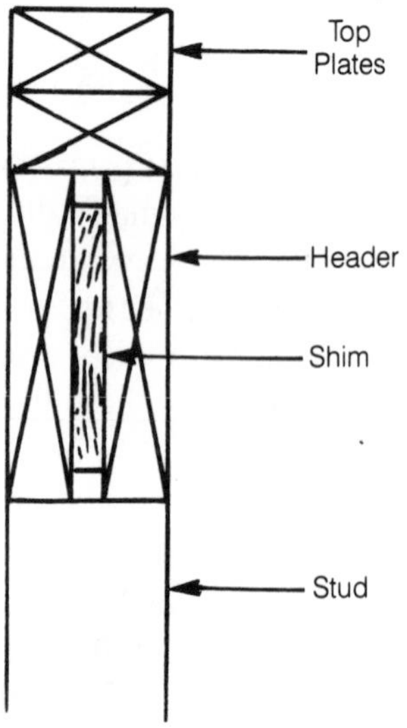
Top
Plates
Header
Shim
Stud

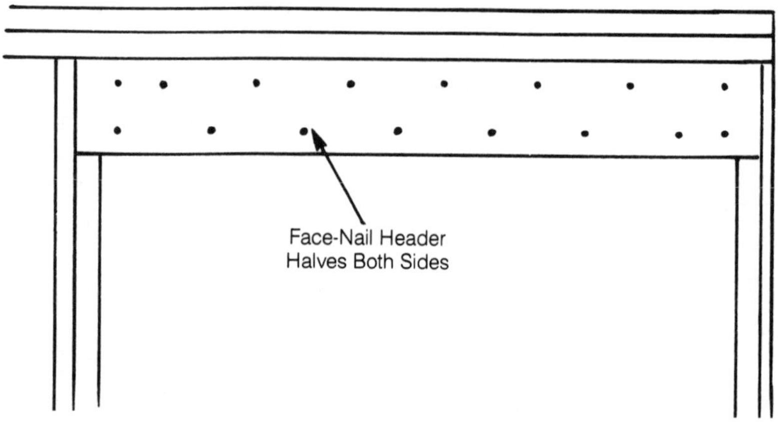
Face-Nail Header
Halves Both Sides

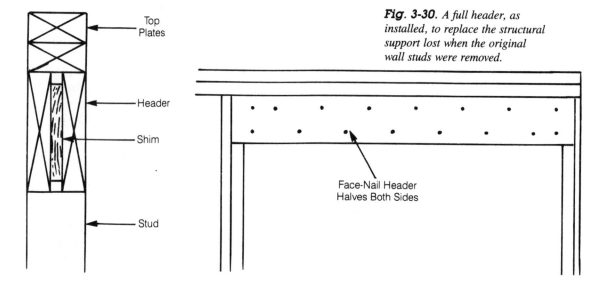

Fig. 3-30. A full header, as installed, to replace the structural support lost when the original wall studs were removed.

this cannot be done, cut away the wall covering/sheathing between two studs, along their centerlines. Then install a pair of studs in the wall, nailed together with 10d nails, and toenail them in place top and bottom with 10d nails (Fig. 3-31). Center the studs on the centerline of the new wall. Replace the wall sheathing, then build the new wall.

The new wall should run at an angle, preferably 90 degrees, to the ceiling joists. Next best is parallel with and directly underneath one joist. If the wall must be located between a pair of joists and parallel with them, attachment is difficult. It might be necessary to cut the ceiling way to expose two voids to each side of the wall location. Nail cross blocking of 2 × 4s between the two joists that will straddle the new wall, face down and flush with the bottom edges of the joists (Fig. 3-32). Then replace the ceiling material.

If the wall will abut a concrete block or poured concrete wall, or rest upon a concrete floor, drill a series of holes about 16 inches apart and insert suitable anchors or shields. The holes should be regularly spaced and drilled on the wall centerlines so that matching holes can be easily drilled later in the framing members.

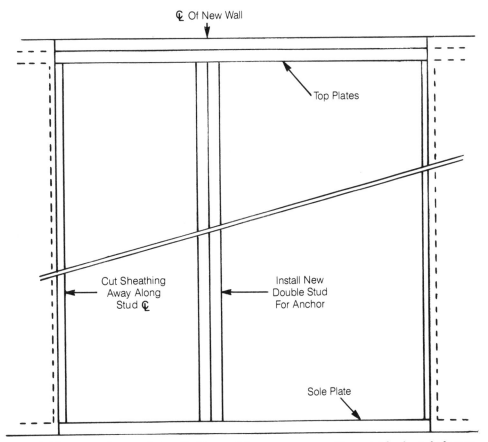

℄ Of New Wall

Top Plates

Cut Sheathing
Away Along
Stud ℄

Install New
Double Stud
For Anchor

Sole Plate

Fig. 3-31. *A pair of new studs installed in a wall frame serve as a nailing point for the end of a new partition wall set at right angles to the existing one.*

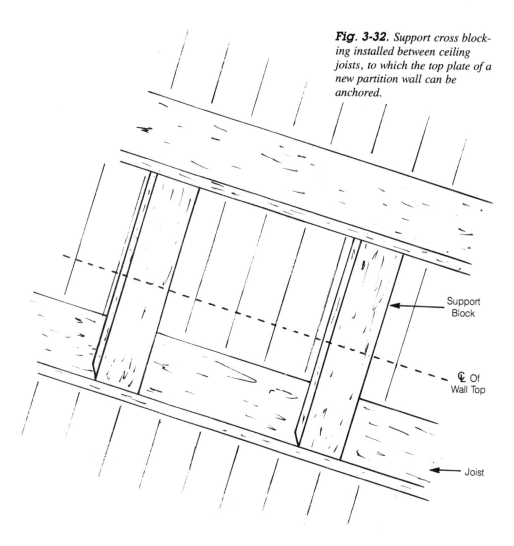

Fig. 3-32. *Support cross block-ing installed between ceiling joists, to which the top plate of a new partition wall can be anchored.*

Support
Block

℄ Of
Wall Top

Joist

Procedure

To begin construction, snap chalklines indicating the position of the wall frame on the floor and ceiling, and on abutting walls as necessary. The lines can indicate the inside or outside edges of the frame, or the centerline, whichever you prefer. If the new wall is open at one or both ends, you can build it as one unit. If it butts against a wall at each end, make it in two sections that will join when you erect it.

Cut a 2 × 4 to the full length of the wall (or use two or more pieces as necessary) and nail it to the ceiling joists or blocking, properly positioned as the uppermost top plate (Fig. 3-33). Use two 12d or 16d nails, depending upon the ceiling thickness, at each nailing point.

Cut two more full-length 2 × 4s for the sole and the lower top plates. (If it is a wall-to-wall installation, make them half length.) Lay them face up on the floor, and starting at

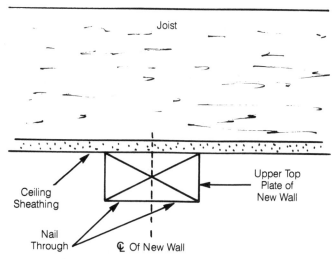

Fig. 3-33. *The upper top plate of a new partition wall is secured first, to either the joist bottoms or to blocking supports, along the centerline of the new wall location.*

one end, mark off the stud centerline locations. The end stud will lie flush with the plate ends, the next in line is 14$^1/_2$ inches from the end of the plates, and the rest follow at 16-inch intervals. If the sole plate is to be lagged to a concrete floor, drill the holes for the lag screws now.

Next, cut the studs to length. If there is any chance of a ceiling sag or a humped floor, measure the length of each stud for its particular location along the wall; they might not all be quite the same. That length will be the distance from the floor to the underside of the upper top plate, minus the thickness of the lower top and the sole plates. Set the plates on edge and a stud length apart on the floor. If the end studs are to be secured to a masonry wall, drill the holes for the lag screws. Nail one end stud in place by driving a pair of 16d nails through each plate into the stud ends. Then follow suit with the remaining studs.

Stand the wall frame, or the half-section, up on the sole plate. Keeping it vertical, work the sole plate into line and tap the lower top plate in under the upper top plate. If you have measured accurately, the fit should be snug. If it is too snug at some point, lower the frame and shave a bit of wood off the top of the lower top plate with a plane. Align the edges of the two top plates and nail them together with two or three 10d nails spaced between each pair of studs. Figure 3-34 shows a typical construction.

Check the end studs for plumbness with a spirit level or plumb bob, and secure them to the adjoining walls as necessary. Check several intermediate studs the same way and align the sole plate correctly, then nail it to the floor with 10d nails. Or, if there are some floor joists that you can nail into, use 16d nails there. Secure the plate to a concrete floor with appropriate lag screws or other anchors. If you are building the wall in halves, construct the second half just as you did the first, cutting the plates to abut those in the first section. When the two halves are secured in place, cover the plate joints at top and bottom

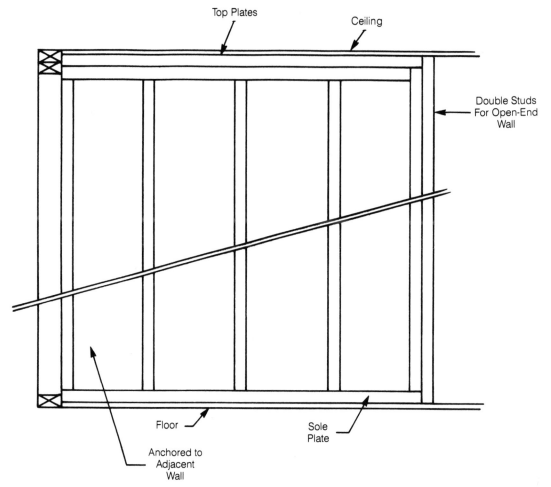

Fig. 3-34. *A typical retrofit installation of a particular wall frame set at right angles and anchored to an existing wall.*

with short lengths of 2-×-4 extending from stud to stud. Nail them to the plates with three 10d nails on each side of the joints to make a solid splice.

Rough openings for a door, window, or other opening can be easily framed into a partition wall. When framing a door, you can make the sole plate full length and later cut away the portion that lies within the door opening. This makes the frame more rigid and easier to handle during installation, and makes it easier to keep the opening in square. You can also cut the sole plate to its separate lengths and frame from it.

Nail the outer, full-height studs in place first. Then nail a double 2-×-4 header between the outer studs with four 16d nails driven through the studs into the header ends on each side. Nail the jack studs to the outer studs next, using 10d nails staggered on 16-inch centers. Toenail three cripple studs in place, one center over the opening and the

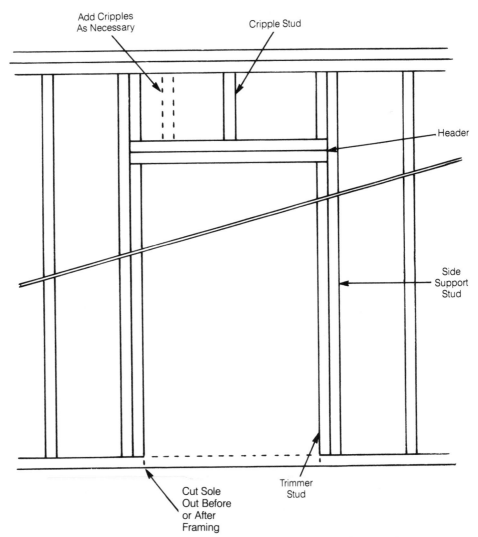

Add Cripples
As Necessary

Cripple Stud

Header

Side
Support
Stud

Cut Sole
Out Before
or After
Framing

Trimmer
Stud

Fig. 3-35. *A common arrangement for framing in a new door opening.*

other two at the sides (Fig. 3-35), using 10d nails. Make sure the studs are plumb and the header level, so that the upper corner angles of the opening are each 90 degrees. Make the size of the rough opening 3 inches higher than the door height (the standard is 6 feet and 8 inches, so the rough opening would be 6 feet and 11 inches) and $2^{1}/_{2}$ inches wider than the door width.

Other openings are done in much the same way. Because the wall is an interior non-load-bearing type, the header can be doubled 2 × 4s attached in the same way as for doors. The sill or bottom members of the opening are likewise doubled 2 × 4s. The easiest construction sequence is: outer full-length studs; top member of the header; upper

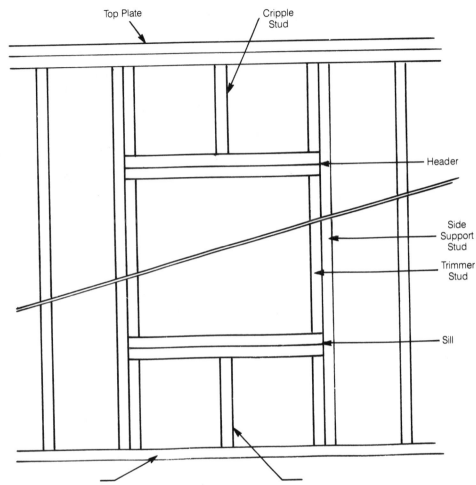

Fig. 3-36. *A common arrangement for framing in a new window opening.*

cripple studs; lower member of the header; bottom jack studs; lower sill member; lower center cripple stud; upper member of the sill; and middle jack studs (Fig. 3-36). Nail the members in place with 10d and 16d nails as appropriate. Generally the size of the rough opening should be about $2^1/2$ inches higher and wider than the finished size. This, however, depends in part on how the opening will be trimmed out, and what will be installed in the opening.

Partition walls that are less than full height—also called stem walls—are constructed in the same way, but are usually free at one or both ends. If the wall is to be merely a backstop for something such as a planter or cabinet that will be attached to both wall and the floor, it can be freestanding. The object attached to it will give it sufficient solidity and stability. A stem wall section alone, however, is too thin and flimsy to stand securely by

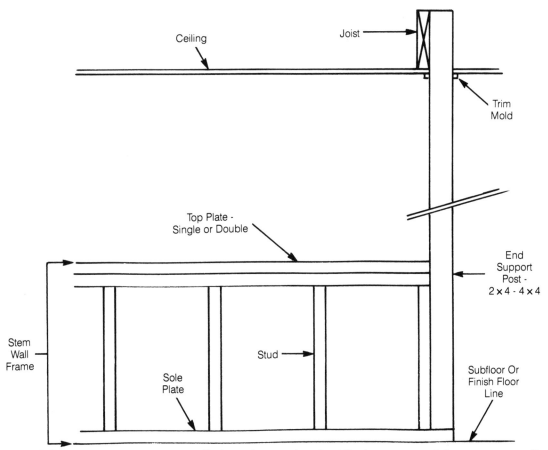

Fig. 3-37. *This framing arrangement affords ample strength and rigidity for an open-ended stem partition wall.*

itself. A post, which might be a single 4 × 4 or two or three 2 × 4s nailed together, should be built into the free ends of stem walls, or within 2 feet or so of the ends. The post should run from floor to ceiling, and be so situated that it can be solidly anchored to ceiling joist or other structural member (Fig. 3-37). The sole plate of the stem wall, as well as an end stud butting against another wall, can be nailed in the usual fashion. An alternative—which works well if the stem wall is no more than 3 or 4 feet high—is to build it of 2-×-6 or 2-×-8 stock, with the studs placed on 24-inch or even 30-inch centers. The thickness of the wall and the rigidity of the components affords reasonable stability.

WALL INSULATION

Thermal insulation may be installed in interior walls as a sound-deadening measure. As an energy-conserving measure—to reduce heating/cooling costs, and to increase comfort—insulation should be present in all exterior walls of a house. It should also be present within interior walls that separate heated/cooled rooms or areas that are often always kept at a temperature differential of 10° Fahrenheit or more.

The walls of older houses almost never meet present-day standards of thermal efficiency, and even newer houses might be inadequately insulated. A level of R-11 should be considered a minimum for both interior and exterior walls of any house in this country. In cold climates, the minimum for exterior walls should be R-17, and some local building codes might require more than this. Indeed, more is better—up to about R-24 in very cold climates. A major remodeling job affords a good opportunity to install new insulation, or replace or augment old insulation.

Installation

If the existing wall covering/sheathing has been stripped away to expose a stud frame, new thermal insulation can readily be installed in the wall cavities. Remove any old, deteriorated insulation. Replace it with fiberglass or rock wool batts or roll material, either faced with foil or kraft paper, or unfaced. Most wall cavities will be about 3¹/₂ inches deep; install 3¹/₂-inch-thick, R-11 insulation. Do not try to compress thicker material into the cavities in an effort to gain greater thermal protection; this doesn't work.

If you install roll insulation, cut the pieces to the full height of the wall cavities, allowing about ¹/₂ inch extra at each end. If you put in batts, place a full 4-foot batt at the bottom of each cavity, snug it down well at the bottom, then cut a piece about 1 inch long to fit into the upper part of the cavity. The fewer seams you can manage, the better. Tuck the material into each cavity and run your hands along the sides against the studs (watch out for splinters) to fluff the material out and set it into the cavity corners and edges. If the insulation is faced, unfold the mounting flanges at each side and staple them to the faces of the studs about every 12 inches. At top and bottom, fold the facing into a flange and staple it at three or four points (Fig. 3-38).

If the stud spacing is nonstandard and variable, as is often the case in very old houses, you will just have to cut and fill as best as you can. Sometimes, especially if the stud spacing is wide, it is easiest to use unfaced insulation and install strips of material horizontally rather than vertically, stacking the pieces up like blocks. Mesh the seam edges together by wiggling the material; friction will hold the pieces in place. Odd-shaped spaces in any kind of framing can be filled in the same way. When all the insulation has been installed, it should be covered with a vapor barrier, whether or not there is one included in the insulation facing.

Methods of Increasing Thermal Efficiency

If you want to increase the thermal efficiency of the wall beyond what can be had from the existing wall construction, there are a few ways to do so. Either method can be employed whether the wall is sheathed or unsheathed. If sheathed, all the door and window trim and all the perimeter moldings must be removed first.

The first method is to fur out the existing framework. Against each wall stud, and around the window and door openings, nail up furring strips. Nominal 1-×-2 strips will

Fig. 3-38. *The ends of faced blanket insulation, cut the ends of the sections a bit long, then carefully fold in and staple.*

not allow enough increase in cavity depth to be worthwhile, but if you put up 2-×-2 strips you can then install 1¹/₂-inch-thick fiberglass insulation and gain a bit less than R-5. These strips can be face-nailed with 10d nails, an easy job. Installing 2-×-4 furring and filling the cavities with 3¹/₂-inch fiberglass will result in an increase of R-11. This is a more difficult job, in that the furring must either be toenailed securely to the existing members with 10d nails, or what amounts to a new wall frame must be constructed flat on the floor and then erected in place against the old one. However, none of the members have to be doubled as they would in a complete partition.

The second method involves nailing panels or sheets of rigid insulation directly to the wall stud faces, or through the existing wall sheathing into the studs. There are several kinds of insulation that can be used for this purpose, with insulating values ranging from approximately R-2 to R-8 per inch of thickness. Several thicknesses are also available; ³/₄-inch is most common. The material can also be doubled for added thermal efficiency, if desired. Some of these products are covered on one side with reflective foil, which acts as a vapor barrier and should be placed facing the interior of the building. The joints are then covered with foil tape to seal them.

If this arrangement is used, insulation contained within the wall cavities should not be fitted with a vapor barrier. Not all kinds of rigid insulation are readily available everywhere, and new products are constantly coming on the market. Your best bet is to find out what is available to you that will best suit your purposes, and follow the manufacturer's instructions for installation. Note that many rigid insulations are flammable, and should be covered after installation with plasterboard sheathing, regardless of what the ultimate finish wall covering will be.

Both of these methods of increasing wall thermal efficiency require some added work to complete the installations. The finished wall extends out further into the room than the original, of course, which means that moldings on adjacent walls must be removed at least partially and retrimmed to fit. All window and door jambs must be extended so that their inside edges are flush with the new wall sheathing surface; window sills must be extended and door thresholds widened. Electrical outlets, wall switches, and wall-mounted lighting fixture boxes must be removed and reset, and the wiring rearranged to suit. Water and drain pipes must be extended or reset to match up with the new wall surface, and any wall-mounted or wall-installed items must be similarly shifted. All in all, this sort of project can turn into a substantial one. In many cases, however, the savings in heating/cooling costs, not to mention the added comfort, more than justifies the effort.

Smooth-surfaced poured concrete, brick, or concrete block walls can be insulated provided they are always dry, or no more than slightly damp from hygyroscopic moisture incursion. Wet walls cannot be successfully insulated or covered, and damage will result if this is attempted. There are a few principal approaches to this type of insulation project. The first step in all of them is to apply a waterproof coating to the entire wall surface. Numerous products are commercially available; ask your building supply dealer which one is best for your purpose.

After allowing a suitable drying period for the waterproofing, you can cut and fit sheets of rigid insulation, and glue them directly to the wall. Fit the seams tightly together and seal them with adhesive or caulk. If there are windows or doors in the wall expanse, extend their jambs and sills to match the thickness of the insulation, and caulk the joints all around. The kind and thickness of insulation to use depends upon the desired R-value, whether or not you want a foil facing, and what is locally available. Use the manufacturer's recommendations as to adhesive, caulk, and joint sealing methods. Plasterboard, hardboard, or plywood sheets can then be glued directly to the insulation surface, if desired, to provide a finished appearance (Fig. 3-39).

Another method commonly used is to nail a top and bottom plate of nominal 1- × -2 or 2- × -2 stock in place at the wall/ceiling joint and the floor/wall joint. Then fill in with furring strips nailed to the wall (this is covered later in the chapter). Space the strips 16 or 24 inches on centers to readily accomodate the later installation of plasterboard sheathing or finish paneling or planking. Then fit standard fiberglass or rock wool blanket insulation between the strips, just as with any other kind of stud wall (Fig. 3-40). Or you can cut and snugly fit rigid insulation panels between the strips. A few daubs of adhesive will hold them in place until a sheathing is put on.

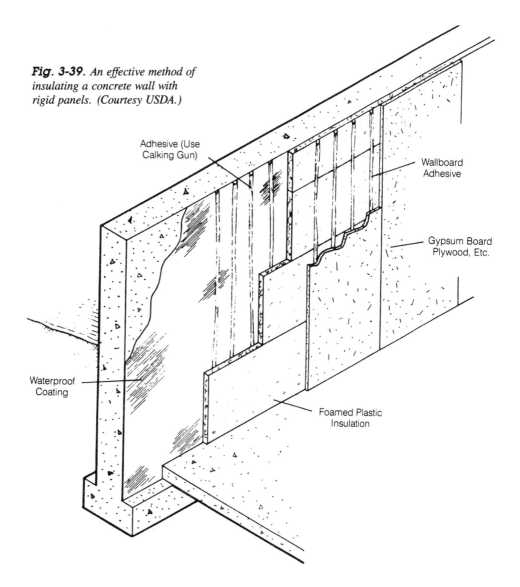

Fig. 3-39. *An effective method of insulating a concrete wall with rigid panels. (Courtesy USDA.)*

Adhesive (Use Calking Gun)

Wallboard Adhesive

Gypsum Board Plywood, Etc.

Waterproof Coating

Foamed Plastic Insulation

A third method is used when a full-depth cavity wall is needed in order to accomodate electrical wiring, piping, heat units or ducts, or other equipment—or when the wall to be covered is rough fieldstone, granite blocks, or some other irregular surface. This involves constructing a new wall frame and anchoring it in place (covered later in the chapter) then installing blanket insulation in the usual manner.

WALL VAPOR BARRIER

To avoid the damaging effects of moisture condensation within the exterior walls of a house, a vapor barrier should be fitted. This is true of houses built in all parts of the conti-

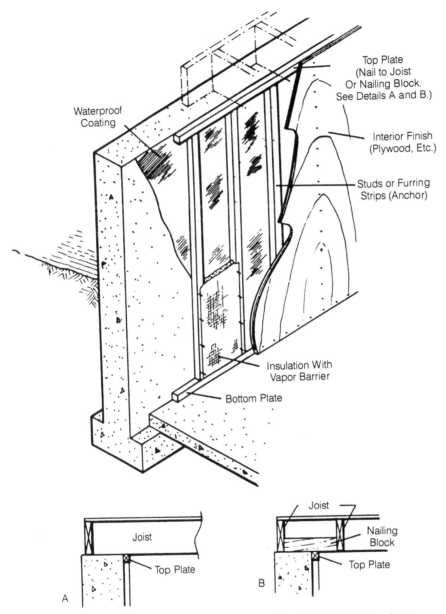

Fig. 3-40. *A concrete wall can be furred out and fitted with blanket insulation in this fashion. (Courtesy USDA.)*

nental United States, with the possible exception of some areas of the South and Southwest. The barrier material must be placed as close as possible to the heated side of the wall; this is usually just beneath the wall sheathing. Only the exterior walls should be so treated. Interior walls—even if separating rooms heated to temperature levels differing by

10° Fahrenheit or more, or separating heated and unheated spaces—are not fitted with vapor barriers. If a wall is fitted or retrofitted with two or more layers of thermal insulation, only the one closest to the heated side of the wall should incorporate a vapor barrier. If, in a retrofit, an insulation layer installed to the exterior side of the wall incorporates a foil or kraft paper barrier, this should be slashed with a razor blade to destroy its effectiveness before the vapor barrier is installed.

Materials

The material most frequently used for the purpose—which is also the most effective, least expensive, and easiest to apply—is 4- or 6-mil polyethylene sheeting. It is readily available at hardware and building and supply outlets in rolls of various sizes. Either the clear or black sheeting can be used.

Fiberglass and rock wool blanket insulation is commonly faced with foil or kraft paper, which acts as a vapor barrier. However, unless installed with extreme care and with all slits, tears, seams, and gaps completely sealed off, the vapor barrier is almost always less than fully effective. Common practice is to install plastic sheeting against the wall studs and beneath the sheathing (Fig. 3-41).

Installation

To install a sheet vapor barrier over wall studs or furring strips, stretch the sheeting out across the wall, right over window and door openings, and staple it near the top of

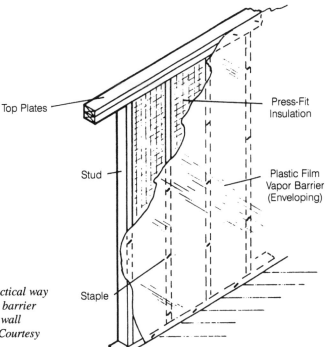

Fig. 3-41. The most practical way to ensure complete vapor barrier protection is to cover the wall frame with plastic film. (Courtesy USDA.)

Top Plates

Stud

Staple

Press-Fit Insulation

Plastic Film Vapor Barrier (Enveloping)

each stud or strip or along the top plate. If possible, use a full, unbroken sheet. If not, overlap the plastic by one full stud cavity (usually 16 or 24 inches). Leave some extra at the top that can be tucked up under the ceiling material, folded out and down over the wall sheathing when installed and, then sandwiched under a ceiling molding, or terminated in some similar way. The object is to bind off the edge of the plastic and seal it down tightly.

Smooth the plastic down the studs or strips and secure it with just one or two staples on each. At the bottom, fold the plastic out so that a few inches of it can be placed under a finish floor layer, tucked up behind a base molding, or some similar arrangement. At electrical outlet boxes and similar small openings, slit the plastic in an X across the opening and tuck the flaps back, keeping the plastic as tight as possible around the opening. Trim the plastic away around door and window openings after securing it along the edges with a few staples. Sometimes it is possible to leave a narrow flap of plastic that can be sandwiched between sheathing and trim, sealing it off. Some workers prefer to tape the cut edges down to accomplish the same purpose. If the completed barrier is accidentally slit or punctured, seal off the damage with tape before installing the wall sheathing. Check all of the staples before installing the sheathing, and drive down any that are not fully flush with the surface of the barrier.

Some forms of rigid insulation are faced with reflective foil that acts as a vapor barrier. This system is a workable one, and no secondary barrier need be installed over the insulation if the sheets are carefully installed and remain undamaged. The seams are sealed off with the manufacturer's recommended caulk or tape. In cases where rigid insulation panels are installed between furring strips, or where the panels themselves do not incorporate a vapor barrier, cover the surface with a layer of plastic sheeting. Either staple it in place or secure it with daubs of adhesive that is compatible with the insulation material.

Most older houses are not fitted with vapor barriers. If you plan to re-cover your existing walls with any rigid material such as sheathing or paneling, you can include a vapor barrier if you wish to. Cut out a small section of the old exterior wall covering, perhaps beside a window or door or just above the floor line, to determine what is in there and how the wall was built. If there is no apparent vapor barrier, or perhaps some tatty old building paper or felt, a new vapor barrier will stop moisture migration and probably cut down air infiltration as well. Before you put up the new wall covering, secure a sheet of plastic to the inside of the exterior walls, right over the old sheathing. Wrap the sheet around corners onto interior walls about 6 to 12 inches or so, and leave flaps at top and bottom that can later be tucked and sandwiched in. Then go ahead with the re-covering.

WALL SOUND CONTROL

Because of the way in which most houses are built and the need to keep costs reasonable, complete soundproofing of walls is almost impossible, even in new residential construction. Certain measures can be taken during a remodeling, however, that will at least minimize sound transfer from room to room through the walls. Keep in mind that good results depend upon similarly treating ceilings and floors as well.

Much of the sound transmission through walls is from airborne sound vibrations that strike the wall sheathing. This is transmitted through the wall space to the opposite wall sheathing and into the next room or area. A small amount of transmission can come from impact noise—objects striking the wall itself—but this is usually minimal. Structure-borne sound is the source of some noise that is transmitted through walls, too. This can occur from pipes in the walls touching upon some part of the structure, and from equipment operating on the floor above or in some other area. The sound vibrations are picked up by structural elements of the building and transmitted throughout, including the wall frame. The walls act as sounding boards. Sound finds its way along numerous pathways, often unsuspected, and travels from one room to another.

There are several ways in which sound transmission through walls can be minimized. Note that where the makeup of the walls is concerned, soundproofing treatments do not necessarily have to be carried out on each side of the offending wall, although this would be more efficient.

The first step is usually the simplest; isolate any machinery or equipment that might be the cause of annoying noise, to reduce the degree of vibration that they transmit into the structure. This was discussed in Chapter 2.

Another possibility is to add to the wall thickness. The heavier and thicker the sheathing/covering, the more sound transmission will be damped. Gypsum wallboard sheathing $5/8$ inch thick will damp more noise than $1/2$-inch material, and $3/4$-inch is even better. While a standard partition wall made up of 2-$\times$-4 studs and $1/2$-inch gypsum wallboard on each side has an STC (sound-transmission class) rating of about 34, two layers on each side will result in a rating of about 39. A layer of $3/4$-inch wood planking, especially if rough-cut or resawn, will improve matters a little more.

Another approach is to cover the walls with soundboard, a fibrous, low-density material that is easily applied and absorbs sound vibrations. The sheets can be attached directly to bare wall studs, or secured right over existing wall sheathing. Another layer of sheathing, generally gypsum wallboard, is then installed over the soundboard. The sheathing should be laminated to the soundboard with adhesive for best results. If the sheathing is nailed to the studs, the STC level will be reduced somewhat in proportion to the number of nails driven. However, sheets can be secured at top and bottom with just a few nails and with little loss of performance. Planking or paneling could also be installed—with or without gypsum wallboard sheathing, if thick enough. A 2-$\times$-4 stud wall frame fitted with soundboard and gypsum wallboard on each side has an STC rating of about 48.

Ordinary blanket thermal insulation can be installed in the wall cavity with good results, if properly done. A thickness of about 3 inches is all that is necessary, because more does not add sound reduction. Use either unfaced batt or roll insulation, or peel the foil or kraft paper off faced types. Fit the material snugly between the studs and tightly into corners and along edges, but do not compress it—friction will hold it in place. When installed in a standard stud wall covered with gypsum wallboard, not much sound reduction results. However, when used in conjunction with other measures, a reduction of STC 6 or 7 is likely.

If a new partition wall can be installed, building it in the staggered-stud method is effective in sound reduction. The fundamentals of construction are the same as for an ordinary partition wall frame, as discussed earlier in the chapter. The difference lies in the placement of the studs. You can use 2-×-4 sole and top plates with 2-×-3 studs, but a more effective arrangement is 2-×-6 plates with 2-×-4 studs (Fig. 3-42). This also makes much stronger, stiffer wall. Install two rows of studs on 16-inch centers. Set the edges of the studs in the first row flush with the plate edges at one side. Position the studs in the second row halfway between those in the first, but flush with the opposite plate edges. Set the bottom plate on a strip of fiberglass sill seal to isolate it from the floor. After the wall frame is finished, seal off the joints on both sides along the floor and ceiling lines with an elastomeric caulk such as silicone rubber.

To further improve the sound-deadening characteristics of this type of construction, you can install unfaced thermal insulation in either of two ways. Use $1^1/2$- to $2^1/2$-inch-thick blanket and install it vertically between the studs on one side of the wall in the usual fashion. If more noise is likely to originate on one side than the other, place the insulation to the noisiest side. Or, install the same material horizontally in long strips, weaving it back and forth through the studs on each side of the wall (Fig. 3-43). This will result in an STC rating of around 46 or 47 when the walls are sheathed with gypsum wallboard. If you install soundboard first and laminate the gypsum wallboard to it, you will add about STC 8.

The most effective sound-deadening construction is a double stud wall, sometimes called a party wall or a sound wall. This consists of two complete 2-×-4 stud frame walls spaced about $1/2$ to 1 inch apart (wider spacing does not increase effectiveness). The two walls must not touch or be connected in any way by anything, and they must be set on sill seal and be thoroughly caulked at all joints. This arrangement—when faced on both sides

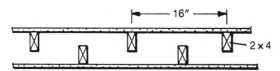

Fig. 3-42. *The staggered stud method of constructing partition walls affords good protection against airborne sound transfer. (Courtesy USDA.)*

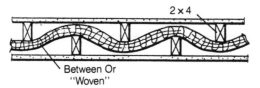

Fig. 3-43. *The addition to the construction shown in the preceding illustration of mineral wool thermal insulation woven between opposing wall studs. This increases sound transfer resistance. (Courtesy USDA.)*

with gypsum wallboard laminated to soundboard and with insulation included in both wall halves—results in an STC of about 56, if carefully constructed (Fig. 3-44). In this particular case the added insulation thickness does help, because the insulation lies in two separated layers.

The double wall arrangement can be used to help soundproof an existing wall, provided the floor space can be spared for the added thickness. Just build an added wall on one side or the other of the existing wall. For best results, strip off the existing sheathing on the side of the wall opposite the location of the new wall, install a layer of insulation, then resheath with gypsum wallboard laminated to soundboard—or at least a double layer

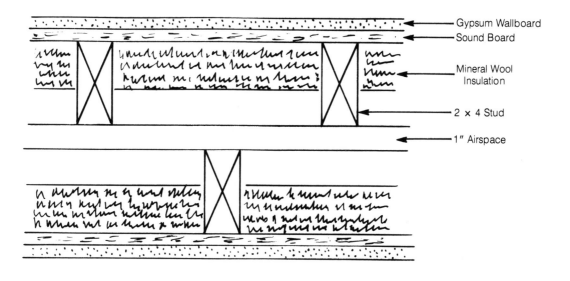

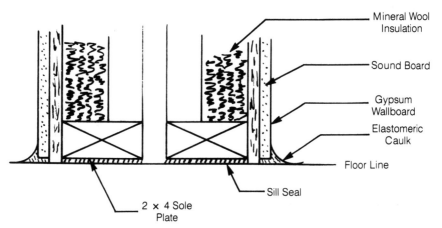

Fig. 3-44. *This party wall construction is about the most resistant to sound transfer that can be managed in wood frame construction.*

of gypsum wallboard. Set the new wall frame about $1/2$ inch from the existing wall, insulate it, and sheath it in the same way.

In many of these arrangements, existing woodwork in the room must be modified and replaced, by extending the jambs of door and other openings and installing new moldings and trim. Electrical outlet boxes must be repositioned and pipes extended, or perhaps relocated. When you install them in new sheathing, cut neat holes and fill the gaps with elastomeric caulk; use plastic boxes in preference to metal ones. Where electrical wires or pipes pass through holes in studs, seal these with caulk at each passage joint. Be sure to use a caulk that is compatible with plastic cable insulation and is approved for the purpose. Wire and pipe runs should be as short as possible, and discontinuous from occupied room to occupied room wherever possible so that they cannot act as sound pathways. Caulk all joints well, including around doors and windows. Solid core doors should always be used in sound conditioned walls, and for greatest effectiveness they should be sealed off with vinyl weatherstripping all around.

INSTALLING NEW WALL SHEATHING

There are several materials that can be used for wall sheathing:

- Fiberboard—should be considered a temporary or "quick-fix" measure, because the material is not durable and has no structural value. There are two exceptions: if fiberboard is installed as an undersheathing for its insulating value (which actually is marginal), and when soundboard—which is a fiberboard product—is installed as an undersheathing for sound-deadening purposes.
- Plywood—in $3/8$-inch or greater thicknesses, makes a good sheathing to which finish paneling or planking, or some of the more unusual finishes such as cedar shingles or clapboards, can be easily fastened. Plywood also lends a substantial amount of strength and rigidity to the structural framework.
- Hardboard—is installed to best effect in a garage or workshop, where solid panels can be interspersed with the perforated type for hanging tools. It can be easily painted and trimmed out with results both attractive and functional.
- Plaster—the traditional wall sheathing. In the past it was standard procedure to lay plaster over wood lath, but today it is applied over metal or gypsum lath. It is an excellent wall sheathing but is infrequently found these days in residences because of the expense involved, and because other materials—especially gypsum wallboard—have supplanted it. Plastering walls is not generally a do-it-yourself project. Experienced professionals are required to lay up a good two- or three-coat wall.
- Gypsum Wallboard—by far the most common wall sheathing today. It is a rigid core of gypsum between heavy paper facings. There are several types. Regular wallboard is the most widely used, and is available in $1/4$-, $5/16$-, $3/8$-, $1/2$-, and $5/8$-inch thicknesses, 4-foot width, and 7- to 14-foot lengths—depending upon thickness. The sheets are backed with gray liner paper and faced with smooth manila paper, and the edges are tapered. The foil-backed wallboard has a bright

finish aluminum foil bonded to the back surface, which serves as a vapor barrier and also, when set against a $3/4$-inch dead air space within the wall cavities, improves insulating performance. It is available only in $3/8$- and $1/2$-inch thicknesses and 4-foot widths. Sheet sizes are 4×8, 4×10, and 4×12 feet.

- Type X wallboard— looks the same as regular wallboard, and is available with or without a foil backing. However, the gypsum core is specially compounded with certain additives and glass fibers for fire resistance. This sheathing is designed for use where fire-resistant frame construction is required, as between living quarters and an attached garage.

- Water-resistant wallboard—also has a specially formulated gypsum core, as well as a chemically-treated face paper, often blue in color. It is recommended for use in bathrooms, kitchens, and utility rooms where moisture might be present, but cannot be used for high-moisture applications like saunas.

- Predecorated wallboard—has a decorative pattern or simulated wood grain applied to the face. A limited number of thicknesses and sheet sizes are available. The sheets are either square edged or bevel edged. The square-edged sheets are usually trimmed out with matching moldings between them, while the bevel-edged sheets are simply abutted. Some types include a flap of facing that is glued down to cover the joint after installation.

- Backing board—is a specialized product that comes in two varieties. One is intended as a backing or underlying sheathing to which other wallboards or acoustical tile can be laminated. The other is a highly water-resistant material for installation in shower stalls as a base for setting ceramic tile.

- Gypsum lath—is intended solely as a base for other wall coverings, particularly plaster. It is typically 16 or 24 inches wide and 4,8, or 12 feet long, and comes in thicknesses of $3/8$ and $1/2$ inch. The edges are rounded and the core is covered with multilayered, tough paper. Plain, perforated, foil-back, and Type X forms are available. The plain lath makes an inexpensive and easy-to-install sheathing for walls that will be covered with a masonry veneer, paneling, or planking.

Installation

To install plywood, hardboard, fiberboard, particleboard (or wafer or chip), or soundboard sheets to a stud wall, simply cut and trim them to fit, then secure them to the studs. Follow the manufacturer's or dealer's recommendations as to attachment methods and fasteners. Nails are most widely used—box or ring-shank 4d or 6d for the hard materials, and often roofing barbs for the soft. Nail spacing is typically 6 to 8 inches around the perimeter and 12 inches at intermediate supports. You can also use construction adhesive, or a combination of both nails (usually on a wider spacing) and glue. Trim the sheets so that each edge lies along the centerline of a stud to afford good nailing and/or adhesion. You can cut fiberboard and soundboard with an ordinary saw blade, but the hard materials, especially particleboard, are best cut with a carbide-tipped blade; ordinary blades will dull very quickly.

Regular gypsum wallboard sheets are installed in much the same way. Start at a convenient wall and trim the first sheet to fit, allowing about $1/4$ to $1/2$ inch for top and bottom clearance. Trim one side edge to fit flush at the wall corner or edge, and the other to align with a stud centerline. Or, to install long sheets horizontally, trim and fit the edges and secure the bottom sheet first, then trim the upper sheet as necessary along the top edge. Secure the sheets with wallboard nails, wallboard screws, or adhesive. For further information on cutting, securing, and finishing tapered-edged wallboard, see the section entitled "Covering Old Wall Sheathing," earlier in this chapter. Note that if the finish wall covering is to be paneling, planking, masonry or other solid veneer, rigid or semi-rigid material, or even a stucco-like texturing, the wallboard joints need not be taped and finished as they must be for paint or wallpaper.

Putting up gypsum lath is a comparatively simple process, because the sheets are much smaller and easy to handle. Joints can gap a little and trimming need not be quite as precise as for wallboard. Start at a wall end and trim the first piece to fit horizontally along the floor line. Then just stack the panels up the wall, trimming the last at the ceiling line. You can also stagger the panels back and forth so that not all the end joints align along the same stud centerline. Secure the panels with wallboard nails spaced 5 inches on centers at both perimeter and intermediate supports (Fig. 3-45). Nail popping is not a concern here, so you don't need to use wallboard screws.

Note: If you are installing Type X wallboard or lath to be in compliance with local fire or building code regulations, be sure to follow all the attendant requirements as to joint treatment, fastening, and finishing.

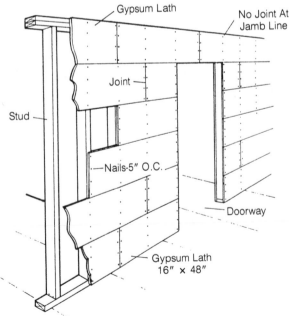

Fig. 3-45. *A typical application of gypsum lath as a backing for plaster or a finish wallcovering. (Courtesy USDA.)*

COVERING MASONRY WALLS

Relatively smooth and regular masonry walls such as concrete block or poured concrete can be covered by direct application of sheathing to the wall surface or by furring the walls out. However, this is not possible with rough-surfaced and irregular masonry walls, such as certain brick or stonework, and is usually not advisable or feasible with ceramic tile. Some kinds of rough-surfaced masonry veneer facings, as well as glued ceramic tile, can be removed and the old sheathing surface re-covered as necessary. Most masonry work, however, is permanent and often part of the structure; obviously this cannot be removed. Other means must be employed to provide a fresh, smooth surface to which a finish wall covering can be applied.

Waterproofing

To cover a smooth masonry surface, first ensure that the entire surface always remains dry, or contains no more than a slight amount of hygroscopic moisture. A masonry wall that weeps or is wet at any time cannot be successfully covered. Coat the entire inner surface with a waterproofing solution if any part of the wall is exposed to the weather and/or lies below grade. Masonry walls or facings that lie entirely within heated and protected portions of the structure need not be treated.

Procedure

After a suitable drying time has elapsed, cut and fit panels of $1/2$-inch-thick gypsum wallboard. Abut the edges tightly, and glue them to the walls with a type of construction adhesive recommended for the purpose. (If you want to insulate the walls with rigid thermal insulation, do this first;) this could be a slow job. It depends upon how well the adhesive works, the atmospheric conditions, whether or not the plasterboard sheets are absolutely flat and have not "set" into a slight bow, and the planeness of the wall surfaces. Individual sheets might have to be kept tight to the walls with braces and props to give the adhesive time to set up and to counteract various inward or outward pressures on the sheets. Where door or window openings are involved, extend the jambs so that their edges are flush with the new wall surface.

An alternative method that involves not much more work and produces better results is to fur the walls out (see Fig. 3-40). Nominal 2-×-2 furring strips are the usual choice for this. After waterproofing the wall surface, if necessary, anchor a sole plate along the floor/wall joint and a top plate along the ceiling/wall joint. Use 12d or 16d nails where fastening to wood. You can anchor the plates to concrete or concrete block with masonry nails, but "shooting" them in place with a gun made for the purpose (rentable) is faster and easier. Then fill in between the plates with studs spaced on 16- or 24-inch centers. Toenail the studs top and bottom with one 8d or 10d nail on each side. Each full-length stud should also be secured to the wall at three or more points along its length. Border all openings with 2-×-2s and extend all jambs. If thermal insulation is called for, install that between the studs. Apply a vapor barrier if necessary. Attach sheathing to the panels or directly apply a finish wall covering.

If the wall surfaces are rough and irregular, you will have to construct a new wall frame. First, determine how far out into the room the outermost rough spots protrude. You can pick out the worst spots by eye. Then hold a floor-to-ceiling length straightedge (a length of straight furring or 2-×-4 will do) against each rough spot and plumb it with a spirit level. Make a mark on the floor at the inner edge of the straightedge. The mark furthest out onto the floor will indicate where the inner edge of the new wall frame must be located. Do this for all of the walls to be covered. Then snap chalk lines on the floor to indicate the outside perimeter of the new walls. Adjust the lines as necessary so that the corners are at 90 degrees, or whatever other angle is necessary to ensure that the wall line is regular and the wall sections are correctly positioned. Once this layout is correct, transfer it to the ceiling by hanging a plumb bob over each end of each line, marking corresponding points on the ceiling or ceiling joists, and snapping new lines there.

Once all the reference points and guidelines are set, you can proceed. If the walls need waterproofing or joints need tuckpointing with mortar, take care of that. Then construct the new wall frames of 2-×-4s and anchor them in place. For details on building new walls, and installing insulation, vapor barrier, sheathing, and finish wall coverings, refer to the appropriate sections.

PAINTING WALLS

Probably the most extensively used decorative wall finishes today are applied coatings: paint on gypsum wallboard or plaster; paint, stain and sealer, or other special coatings on concrete block or poured concrete; wax, stain-wax, varnish, stain-varnish, or stain with varnish top coats on wood. Preparation for these applied finishes is not difficult, application is easy by comparison with other wall finishes, and the cost is low. Refinishing in order to freshen up the appearance or change the decorating scheme is simple enough, and can be done as often as desired without major impact to the household routine.

Preparations

First remove any furnishings, paintings, or other objects that might be in the way. If you plan to paint with a roller, completely cover any objects close to the walls, because the paint is likely to spatter. Turn off the electrical circuits that feed convenience outlets, wall switches, and wall-mounted lighting fixtures, and remove the fuses or tape from the circuit breaker handles. Take down wall lighting fixtures, and remove all switch and outlet cover plates.

If you have a steady hand, it usually is not necessary to mask off the edges of door and window casings, base or ceiling moldings, or other trimwork. However, if an adjacent surface is unpainted or a different material, masking might be in order. This could be the case, for example, along a chair rail above a wood wainscoting, or in a corner where an adjacent wall is natural wood. Many such break lines can be protected with just a single strip of 3/4- or 1-inch masking tape. If more protection is needed, such as to cover an entire wainscot, secure sheets of newspaper or masking paper to the break line with masking tape. Cover the floor area near the walls with old sheets, plastic film drop cloths (available at very little cost at any paint store), or whatever is handy.

Preparing the Surface for Painting

An important point to remember in preparing wall surfaces for paint is that the glossier the paint, the more it will reveal any defects in the surface. Bumps, depressions, ridges, small holes, scuffs, and practically any irregularities will stand right out—especially under artificial light—on a high-gloss surface because of the shadows and reflections cast. Think of how even tiny blemishes and imperfections show up on the polished hood of your car. And the lighter and brighter the color, the worse they will appear. A perfectly flat paint will reveal surface imperfections the least, with barbecue black probably being the most forgiving of all.

On the other hand, a glossy surface will clean the best and is the least suspectible to retaining soil, stains, or grease film (which is present in every house). A flat finish is the most susceptible and the hardest to keep looking fresh and clean, and it requires more frequent refinishing. Semi-gloss paints are the usual compromise, and for a successful finish good preparation of the wall surface is the main key.

If the wall surface is new gypsum wallboard, inspect all of the taped joints and the covered nails or screws. The joint compound should be sanded perfectly flush with the wallboard surface. It should also be flat and ridgeless, free of pinholes, and feathered out to the edges without scuffs on the paper surface. A good way to tell if the surface is true, and to spot any imperfections, is to place a strong light right next to the wall, shining along it. Face the light, with your cheek right next to the surface, and sight along the wall. Imperfections will show up as shadows or silhouettes against the otherwise flat surface. Repair them with more compound or vinyl paste spackle, with sanding, or whatever else is necessary. Also, check the surface in general for any damage that might have occurred, and make needed repairs. Finally, wipe down or vacuum the walls thoroughly to clean off any dust accumulations.

If the wall is new plaster, follow the same routine for finding damage or imperfections, and make repairs. Note that coatings should not be applied to a new plaster surface until it has *completely* cured and there is no longer any moisture present. This might require several months.

Much of the wood wall covering produced today, especially plywood paneling, is factory prefinished and needs no further attention. However, some is not, particularly wood planking, and this should be finished as soon as possible after it is installed. If the surface is rough-cut or resawn, preparation consists only of a thorough vacuuming. Planed surfaces seldom need any attention beyond light scuff sanding with fine paper at the rough spots surrounding knots or where a planer blade has burred an edge. Nail holes or other small defects should not be filled at this stage. However, if the finish is to be paint and the wood contains resinous knots, seal the knots by spraying or brushing on two coats of a sealer intended for this purpose, so the resins and/or coloration will not bleed through the new finish.

If the wall covering is gypsum wallboard or plaster that has previously been painted, wash it thoroughly with warm water and a mild detergent and remove all grime and grease film. Attack heavy accumulations of dirt, grease spots, and stains individually with warm water and a strong detergent. If the finish is glossy, treat the entire wall with a commercial deglosser or scuff-sand it lightly with fine sandpaper. Also spot-sand off any dust nibs, paint dribbles, or small particles caught in the old paint, smoothing the surface wherever necessary. If there are stains that will not come off, such as indelible marker ink, crayon, lipstick, grease, or ball-point pen scribblings, these must be sealed off or they will bleed right through the new finish. Shellac, which you can spot-apply with a spray can, is a good sealer if the topcoat will be an oil-based paint. A primer-sealer made for the purpose can also be used under any kind of topcoat.

If the wall covering is wood already coated with an applied finish and it is to be refinished, first check over the entire surface for defects. If the finish is clear and it is to be recoated with a clear finish, sand out small nicks or scratches with very fine sandpaper. Regardless of the existing finish, if the new coating will be paint, fill all nicks, scratches, holes, and other defects with wood putty or vinyl paste spackle. In any case, wash the walls with mild detergent and warm water. If the old finish is high-gloss, treat it with a commercial deglosser or lightly sand the entire surface with fine sandpaper to cut the gloss. Semi-gloss or flat surfaces need no such treatment. However, if the old finish is rough and full of dust nibs or bits of debris from a careless painting job, sand the entire surface smooth with fine sandpaper.

If the surface to be painted is new, unfinished concrete block or poured concrete, it should be allowed to age about six weeks before cement paint, vinyl paint, or latex paint is applied, and at least three months (but preferably a year) before an oil-based paint or stain is applied. If the wall has rough spots and/or bits of splattered-on mortar or dribbles of concrete that oozed out around the forms, smooth the surface off. Where the surface is very smooth it should be roughened up, especially if a cement paint is to be applied. You can accomplish both purposes at once by rubbing the wall with the face of a brick or chunks broken from a concrete block. Also, remove all traces of efflorescence—that white, powdery substance that sometimes forms on concrete surfaces or around block joints. Most of this will disappear if the walls are rubbed. Otherwise, it can be cleaned away with a stiff wire brush. If there is some stubborn residue left, you will have to resort to washing the walls with a solution of commercial muriatic acid (available at paint stores and masonry supply houses) and water, applied with a stiff-bristle brush. Use according to directions, and follow up with a copious rinse of clear water. Allow plenty of time for thorough drying before applying the paint. If the walls need not be washed, vacuum them thoroughly to remove all the dust. If there are stains or grease spots on the wall, clean them off as much as possible. Try a strong solution of trisodium phosphate (TSP) first. If that doesn't work, try appropriate solvents like alcohol, mineral spirits, lacquer thinner, or acetone. (Note: Be careful when using acetone; it is highly flammable with toxic fumes). Whatever remains, spot-prime with a sealer intended for the purpose. Patch any holes or cracks with an epoxy concrete patching compound or cement.

If the block or concrete has already been painted, check the surface to make sure that none of the old coating is lifting or chipping off, and scrape any such areas you find. If the surface is very smooth and glossy, treat it with a deglosser. Otherwise, just wash it with warm water and mild detergent. Patch any holes or cracks with an epoxy concrete patching compound or cement.

Brick walls are treated in much the same way as concrete or concrete block. They should not be painted until the mortar joints have had ample time to cure, as with poured concrete. Smooth any mortar crumbs out of and away from around all the mortar joints, and clean off the brick faces as well. If the finish is to be a clear sealer, remove all traces of mortar and/or efflorescence from the brick faces by rubbing, wire-brushing, or scrubbing with muriatic acid. If paint will be used, traces of mortar on the bricks are of no concern, but efflorescence must be removed. If you wash new brick, allow at least a couple of weeks for drying before applying a coating. If the brick has previously been painted, just wash the surface with warm water and mild detergent to remove dust and grime, paying particular attention to the mortar joints. A stiff-bristled scrub brush is best for this.

Choosing the Paint

The kind of applied coating you select, apart from the myriad of different colors and tones that are available, depends upon the wall surface, the service conditions, and the desired appearance of the finish. There are many brands and types of paints available at any given moment, and formulations are continually being changed. They come and go in the marketplace all the time, and certainly not all paints are available in any one location at any time. Thus, you will have to check your local paint supply outlets, study the situation, and be guided by dealers' and manufacturers' recommendations as to which specific products might be best for your purposes.

Although there are other types of paint that might be used, interior latex paint is the kind most often applied to interior plaster or gypsum wallboard sheathing. The walls in living and dining rooms, bedrooms, dens and libraries, and other such rooms are unlikely to become soiled and need only infrequent cleaning. These are usually treated with a flat latex. The "scrubbable" or "washable" types that have recently been introduced can be cleaned more often and more vigorously with fewer ill effects than plain flats. In kitchens, baths, laundries, nurseries, and similar rooms, a semi-gloss latex, which more readily sheds soil and grease, will hold up under repeated washings. High-gloss coatings are seldom used on walls, but they can be. The usual choice is an oil-base high-gloss enamel or a polyurethane enamel.

Wood paneling or planking can be coated with either flat or semi-gloss interior latex or acrylic latex paint, with oil-based enamel, or with semi- or high-gloss polyurethane enamel. For a clear or natural finish, the usual choices are polyurethane varnish or interior varnish for a hard coating. These are available in semi-gloss, satin, and high-gloss.

Shellac can also be used for a high-gloss finish, but is not a tough, durable coating like the varnishes and is susceptible to water-spotting and other problems. Protective dull finishes are also popular: paste wax, resin/oil finish, or wood sealer. Transparent stains are often used, usually of the oil-based variety. Although the wood-tone browns and reddish browns are most common, many bright colors are also available and can provide startling and emphatic decorative effects. Stains are given a top-coat varnish or wax. Semi-transparent and opaque stains can also be applied, but they largely or completely hide the grain and figure of the wood.

Brick walls can be treated with a clear sealer made for the purpose, or with polyurethane or interior varnish in either satin or gloss finish. You can also paint interior brick with polyurethane or oil-based enamel, epoxy paint, rubber-based paint, masonry paint, or portland cement paint.

Dry poured concrete or concrete block can be treated with an oil-based semi-transparent or opaque stain, then coated with polyurethane varnish. Or, you can apply any of the paints mentioned for brick. If the wall surface is damp, however, or has an alkali problem, only certain waterproofing compounds, portland cement paint, and masonry paints made for the purpose can be successfully applied. This is a matter for discussion with your paint supplier.

Note that in most finish paint applications on a new, raw surface, a primer or undercoat must be applied as a base for the finish coats. The exact kind of primer varies with the type of finish coating. Follow the manufacturer's and/or the dealer's recommendations as to what product should be used. Whenever possible, it is wise to use primers and finish coatings made by the same manufacturer.

If you plan to apply any of the hard coatings over an old finish coating, be aware of a major potential problem: the compatibility of the two coatings. It sometimes happens that a new coating will partially dissolve and lift the old one. A lacquer, for example, will instantly react with an oil-based paint, and any water-soluble paint will interact with shellac. Whenever possible, refinish an old coating with the same kind of finish. If you don't know what the old finish is, try a small test spot in some hidden corner. You will need to cover at least a square foot of area with the new finish to find out if it will fail to bond properly.

Materials

Applications of all paints, unless otherwise noted in the manufacturer's instructions, can be made by roller, pad, bristle brush, or foam brush applicator, a throwaway wedge of plastic foam stuck on a handle that looks and works like a brush, but in many instances does a better job more easily. For large areas of wall, a roller is the fastest and easiest to use: brush marks are eliminated, and lap marks are usually not a concern. Pads also work well. For smaller areas, bristle or foam brushes are often the most convenient, and many painters prefer to use bristle brushes for everything. Coatings can also be applied with a spray outfit, but this is a dusty, messy job that isn't usually recommended for the ordinary interior redecorating and remodeling jobs.

Procedure

How do you paint a wall? Actually, it's easier to do than to explain. Experience is the best teacher, and just a little bit will see you well on your way to being a competent painter. A few tips, though, might help. Use only top-quality coatings and applicators. The cheap stuff can cause problems and can result in a below-par job. Also, select the right coatings, with brushes/rollers/pads to suit both the coatings and the surfaces to be coated. Try to do the job when the atmospheric conditions are ideal range. This will afford you the best finish with the least effort. In general, a temperature of 65° to 75° Farenheit and a relative humidity of 40 to 60 percent is just about right, for both paint and painter. The further above or below those ranges you get, the more difficult application becomes, and the greater the likelihood of a less than satisfactory finish.

If you plan to use a roller or pad, first coat a strip about 2 inches wide around all the perimeters at break lines, such as beside window and door casings and along the moldings. Use a 1½-inch (or smaller) brush for this, or a tiny trimming pad or roller, masking as necessary. Also paint out any small and cramped places that might not be accessible to a full-sized pad or roller. If you paint with a brush, use a 1½- to 2-inch one for the tough spots, and 3- to 4-inch one for the wide open spaces.

Begin at a corner and work left to right (or vice versa if you are a lefty) so your eyes are mostly in front of the area you just coated and you can catch missed spots and smooth out dribbles or spatters. Use vertical strokes, and strive for as little overlap as possible. Move along the wall in strips, but make the borders of the strips variable, not a straight line from floor to ceiling, to minimize overlap marks. Try not to rework freshly laid paint; if you do after the surface has started to tack over, you will leave applicator marks and streaks. With most coatings, it is best to work from uncoated portions and feather back into freshly coated areas, rather than from fresh coating onto the dry wall, especially if you are using a brush.

When brushing, use even strokes of variable length, spreading the paint from the tip of the brush. When rolling, run the roller vertically once, then diagonally once in slightly overlapping strokes for good coverage and blending. Whatever the applicator, don't overload it with paint. You will soon discover how much is right, and that depends upon the consistency of the coating. Never stop in the middle of a strip or out in an open wall expanse. When you quit for lunch, break the coating above a door or window in line with one of the vertical casings, or at a corner or at some other natural break point.

TEXTURING WALLS

Applied coatings such as paint or varnish dry to a thin, smooth, even surfacing that may range from dull or flat to a highly reflective gloss finish, depending upon the formulation. Texture coatings, on the other hand, leave a thick, rough, flat finish.

Types of Coatings

Most texture coatings come in neutral colors such as white, buff, or eggshell, although a few colors are available and some texture coatings can be tinted. There are several kinds of texture coatings available, but for practical purposes they can be separated into two types. The first has a relatively thin consistency, like a heavy enamel paint. In home remodeling projects, it is usually applied with a brush. The other type is very thick and sticky, like overdone oatmeal. Some of these paints can be applied with a stiff-bristled brush, while others must be spread on with a plasterer's trowel.

The lighter-weight texture coatings are often called texture paints. Some of these, like sand paint, are meant to be just spread on and left to dry to the resulting characteristic texture. The heavier coatings are compounds of various sorts that are meant to be applied, then immediately reworked into whatever surface texture or pattern is desired before the material cures. There is a third kind of material that can be used for texturing: stucco. This portland cement or lime cement plaster was once widely used, but because of the complexity and high cost of its application (this is a job for professionals), it has been almost entirely replaced for interior decorating purposes by the easy-to-apply and inexpensive texturing compounds and paints.

Texture coatings can be applied to any clean, dry, solid surface such as plaster, gypsum wallboard, poured concrete or concrete block, brick, or plywood. They can also be put on over old coatings like paint or varnish, if they are well bonded. They should not be applied over wallpaper because the moisture in the coating can loosen the glue and lift the paper from the sheathing. One of the advantages of texture coatings is that they will cover well and hide minor surface defects. The thinner the coating, the more will show through or be replicated on the new surface. The heavier the coating, the more will be hidden. Also, the smoother and more delicate the texturing, the fewer the defects that will be hidden, while increasingly intensive and complex texturing patterns will obscure more of the underlying surface, so defects become undetectable in the overall pattern.

Preparing the Surface

To prepare for a wall texturing job, wash the surfaces thoroughly with warm water and a mild detergent. Clean away grease spots. If there are stains or marks, such as from a ball point pen, crayons, or lipstick, seal them off with a coat, or preferably two, of spot sealer. If the surface is unfinished wood, or has been coated with an oil-based stain and has no top-coat (sealer or varnish, for example), or if wallpaper has been stripped from the sheathing, prime the entire surface after cleaning with one coat of white, oil-based primer paint. If you will be applying one of the thinner varieties of texture coating, fill any cracks, holes, or other defects greater than about $1/16$ inch with vinyl paste spackle. The heavier compounds have better crack-filling and bridging capabilities, so only defects and joint cracks of about $1/4$ inch need be filled. This often can be done by spot-repairing the defects first with the coating material itself. Regardless of the kind of coating you use, be sure to follow all the manufacturer's instructions for preparation. There are some detail differences depending upon the material used and the nature of the surface being covered.

Procedure

To apply a texture coating, simply spread it on the prepared wall surfaces with a brush, trowel, or wide-blade putty or joint knife, according to instructions. Allow a thin coating to dry for several days before hanging pictures or setting anything against it. Thick coatings are usually applied in small sections, depending upon how quickly the material begins to cure.

Spread the material in a coat of fairly even thickness, which can usually be anywhere from $1/8$ inch to as much as $3/8$ inch. However, the thicker the coating the less wall surface you can cover per gallon of material. A thickness of $1/8$ to $3/16$ inch is ample for most texturing patterns, $1/4$ inch for more pronounced ones. And, of course, the coating can be used to level out depressions, dips, or defects in the surface, so it might be thicker in some places than in others. This does not affect the integrity of the coating. There is the danger in getting some coatings too thick, however, that as they cure they shrink inward and form cracks around the area. To remedy this, allow thorough drying, then carefully patch the spot with a bit more of the same material; it should blend right in.

The fun part of the job is the texturing itself. There are dozens of pattern possibilities. The final effect is entirely up to you, but give it some thought before you begin so you will know just how to proceed and what the result will be. For example, you can score the wet

Fig. 3-46. *A scored or striated texture, which can be formed into any number of straight or curvilinear patterns.*

surface vertically, horizontally, diagonally, or in any number of patterns by drawing a whisk broom, a stiff-bristled brush, or some similar object across it (Fig. 3-46). You can comb it with a steel comb made for the purpose. You can stipple it by running a paint roller across the surface; the nap of the roller and the amount of pressure you use will determine the stipple pattern (Fig. 3-47). If you pat a sponge against it, you will pull the whole surface up into tiny mountain ranges. A damp sponge produces one pattern, a dry one another, a holey sponge yet another. A flat block of wood or a flat steel trowel pressed against the surface and then lifted straight away will produce somewhat different little mountains (Fig. 3-48). Or you can swing either of those tools in flat arcs across the surface and leave multiple intersecting semicircular ridges. You can make repetitive circles by gently pressing the smooth bottom of a water glass gently against the surface, twisting it slightly, and removing it (clean the bottom off each time). A patterned glass bottom will leave its particular imprint, and with cookie cutters you can make all sorts of patterns. With any of the patterns that leave raised ridges, points, or borders, you can go back over the textured surface and knock those edges down with the edge of a plasterer's trowel to create another look.

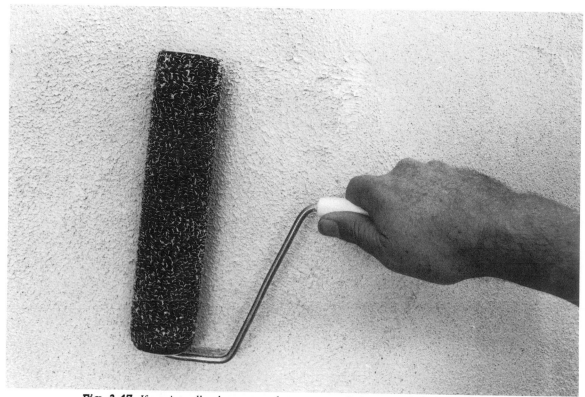

Fig. 3-47. *If a paint roller does not produce a pattern distinct enough for your purposes, you can use a special texturing roller.*

Fig. 3-48*. Patting with a flat block of wood or a steel trowel will produce many slightly varied textures.*

To further change the appearance of a textured wall, apply a little paint after the material has thoroughly cured. Use an interior latex paint in a color or tone that complements the texture coating. Apply the paint with a fine-bristled brush held almost flat to the surface, with only a small amount of paint loaded on the brush tip. Draw the brush smoothly across the surface so that the paint only coats the protruding portions of the texture, leaving the depressions the original color. The effect can be quite dramatic, or quite subtle, depending upon your color choices and the texture pattern.

In short, you can use whatever your imagination can dream up to make and modify impressions on the surface, within the time limitations of the coating material. And you can be sure that you will never see another wall surface just like yours—it will be unique.

PAPERING WALLS

"Wallpaper" has come to be a handy generic term for a number of different wall coverings. Apart from the thousands of color and pattern combinations you can choose from—even a small shop will have a hundred or so books of samples—there are several basic materials to consider. Inexpensive wallpapers are just that—thin paper of relatively low quality with a printed design. Paper wall coverings progress through a substantial range of weights, quality levels, and durability. There are many coated papers also, which are usually washable. Some papers are not paper at all, but vinyl. Others have decorative foil facing. There are many fabric wall coverings, some of heavy plastic, and a few special systems for covering masonry or ceramic tile walls. Some papers are plain, but most are patterned by roller printing, screen printing, engraving, lithographing, or photographic means. Any of these may be smooth, textured, or embossed, or have other decorative materials such as grasses or fibers bonded to them. The choices and varieties seem limitless.

Estimating Material Requirements

Wallpapers are usually sold by the single roll or the double roll. The width varies, as does the roll length, but most rolls contain about 36 square feet. Whenever possible, buy pretrimmed paper in double rolls, because this saves waste. Determining how much paper you will need can be a bit tricky. Some papers need not be matched—that is, there is no repeating pattern that must be matched together at each strip edge. Papers that must be matched require a greater quantity of material, depending upon the distance between pattern repetitions.

The rule of thumb states that you will use 30 square feet of every 36-square-foot single roll. The extra 6 feet is used for trimming, waste, and matching. Measure the walls, including doors and windows, and determine the gross square footage. A room 8 by 12 feet with 8-foot walls, for example, would have a wall area of 320 square feet. Divide this by 30 (usable square feet in a single roll), which in this example, leaves 10.66. Thus, you would need $10^2/3$ single rolls. From this, subtract 1 single roll for every two window or door openings. Assuming four openings, you would then need $8^2/3$ single rolls. Round this up to 9 single rolls, and always add 1 single roll for good measure—running short of paper is a major disaster. For this example, buy 10 single rolls or 5 double rolls for the room.

This method is obviously an inexact one. A better method is to determine the gross wall area, the number of doors and their area, and the number of windows and their area. It would do no harm to make sketches of each wall of the room with all the dimensions included. Then select the specific wallpaper that you are going to use, and make a note of the roll width, square footage, and pattern repeat lengths. Sit down with the wallpaper supplier and work out the exact number of single or double rolls needed to do the job, plus a little extra for insurance.

Equipment

You will need some special equipment for wallpapering (Fig. 3-49), but none of it is very expensive. The best bet is a small kit that includes a paste brush, a smoothing brush, a seam roller, and a razor knife that uses replaceable industrial-grade single-edged razor blades. Instead of, or in addition to, the smoothing brush, you might find a 12-inch-long Styrofoam smoothing block faced with suede fabric on a resilient plastic base a very useful tool. And, for trimming in tight spots, a small X-Acto or similar knife with a long, pointed

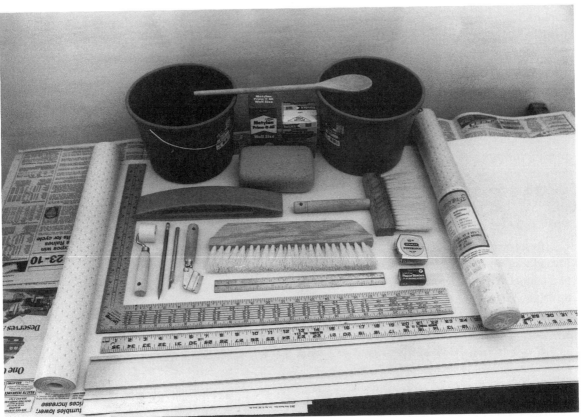

Fig. 3-49. All the tools and supplies needed for hanging wallpaper. A pair of large shears could be added, although many folks prefer to make all cuts with a razor knife.

blade is almost a must. You will also need a large wallpaper sponge and a water bucket. If you will mix your own wallpaper paste you will need a second bucket or a large mixing bowl; a wooden paddle or a big wooden kitchen spoon makes a good mixer. If you will be using a prepared adhesive, the paste brush that ordinarily comes in a kit might not do the job; you might prefer a 4-inch, stiff bristle paint brush instead. For cutting and measuring you will need a tape measure and a 12-inch ruler, a long straightedge, long shears or scissors, small scissors, a carpenter's framing square, a pencil, and a plumb bob. And finally, you will need a platform of some sort to work on. A typical arrangement utilizes either a pair of sawhorses or a long table, topped with a sheet of smooth ³/₄-inch plywood about 2 feet wide and 6 to 8 feet long, or a smooth hollow-core door (without the hardware). Add a stack of old newspapers, and you're ready.

Adhesives

A variety of adhesives is available for hanging wall coverings, and your choice depends partly upon the nature of the wall covering. You will need wall sizing, which is applied first. Light- to medium-weight wall coverings are usually hung with ordinary flour paste, which comes in small packets to be mixed with cold water. Various additives can be mixed with these pastes to improve sticking power. Heavy papers and fabrics usually are hung with a thick, heavy-duty adhesive. The nature of the wallpaper surface also determines what adhesives should be used, and in what sequence. For example, a nonporous surface being covered with a heavy material might first require an initial coating of a latex primer-sealer, plus a coating of sizing. Your supplier can solve the adhesives puzzle for you; be guided by his experience as to which products have worked best in various situations. Some papers are prepasted and theoretically only have to be dipped in water; special trays are available for this purpose. However, many paperhangers prefer to also apply a liberal dose of paste to such paper, prepasted or not. Again, check with your supplier.

Preparations

To set up for a papering job, first clear the room. Put down dropcloths or sheet plastic around your work area and along the walls. If possible, set up your work table in the center of the room to be papered. Turn off the electrical circuits and tape the circuit breaker handles closed or remove the fuses. Take down wall lighting fixtures and remove switch and outlet covers. Also, remove any other trim frames—around heat registers, for example— that will come off easily. Wash down the wall surfaces and make sure they are free of dust, grime, and grease, and do any other wall prep work that might be recommended by the manufacturer or the supplier.

Prime and/or size the walls as required, and allow the necessary length of time to elapse before starting to hang the covering. The job will be easier if you can accommodate your work schedule to suitable weather conditions. A temperature range of 65° to 80° Fahrenheit and a humidity range of about 40 to 70 percent is fine. Working in temperatures below 55° is not a good idea, and hot, dry conditions—especially with a breeze blowing in the windows—makes the job very difficult.

Procedure

There seem to be as many ways to hang paper as there are paperhangers; almost everybody has a few pet tricks and ideas picked up through experience. However, the fundamentals remain the same even if there are differences of opinion concerning the details. The best place to start is along a door or a tall window whose top is close to the ceiling. If there is some mismatching or misalignment after you have gone all around the room and back to the starting point, this is where it will show the least. Some folks, though, prefer to start in a corner. If you are right-handed, work from left to right; lefties go the other way.

Place a straightedge along the door casing up to the ceiling and make a mark at the ceiling. From that mark (or a corner), measure out along the wall-ceiling line the width of your wallpaper minus 1 inch, and make another mark. Drop a plumb bob from that mark to the floor and make another mark. It is usual procedure to connect these two marks with a floor-to-ceiling snapped chalkline, but this can cause problems—not the line, but the chalk. The red or blue chalk commonly used to snap lines will smear in the paste and can become embedded in many varieties of wallpaper, leaving stains that won't come out. To be safe, hold a plain string taut between the two marks and carefully make several more marks on the wall under the string. Then connect the marks with a long straightedge and a light pencil line (Fig. 3-50). This line will be your starting point.

Measure the distance from floor to ceiling and add 2 inches. Cover your work table with a layer of newspapers, roll out a length of wallpaper, and cut it to this measurement. Square off the cut with a framing square, mark the cut line with a pencil, and lop off the

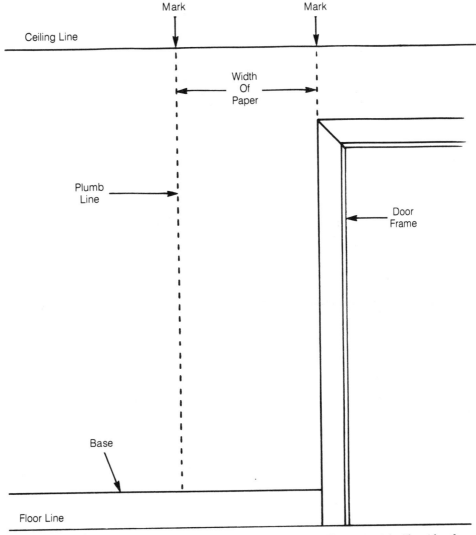

Fig. 3-50. *One method of laying out the starting line for a wallpapering job. The side of a tall window or a wall corner could be used.*

piece with shears. If the wallpaper is pretrimmed (which saves a lot of work and results in better seams), you're ready to go. If not, trim off the selvage edges with a straightedge and a razor knife.

The next step is adhesive; prepare that if necessary. For prepasted papers, soak the paper according to directions. If you prefer to apply paste anyway, do not soak. For applying paste to the paper, lay the paper face down on the table (probably with one end trailing off the end) and brush on an even coating of paste. For directional patterns (which must be hung right end up) start pasting at the bottom of the strip. When the strip is about two-thirds covered, loop the bottom end back on itself, paste to paste, in a loose U shape. Be careful not to make a crease or fold. Slide the strip fully onto the table and finish the paste coating. If your adhesive is the type that is applied to the wall rather than the paper (a system that is easier in some respects) coat the wall with adhesive to a point slightly beyond where the strip edges will lie.

Pick up the strip of paper with a thumb and forefinger at each top corner, climb onto the stool or ladder (already set in the proper place, please), and align the edge of the strip with your starting guideline. Press the paper gently against the wall with the heel of your hand, leaving a flap of about 1 inch extending above the bottom of the molding, or lapped against the ceiling (Fig. 3-51). Smooth the paper down along the guideline, making sure that it is exactly aligned.

Meanwhile, hold about half the width of the strip off the wall with your other hand. If it does not line up properly, peel it off and start again. Once you have good alignment for 2 feet or so along the guideline, bring the other side edge into contact with the wall at the top. Smooth the paper across the top end with your hand, then sweep across and downward lightly to smooth more of the strip in place (Fig. 3-52). With a fingernail, crease the paper at the top by crimping it into the wall-ceiling joint or along the inner edge of the molding.

Once the sheet is secure, smooth the whole top part with a smoothing brush or block and work out any air bubbles. Then pick out the bottom corners of the strip with your thumbs and forefingers and peel the loop of paper apart by pulling gently and straight downward (Fig. 3-53). Let this bottom section hang free, and smooth it from the center of the strip downward and outward to each side in repeated short strokes (Fig. 3-54) until you reach the bottom. Then crease the paper along the joint between the wall and the base molding. (Note: With many heavy wall coverings creasing is either ineffective or impossible.)

The side edge of paper at the door or window casing is now folded out for approximately 1 inch along the edge of the casing. Make a cut with small scissors at the top of the casing and fold that flap onto the wall, into the paste. Then trim from the top of the casing to the base molding, along the joint between the wall and the inner edge of the casing. Use the razor knife held at about a 45-degree angle, both to the wall and to the direction of travel of the blade. Cut with only the corner of the blade (Fig. 3-55). Trim slowly and watch that the blade doesn't wander. It helps if you can crease the paper into the joint with a fingernail, just ahead of the knife. Then trim across the top and bottom in the same way.

Fig. 3-51. *The first strip of wallpaper set to the starting plumb line. The starting point is the tall window to the right.*

Go back over the whole strip with the smoothing brush or block and make sure the paper is flat and there are no air bubbles underneath. Then run along all the edges with a seam roller (Fig. 3-56), using moderate pressure; too much will force all the adhesive out and you'll have a starved edge that will eventually lift off.

Cut the next strip and hang it in the same way. (Some paperhangers prefer to cut all the strips to length first.) If there is a repeating pattern, make sure to cut the strip long

Fig. 3-52. *The wallpaper strip adjusted to exactly follow the plumb starting line. The upper portion is smoothed back toward the window casing at the right.*

enough to match the pattern correctly to the first strip. After checking the match, you can trim to length, leaving about 1 inch extra at top and bottom. Abut the edge of the second strip exactly against the first, with no overlap anywhere and matched as necessary. Some workers prefer to overlap each seam about 1/4 inch, but the appearance is not as clean and

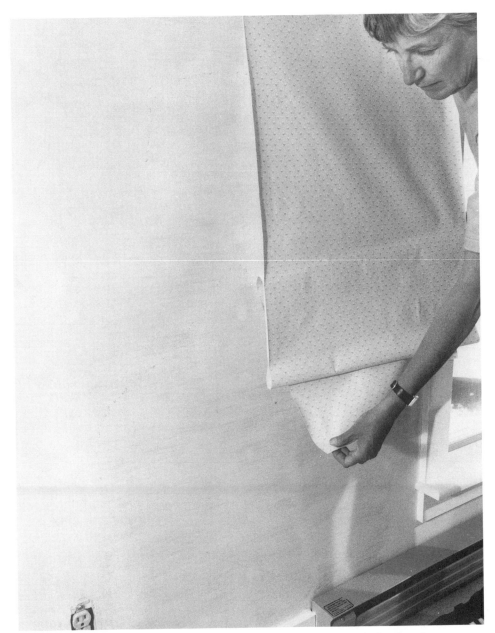

Fig. 3-53. *With the upper third of the wallpaper strip smoothed and stuck,*
the bottom loop can be peeled down into place.

overlapping will not work with matched papers. Trim this strip in the same way as the
first.

Here is another option; trim the first sheet only after the second has been put up, the
second after the third has been hung, and so on—which can allow a better trim line. Make

Fig. 3-54. *A smoothing block used to set the lower portion of the wallpaper strip exactly to the plumb line. It is then smoothed back to the window casing.*

sure that there is plenty of adhesive under the edges along the vertical seam. If you are in doubt, lift the paper away slightly and squidge a little more underneath with a fingertip. Roll the seams with moderate pressure. Then wash the surface of the paper at least twice with clean water and a sponge. Fabrics are an exception: with those, you have to be careful not to smear adhesive, and spot clean whatever does get on the surface.

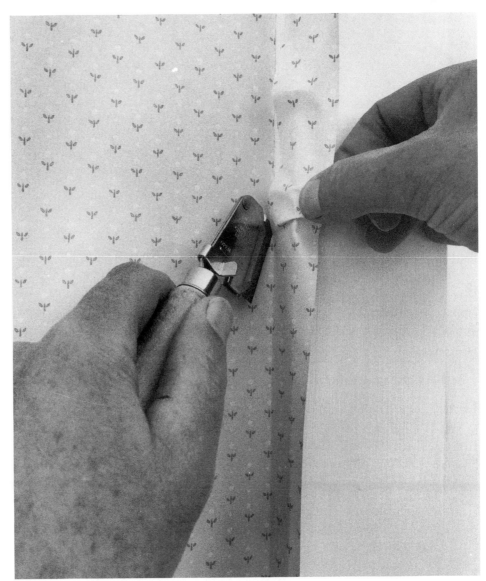

Fig. 3-55. *Here the right edge of the paper is trimmed to the window casing with a razor knife.*

Continue putting up the strips consecutively. Paper right over outlet and switch boxes and trim out the openings afterwards. Match up and hang short pieces below windows and above doors and windows in the same way as full strips. In many instances you will have to trim out large L shaped strips. Rough-cut these pieces first with shears, leaving about 1 inch extra all around, and then hang and trim them. Trimming around intricate shapes takes some ingenuity. The usual procedure is to make a small relief cut with scissors or a razor knife, press a little bit of paper into place, trim off the scrap closely with an X-Acto (Fig. 3-57), make another relief cut, and so on until you have trimmed to the right contour.

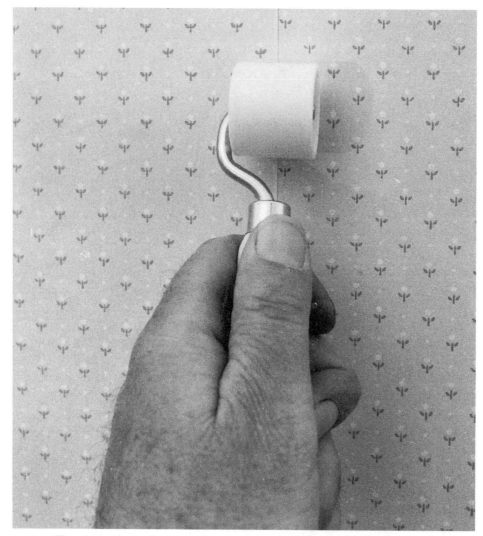

Fig. 3-56*. The vertical seams between wallpaper strips should be rolled gently
to make sure they are stuck down tight.*

With many wall coverings it is not advisable to wrap a strip into an inside corner.
When the adhesive and the wall covering dries, the covering will shrink and lift away
across the corner for as much as a half inch or so. When you come to an inside corner,
trim the strip lengthwise so that only about 1 inch wraps onto the adjacent wall. Run over
this narrow strip with a seam roller and make sure the material is tucked tightly into the
corner. Then match and fit the next sheet right into the corner, overlapping the 1-inch strip
but keeping the full strip plumb. Outside corners of 45 degrees or less seldom present any
problems with shrinkage. Sharper corners might, depending upon the wall covering material. Here you can use the same system. Wrap only about 1 inch of material around the

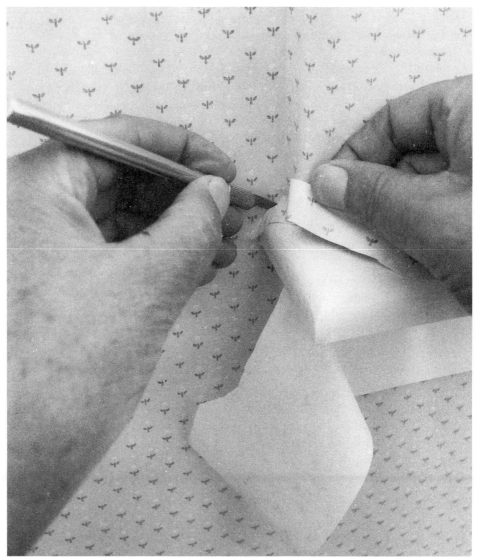

Fig. 3-57. *Small, tricky trim cuts, are most easily done with a small, maneuverable, and very sharp modeling knife.*

corner, then overlap the next full strip onto the 1-inch strip and trim its edge flush with the corner.

If you should come to a meeting point between the edges of two strips that are out of vertical alignment—as when you have papered all the way around and back to the starting point—let one strip overlap the other by 1 inch or so and paste it down. Center a straight-edge about midway of the overlap, hold it firmly, and cut through both layers of paper with your razor knife. Peel away the scrap end on top, then lift the paper edge with the point of a knife and remove the piece of scrap underneath. Press the paper back down and go over the edges with a seam roller, and you should have a perfect seam.

Sometimes making a perfect match at the final meeting point is impossible because the total distance around the room is not equal to an exact number of full wallpaper strip widths. The last piece, which should lie above a door or window, must be only a portion of the full width of a strip. In that case, by juggling smaller pieces and pasting up two or three small sections, you might be able to work out an approximate match by blending different parts of the pattern or by pasting in an extra few inches of blank, patternless background. Cut and fit the pieces dry on the worktable, experimenting until you get a suitable arrangement. The pieces might appear to match when seen from a short distance away, when they actually do not. What you are trying to arrange is an optical illusion, and with a little experimenting this often works out nicely. If necessary, this sort of "piecing in" can also be done in corners or other unobtrusive spots.

As a final step, after all the wall covering has been hung, inspect the walls to make sure all the trimming got done and all the seams are flat and tight. If you find any lifted edges, ease a bit of paste under them with a toothpick and press them down. Check the base and ceiling moldings and edges for smears of adhesive, and make sure all the surfaces are well washed down. Then replace electrical and plumbing hardware, along with any other items that were removed, and clean up the area.

PANELED AND PLANKED WALLS

There are numerous panels and plank wallcoverings from which you can choose, with a tremendous array of decorative possibilities. The fastest and easiest way—aside from paint—to finish a room is to put up factory prefinished paneling. Although it takes a bit longer, unfinished paneling or either finished or unfinished planking also present a rich custom-finished appearance. Both can be applied over a variety of existing surfaces, even those in poor condition. Some of these products also add strength to the structure and slightly increase the thermal efficiency of exterior walls. There are several types of products to consider.

Options

Hardboard panels come in a variety of finishes, including patterns and simulated wood grain or planking. Also available are panels coated with baked-on modified melamine finish—again in solid colors and in patterns—that are waterproof when properly sealed and suitable for kitchen and bathroom walls, and even shower stalls. Most of these products are $1/4$ inch thick and made in 4-$\times$-8 foot sheets. Some have matching molding systems that join and cover sheet corners, edges, and tops.

Hardwood plywood panels, which are composed of three or more layers of wood including a face veneer, may be finished or unfinished. Most readily available varieties designed for home installation are finished. Some are made to resemble planking (Fig. 3-58); most are plain-faced with solid veneer. This veneer is made up of smaller pieces matched together in various patterns, including diamond and herringbone. The least

Fig. 3-58. A typical and elegant wall paneling installation of curly birch. Note the cornice arrangement. (Courtesy Hardwood Plywood Manufacturers Association.)

expensive are faced plain with common veneers, such as birch, oak, lauan, ash, pine, and aspen. The most expensive are those made in complex piece matches using exotic woods like rosewood, teak, limba, prima vera, or bubinga, or more common woods with unusual grain or figure such as walnut burl. There are about 150 different wood species regularly used as face veneers for hardwood plywood panels, each of which can be cut in one of three ways to present a different grain and figure appearance. Together with the several matching possibilities and the various finishes, there are something like 18,000 combinations possible. Standard panels are considered as $1/4$ inch (or sometimes $3/4$ inch) thick, 4 feet wide, and 8 feet long, but numerous other sizes are sometimes available or can be made up specially.

Wall planking presents many additional options. It is packaged random-length, tongue-and-grooved, and end-matched in thicknesses usually ranging from $1/4$ inch to

Fig. 3-59. *A typical vertical wall plank installation. Diagonal, horizontal, and geometric design installations are also popular. (Courtesy Wood Moulding & Millwork Producers Association.)*

$1/2$ inch. It is readily available at most building supply centers, either in stock or by order. This very popular material (Fig. 3-59) offers lots of decorative possibilities and is easy to install. Pine, cedar, and redwood are perhaps the most common, but you can also choose among walnut, cherry, pecan, black ash, and oak. Some of these products are factory prefinished clear or in various color stains, many are unfinished.

Wall planking in $3/4$-inch thickness might also be found through building supply dealers. It comes in varying lengths, widths, and a few finishes—but usually unfinished. This planking is generally tongue-and-grooved, but not end matched because it is intended to be put up full length vertically, floor to ceiling. Some, typically pine or cedar, might be surface-milled into various patterns.

For greater variety or more unusual woods, you can turn to the specialty wood dealers for an amazing selection of wood species. They are unfinished and in various widths and lengths, but usually 3/4 inch thick and often tongue and grooved. Readily available species include white and red oak, white ash, rock maple, curley and bird's-eye maple, black cherry, black willow, butternut, pecan, cypress, elm, walnut, teak, mahogany, poplar, pine, cedar, and—when available—exotics like Thai boxwood, padouk, or kempas. Note that many of the exotics, while they make an unusual and distinctive (and generally expensive) wall covering, must often be pieced out from relatively small and random lengths and widths.

Estimating Material Requirements

Because most plank and panel wall coverings are rather expensive, you will need to figure your material requirements with care and accuracy to minimize the amount of leftovers. On the other hand, it's usually wise to include a small allowance for spoilage and waste. If the material you select is readily available from local stock, you can purchase a bit less than your estimated needs, install that, and then go back to your supplier for just enough to finish the job. If the material is unusual or a special-order item, plan on buying some extra at the outset.

When estimating paneling requirements, start with the known size of the sheets (usually 4- × -8 feet) and plan the actual layout as the panels will fit on the wall, piece by piece. At the same time, determine how and where you can position to best effect the smaller pieces that will have to be cut from full panels. You might also have to figure out where various panel cuts will have to be made in order to match side edges together for correct appearance. This is sometimes the case, for example, with panels that are scored vertically to simulate planks. Once your layout is made up, just total the number of full panels required, then add one extra for a small job, perhaps two for a large room.

Plank estimates are made a little differently. If the material is packaged wallcovering stock, each package will cover a certain number of square feet of wall area. If the material is end-matched—as is usually the case—and packed in random lengths, there will not be much waste per package. Measure the wall surfaces to be covered by breaking them into rectangles—open areas, above doors, above and below windows. Do not subtract for small openings in the wall, such as heat registers. Determine the area of each rectangle, round up to the next highest full square foot, and add them up. Then, if you will be installing the planks vertically or horizontally, add 10 percent, divide the result by the number of square feet in each package of material, and round the answer up to the next highest full package. If you will be installing the planking diagonally, add 15 percent to the area total; for intricate geometric patterns, add 20 percent.

Unpackaged random-length and random-width stock is usually sold by the board foot, whether intended specifically for wall covering or for general purposes. This material is seldom end-matched, however, so there might be more cutting waste depending upon available lengths and the wall surface to be covered. Calculate the area of the walls closely, then add about 20 percent for spoilage and waste. The number of square feet of wall area will be the number of board feet of material required, assuming 3/4-inch-thick material. If it is not, the dealer can make the necessary translation for you.

Installing Paneling

Putting up wall paneling is not a difficult job, but does require some patience and attention to detail, particularly in making accurate measurements and clean cuts. If $3/4$ inch thick, the panels can be fastened directly to an open wall frame, because the material is strong and rigid and actually adds strength to the structure. If less than $3/4$ inch thick, install it over gypsum wallboard or lath sheathing. Panels can be laid up directly against plaster, wallboard, fiberboard, particleboard, plywood sheathing, or rigid insulation panels. You can secure it with nails into the studs, or glue it to clean, dry surfaces with a compatible adhesive. Gluing to wallpaper, however, isn't a good idea—if the paper should come loose, so will the paneling. If the surfaces are smooth, dust-free, and always dry, paneling can also be glued to poured concrete or concrete block walls. A better method is to waterproof the surfaces and fur them out, cover the furring with sheathing (see "Covering Masonry Walls", earlier in this chapter) and then either glue or nail the paneling in place. Whatever the material, it should be left spread around or loosely stacked to acclimatize for at least 48 hours, preferably a week, in the room where it will be installed.

To put up the paneling, start at an inside corner and set the first panel into the corner, just as with sheathing panels. If the corner is not plumb, scribe or measure the irregularity and trim the panel to fit. The panel edge away from the corner should be plumb. Trim the panel to floor-to-ceiling height, minus about $1/2$ inch. When you position the panel, shim it $1/4$ inch off the floor with a piece of scrap.

To fasten with nails, use ring-shank panel nails of appropriate length and color. If glueing, the usual practice is to run a $1/8$-inch-thick bead of panel adhesive all around the perimeter, about 1 inch in from the edges, and zigzag beads vertically at about 16-inch intervals. Position the panel slightly away from the wall, press it in place, then pull it away again. Allow the adhesive to tack for about 2 minutes, then reset the panel. Note, though, that this procedure might vary depending upon the adhesive, so follow the manufacturer's instructions. Paneling can also be both nailed and glued. Many installers nail the panel corners, or nail across both top and bottom and at a few intermediate points, to secure the panel while the adhesive cures.

When you trim panels, leave about $1/4$ inch of clearance for expansion wherever the joints or gaps will be later concealed by moldings or trim. Otherwise, abut the panel edges tightly. Cut with a fine-toothed crosscut handsaw (best bet is a panel saw intended for the purpose) or a power circular saw fitted with a fine-toothed plywood blade. Some major cuts can also be made on a large table saw fitted with the same type of blade. Make such cuts with the panel face up. You can also cut with a sabre saw; in that case, place the panel face down. If necessary, when matching panel edges together you can make the cuts a tiny bit wide of the mark, then pare the edges down a little at a time with a block plane to get a perfect fit.

In some cases it is necessary to drive nails in spots where they will show. If this occurs, simply countersink the nail head a tiny bit, and fill the hole with a matching color putty stick. If the paneling is unfinished, do the filling after the final coat of finish has been applied. If you are glueing panels up and they don't stick tightly to the wall because of a bow or a depression, or because they are warped, wedge the panel in place: Set a 2-×-4 flat against the panel. Wedge a length of 1-×-4 between the center of the 2-×-4 and your table saw or a heavy piece of furniture and jam the panel flat.

Installing Planking

Thin wall planking is usually glued in place, and you can drive an occasional panel nail here and there to help secure it. This method works well for applying the material to any smooth, clean surface, and allows the pieces to be installed in any orientation—including geometric designs. Two or three wavy 1/8-inch beads of adhesive, depending upon the plank width, are sufficient. The tongues and grooves, along with occasional nails and the tackiness of panel adhesive, will hold the pieces firmly as you work along. If the wood is hard or you drive nails near the plank ends or edges, drill slightly undersized pilot holes first. Ring-shank panel nails are the best choice here.

Allow expansion gaps at floor and ceiling and next to window and door jambs, but otherwise abut the pieces tightly. Set the first piece, or row of pieces, plumb at a corner. They should be level along the floor line, or otherwise oriented to the correct angle of your design. Subsequent rows will follow this first one, but alignment is not automatic. Pieces will vary just slightly in width, and some will be a bit crooked. Check the plumb, level, or angle lie about every three or four rows to make sure you are on target. If not, make the necessary adjustments by shifting the tongues just a bit in the mating grooves. A small amount of creep in joint widths and alignment can make quite a difference over a span of several feet. It will work against you if you don't check often, but for you if you need to compensate in the overall alignment.

Make your cuts with a fine-toothed saw; a table saw is the best for close control. With end-matched planking, position your piece so that a cut end will come at the ceiling line and the cut-off piece can be used, cut end down, at the floor to start one of the next rows. Do not mate two butt ends in the field. If this becomes necessary, fashion end-matching rabbets to resemble the factory cuts.

Full-length, 3/4-inch-thick planking is installed differently. It must be securely attached to studding, blocking, or furring strips with nails—adhesive won't hold it in place if it decides to warp. The planking can be installed vertically, horizontally, or diagonally, so long as there is something solid to fasten it to. If set horizontally or diagonally, the planks can be nailed right to the wall studs, with end joints meeting at the centerline of a stud. For vertical placement, there must be at least two rows of horizontal blocking installed between the studs, spaced equidistant from floor to ceiling, to which the planks can be nailed (Fig. 3-60). If there is no blocking present, you can run four lengths of 1-×-4 furring strips across the walls and fasten them to the studs with 12d nails—one each at floor and ceiling and two or more spaced in between.

To install this kind of planking, start plumb in a corner, set to a diagonal, or level along the floor line. At points where the wood will be covered by base and ceiling moldings or window or door casings, you can top-nail the planks directly to the supports. Use either finish or box nails of appropriate length—8d or 10d in most cases. Elsewhere, blind-nail by driving the nails angled through the plank edges. If the planks are tongue-and-grooved, set the first plank groove to the corner, or headed down, so you can blind-nail at the tongue corners. If the planks are square-edged, only one side of each plank can be blind-nailed at the intermediate points. You will have to top-nail at these points, set the nail heads and fill later with matching putty. Often it is possible to drive the nail on an angle right at the edge corner, which makes it less obvious. Leave small expansion gaps top and bottom and at jambs, but otherwise butt the planks as tightly as you can.

If planks are squared-edged, some shrinkage can be expected regardless of the wood species. This will open gaps between planks that might become obvious. Painting the sheathing surface in a color similar to that of the wood will render practically invisible any narrow cracks that occur. Another way around the problem is to install the planking in board-and-batten style (Fig. 3-61). Set the planks about 1 inch apart, then cover each space with a narrow batten of the same wood, nailed to the planks. Or, use the board-on-board system: Space the planks of the first layer apart about 1½ inches less than the width of the planks, then nail the second row to the first to cover the spaces (Fig. 3-62). Either way, shrinkage is negligible.

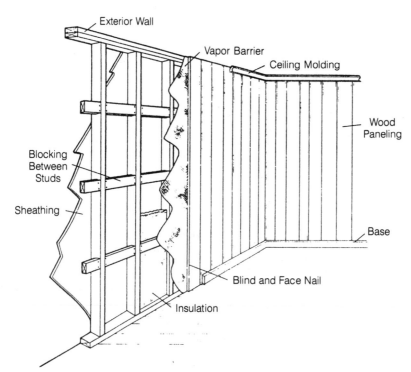

Fig. 3-60. *Blocking should be installed between the wall studs to provide a nailing base for paneling/planking installations. (Courtesy USDA.)*

Wall Sheathing or Backing

Fig. 3-61. *A board-and-batten arrangement, top view. The thickness and width of the pieces can be whatever appeals.*

Wall Sheathing or Backing

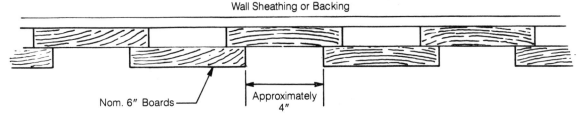

Fig. 3-62. *A board-on-board arrangement makes an attractive wallcovering.*

MASONRY VENEERED WALLS

Inside walls constructed of various kinds of masonry units have long been a popular interior decorating feature. They are usually built as an integral part of the structure during initial construction. It is also possible, and often not difficult, to create the same look during a remodeling project by veneering existing inside wall surfaces. Among the materials that can be used are new brick, recycled or old handmade brick, natural stone, prepared marble, granite, terrazzo tiles, or any of several synthetic brick, slate, or stone facings available. These materials can be used to completely cover an existing wall surface, or in a wainscot arrangement 3 or 4 feet high.

Considerations.

Where full-sized masonry units like brick or natural stone are used—especially to make up a full-height wall of considerable length—the weight of the material is the biggest concern. A single-thickness brick wall 8 feet high and 12 feet long, for example, would weigh in at around 4000 pounds, or about 330 pounds per linear foot. A stone wall would weigh even more. For this reason, such walls based on the wood floor of an ordinary wood framed house, or even on an unreinforced poured concrete basement or garage floor, could cause problems. A wainscot about 3 feet high would weigh approximately 125 pounds per linear foot, possibly more. Even this construction could cause difficulties.

This is not to say that such a job cannot be done. Construction of the wall itself follows the usual procedures of bricklaying and stone masonry. A low wainscot wall of brick or cut stone can sometimes be freestanding, but in most cases—and always with full-height veneer—the units must be tied to the existing wall for stability. There are many patterns and course arrangements (called *bonds*) possible with brick walls, and with stonework. The many intricacies of bricklaying and stone masonry are beyond the scope of this book, and the business of building these kinds of veneer walls is complex, although the work itself is not terribly difficult. If you have your heart set on such a project, you should first consult with an architect, engineer, or builder to see if your plans are practical, or if not, what can be done to make them so. If you are not familiar with masonry work, study up on the subject before diving into your project, or hire a professional mason to do the job.

Putting up a veneer of stone tile or synthetic material (Fig. 3-63) is another matter. These products are lightweight—typically about 2 to 6 pounds per square foot—and are designed for the purpose, often with the do-it-yourselfer in mind. There are several kinds of $1/4$-inch-thick marble tiles bonded to fiberglass backers, $3/8$-inch-thick granite tiles, and terrazzo tiles—all in various sizes—that make handsome wall coverings. The synthetic materials such as brick, random or rubble stone, ashlar stone, and slate are also thin and come as individual pieces or in small sheets for easy application.

The stone tiles are glued to the sheathing with mastic, and may be tightly butted or set slightly apart with grouted joints. The synthetics are likewise applied with special adhesives, many of which are formulated by the manufacturers for their own particular products. Simulated mortar joints may be a part of the material, or individual joints may be made up with the same mastic used to adhere the pieces to the wall surface. Stone tile materials are not off-the-shelf items and there is only a handful of manufacturers, but any good masonry or tile supply dealer or interior decorating shop should be able to obtain both information and material for you. The synthetic veneers, on the other hand, are widely distributed all over the country through lumberyards and building supply houses.

The stone veneers can be applied to any flat, smooth, rigid, stable, clean, and dry surface as gypsum wallboard, plywood, plaster, planking or paneling, poured concrete, or concrete block. The surface must be plane within a variance of no more than about $1/16$ inch in 3 feet in any direction; this is particularly important with the larger tile sizes. The synthetic materials can be applied to the same subsurfaces, but are more forgiving as to the regularity of the surface. This is because they are secured with thick, very sticky mastic that levels the surface and fills defects as the flat-backed pieces are pressed into place. Some of the synthetics can even be successfully applied over old ceramic tile— grout lines and all.

Estimating Material Requirements

To estimate your needs, first select the material and determine the tile or piece size or the area of coverage per package. Then work out just how you are going to lay them up on the wall. It's not a bad idea, especially with the expensive materials, to actually lay out the installation right on the wall with chalk lines, so you know where each piece will fit and where the trim cuts will have to be made. This will also show you where, or if, cut-off pieces can be used. Then count up the number of units needed and add several more, or a package, to cover damage and mistakes. The alternative to this method—easier if the room is large but chancier if you miscalculate—is to determine the square footage of the wall

Fig. 3-63. *Synthetic brick veneer is easy to apply, attractive, and very durable. Note the heat shield attached to the stovepipe.*

surface and add 20 to 30 percent. In any event, all the required material should be obtained at once to avoid problems with mismatch of color or other characteristics stemming from different production runs.

Procedure

To veneer a wall with any of these materials, first remove all electrical components, moldings, trim casings, and any other wall-mounted items. Vacuum or wash the wall as necessary to remove dust, grease, and soil. Start at any convenient point with a level or plumb line, depending upon how the material will be installed. Layout details are much the same as for wall tile, detailed in "Installing Ceramic Tile" late in this chapter. Then simply follow the instructions provided by the manufacturers of both the adhesive or mastic and the veneer materials. Keep a constant check on you vertical and horizontal edge lines to make sure none of them run off. Application details will vary from product to product. To cut the stone tiles, either rent a masonry saw or have the cuts made by a masonry contractor. Synthetic materials vary in their makeup, but can usually be cut with a masonry saw or scribed and broken or nibbled the same as ceramic tile.

Many manufacturers of the synthetic materials recommend that the veneer surface be sealed with a special sealer, which protects the surface and facilitates cleaning. Brick and natural stone can also be sealed with varnish or a masonry sealer. Granite, marble, and terrazzo tiles are factory finished; only the specific waxes or sealers recommended by the manufacturers should be used on them. These coatings should be applied only after any exposed adhesive, grout, or mortar has thoroughly cured.

INSTALLING NONCERAMIC TILES

There are several varieties of nonceramic tile that can be applied as finish wall covering. Stone tile and resilient floor tile are two examples. Any of the large family of resilient tiles, particularly the vinyl types, can be applied as wall covering and are effective as wainscoting in nurseries and bathrooms and as kitchen counter backsplashes. Strips of vinyl sheet flooring can be used for the same purpose. There is a small selection of plastic wall tiles available, typically in $4\frac{1}{4}$-inch square, that somewhat resemble ceramic tiles and come in several colors and patterns. These are mostly installed in bathrooms and sometimes in kitchens, but are not recommended for kitchen counter backsplashes.

Metal tiles work nicely in all kitchen locations (Fig. 3-64) and particularly as a backsplash, and can be employed for other purposes as well. They are also $4\frac{1}{4}$ inches square, and available in copper or coppertone, brushed aluminum, and brushed stainless steel.

Mirror wall tiles come in various sizes and shapes and a few color tints. Some are internally decorated with patterns or motifs, others are plain, or square edged, or bevel edged. They are intended primarily for decorative accents, but can be applied in any desired manner, including full walls. There are also two or three different kinds of cork tiles (availability varies) that can be applied to cover full walls, as accents, or as bulletin boards.

Fig. 3-64. *Metal tiles are simple to put up and make an excellent counter backsplash. These are brushed aluminum, each tile turned 90 degrees to the next.*

All of these materials can be applied to any smooth, dry, clean surface. The most difficult to install are the stone tiles (refer to the previous section). To apply sheet vinyl flooring or resilient floor tiles to walls, follow the same procedures as applying them to floors, modified as necessary to suit the job conditions. Note that there are several different kinds of aluminum and stainless steel trim moldings available that can be used along with sheet and tile floorings as caps, seam covers, inside and outside corners, and edgings. Consult your building supplies or hardware dealer.

The remaining kinds of nonceramic tiles are simple to apply. Because they are so lightweight they can be put over old wallpaper or painted or varnished surfaces, provided they are well bonded. The one exception is large mirror tiles, which are heavy enough that they must be securely anchored to a solid surface. They all can also be installed over defects in the old surface; on surfaces slightly out of true; or over joints, cracks, or decorative grooving such as might be found on wall planking, provided the tile corners can be kept pretty much in the same plane.

Plastic Wall Tile

Plastic wall tile, when applied over a substantial area, is set in adhesive. Use one that is compatible with the plastic and formulated for your purpose. To begin, set up a level or a plumb working line from which to start. If the area is large, establishing quadrants or regular sections is a good idea, so that you can complete one area before going on to the next. Calculate the rows of tiles so that trimmed tiles can be positioned on each horizontal row, butted against a molding or wall. Vertical rows can be set the same way, or you can start with a full-sized tile at the bottom and trim the top tile of each row as necessary to fit against the ceiling or ceiling molding. You can also simply stop the tiles at any intermediate point on the wall and apply some other finish wall covering above them. You can cut these tiles with a fine-toothed hand saw or a sharp utility knife, then smooth the cut edges as necessary with a file.

Spread adhesive on the wall with the recommended applicator, and press the tiles in place. Move them around as little as possible. Continually check the alignment of the rows. The usual placement is in a stack bond—one tile above another, but you can set them in a half-overlap, diagonal, or other pattern if you wish. You can also grout the joints, but this is usually less than satisfactory. A cementitious grout does not adhere well to the plastic, especially if the tiles have rounded edges, and will eventually break free. An epoxy grout fares better, but must be compatible with the plastic. Grout lines made up of carefully applied silicone, acrylic latex, or butyl caulk will adhere well, but be ready for a long, tedious job.

Metal Tile

The same application procedures for plastic wall tile apply for metal tile, except that the tiles are individually secured to the wall with $3/8$-inch squares of double-stick plastic foam, about $1/8$ inch thick, which are supplied with the tiles. The wall surface must be completely free of dust and grease film for the tape to adhere properly. After removing the protective paper from one side, stick a square to each backside corner of the tile. Then remove the paper on the faces of the squares, position the tile carefully and just shy of the wall surface, and press it gently in place. Once set, the tile can't be adjusted. If the set is wrong, you'll have to remove it, replace the mounts, and start over. Running out of mounts poses no problems, because this material is sold in rolls as double-stick foam mounting tape.

Metal tiles are usually set in a stack bond. If the faces are brushed, you can turn each tile 90 degrees to the previous one and alternate the orientation from row to row. This results in a checkerboard appearance because of the dissimilarity in light reflection. That appearance can be further changed by using a half-overlap patterns. Aluminum tiles can be cut easily with sharp scissors or shears, but other metals might require tin snips, depending upon their thickness.

Mirror Tile

The smaller sizes of mirror tiles may also be mounted to the wall with double-stick tape or mounts, or possibly with daubs of special adhesive spotted on the back surface. Larger and heavier pieces might have to be secured with screws driven through predrilled holes, or with mirror-hanging clips. Follow the manufacturer's recommendations.

Cork

You can put up small areas of cork tile by daubing the backside with construction adhesive. However, many of these products are fragile, and if bumped or scraped might break apart and the unglued portions break off the wall. It is best to secure cork tile solidly to the wall. First outline the area to be covered (if it is not to be a whole wall). Then brush a heavy-duty wallcovering adhesive—the kind used for hanging heavyweight vinyl or textile wall coverings—over the wall surface, staying just within the outline. Work in small sections of only a few square feet at a time. Let the adhesive tack up for a moment or two, and set the tiles in place, pressing them down gently with your hand. You can adjust them slightly as you do so. Allow the tiles to set for a few moments, then roll the surface with a pastry rolling pin, maintaining an even and fairly light pressure, just enough to ensure that the tile backs become fully and evenly bedded in the adhesive. Allow at least a couple of days for curing before applying any sealer or using the tile for bulletin board purposes.

INSTALLING CERAMIC TILE

Much of the information concerning ceramic wall tile and the installation thereof is the same as for ceramic floor tile. Refer to the section entitled "Ceramic Tile Flooring" in Chapter 2.

The same kinds of tile used for flooring can also be installed on walls, although in fact only the glazed and matte finish ceramic mosaic tiles are commonly used. A tile group specifically for walls, called glazed wall tile, is the most popular choice for this purpose. There is a tremendous variety available in many colors and patterns—plain, scored, imprinted, and textured surfaces; brilliant, matte, and semimatte or crystalline finishes; and sculptured or hand-painted designs. Although most residential wall tile installations are found in kitchens and bathrooms (Fig. 3-65), they certainly need not be confined to those rooms.

Along with regular field tiles, you can select many special shapes to trim out an installation. These include bead tiles; various sizes of bullnose, cap, and cove tiles; and base and curb tiles. The most common field tile size is 4 $1/4$ inches square, but many other sizes and shapes are available. Tile thickness is usually $1/4$ inch, although some are either $3/16$ or $5/16$ inch thick.

Wall tiles can be applied directly over almost any clean, dry, flat, smooth, plane surface that is tight, sound, and rigid. This includes gypsum wallboard, plywood, wood or hardboard paneling, wood planking, poured concrete, brick, concrete block, plastic laminates, or even old ceramic tile. With proper preparation, wall tile can also be applied to

Fig. 3-65. *A typical bathroom wall tile installation. (Courtesy Ceramic Tile Institute, Los Angeles.)*

painted or varnished surfaces. It should not, however, be applied over wallpaper. In places where moisture is prevalent, such as bathrooms, laundry rooms, or behind kitchen counters, ceramic wall tile should be applied over water-resistant gypsum wallboard or a special waterproof tile backer board. Although water-resistant gypsum wallboard is sometimes used as a backing in shower stalls, the backer board is a much better choice, and imperative in steam rooms and similar locations.

Preparing the Surface

In most cases, preparation of the existing wall surface is a simple matter. Turn off the electrical circuits and tape off the circuit breaker handles or remove the fuses, then remove electrical outlets, wall switches, and wall-mounted lighting fixtures. The outlet boxes might have to be removed and reset later, or possibly they can be left in place and sleeved out with box extensions; this depends upon the kind of boxes and the way they were originally installed. Remove plumbing fixtures and cap off pipes as necessary. Existing wood base or ceiling moldings might or might not have to be taken out, depending upon the tiling job.

Papered walls should be stripped and thoroughly washed. Other surfaces should be dusted or vacuumed, or washed free of grime or grease film. Paint or varnish finishes should be sanded lightly to roughen them up slightly if glossy or semi-glossy; flat finishes need only be cleaned. Treat waxed surfaces with a wax stripper, then wash. Old ceramic tile must be cleaned of soap scum, mineral buildup, and dirt, and the glaze should also be broken. The easiest way to do this is with a sanding disc mounted in an electric drill or sander/polisher. Wear safety goggles and a dust mask while doing this, because the dust you raise will be partly composed of glass and mineral particles. Smooth off nibs, little bumps, and rough spots on poured concrete or masonry walls by rubbing the surface with a chunk of brick or concrete block. Clean off efflorescence by scrubbing with water and a brush, using a solution of muriatic acid, if necessary. Clean off all dust and dirt, and attack grease spots with detergent, trisodium phosphate, or solvents as needed.

Repair damage and defects if substantial enough to interfere with the lie of the tiles. Holes, cracks, open seams, grouted or mortared joints, low spots and depressions, and similar faults can be repaired or filled with appropriate patching compounds. Depending upon exactly what type of bonding agent you use, as well as the particular product and the kind of surface to be covered, you might have to apply a primer or a sealer to the surface before setting the tile.

Wall tile can be bought by the case, by the square foot, or by the piece. If you are only going to cover a small area, lay out the entire tile pattern, either in place or to rough scale on a sheet of paper. Count up the number of individual tiles, plus any different trim pieces, that you will need, and several extra pieces to the total for insurance.

For larger areas, simply compute the total square footage to be covered and add about 10 percent for breakage and waste. Purchase that many square feet of tile. Or, translate the total area in cases of tile and round up to the next highest number of cases that will be needed.

If the tile layout is complex and will require a lot of cuts, as do diagonal patterns or fancy geometrics or designs, it's a good idea to estimate the number of cuts that will be needed and order about 10 percent extra. As for the molded shapes like caps and coves, measure the linear footage needed of each type and divide that by the length of the particular tiles involved, then add a few extra.

Like floor tile, wall tile can be set with the deep-bed method—which incorporates a thick one- or two-coat setting bed of portland cement mortar—or with the thin-set method. The thin-set installation is much more advantageous for the do-it-yourselfer, and it is also more sensible and less costly in residential work. An organic adhesive or mastic is usually used with the thin-set method. For best results, the wall surface should be plane to within $1/8$ inch in 8 feet. The adhesive can be applied to almost any kind of surface.

Adhesive

There are two types of organic adhesives. One is solvent-based and intended for installations that are subject to considerable moisture and require prolonged water resistance, such as shower or tub enclosures. The other type is latex-based and is used in dry locations or areas that are only occasionally subjected to moisture, such as a countertop or backsplash. Both types are further divided into two categories, one for floor tile and the other for wall setting.

In special situations where an organic adhesive might be unsatisfactory, a dry-set, latex-portland cement, or epoxy thin-set mortar should be used, the choice depending upon installation and service conditions. In some cases these mortars might first requre the installation of a mortar bed, especially if the backing is rough or uneven. Your supplier can help you select the best products for your purposes; follow the manufacturer's application instructions exactly.

Layout

Begin the job by setting up the layout guidelines. You will need two—one vertical, and one horizontal (Fig. 3-66). Wall tile is almost always set working from the bottom up, so the horizontal line will be at the bottom of the field. If the bottom edge of tile will be partly hidden by a molding to be applied later, or if the bottom row of tiles will lie in the open, all you have to do is strike a level line at the appropriate point.

However, if you are setting the tile to an existing line such as a countertop or a base molding that happens to be off level, you have a problem. Some sort of adjustment will have to be made, or the tile joint lines will all be askew. One possibility is to plan for a full tile at the lowest point of the off-level line and trim slightly increasing amounts off each tile in the bottom row. This requires a masonry saw for good results. If the problem is a molding, perhaps it can be removed and reset level, or trimmed across top or bottom to compensate for the difference. Alternatively the uneven bottom edge joint of the tiles could be hidden by an added cap molding, or on a countertop by an added cove molding.

The vertical line should be plumb, at right angles to the level line. Strike this line at the center of the area to be covered with tile. Then measure from this centerline to one end

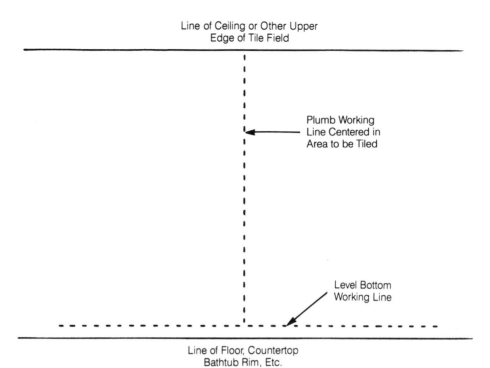

Line of Ceiling or Other Upper
Edge of Tile Field

Plumb Working
Line Centered in
Area to be Tiled

Level Bottom
Working Line

Line of Floor, Countertop
Bathtub Rim, Etc.

Fig. 3-66. *Layout lines for a ceramic wall tile installation.*

of the area and see whether or not the row will end in a half tile or more. If it will not, adjust the centerline and strike another parallel vertical line at a point that will allow each end of the row to end with a half tile or more. This will be the vertical working line. If the tile will stop at some point on the open wall rather than at the ceiling or a cabinet bottom, strike a horizontal line at this point, too. Then you will know just how far to extend the adhesive coverage.

You can set the tiles in a half-overlap or running bond, but the stack bond (one above another) is more common. Either way, set the tiles pyramid fashion. If the area is large, spread the adhesive in sections small enough to finish before the open or working time of the adhesive elapses. If the bottom row of tiles is in the open and the tile bottoms will not butt against something (a molding, for instance), it's a good idea to nail up a temporary batten. A straight piece of lath or 1- × -2 set along the level line will act as a backstop to set the tiles against, and will hold them aligned and in place as you set the tiles above.

Procedure

Use the recommended adhesive spreading method, and coat the first wall section. For a half-overlap bond, set the first tile centered on the vertical line. For a stack bond, set the first tile with one side aligned against the vertical working line. Then start building up the pyramid in the sequences shown in Fig. 3-67. Fill in the diminishing corner areas in

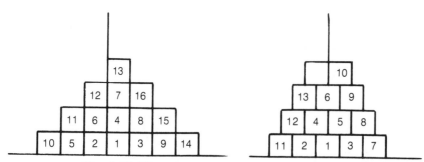

Fig. 3-67. *Tile setting sequences for ceramic wall tiles.*

the same relative sequence. If you will have a substantial number of tiles to cut at row ends and you are certain of their dimensions, it's a good idea to trim them to size before you begin and lay them aside, ready to set. If you are uncertain as to how the dimensions might work out, or if some intricate fitting that calls for tile nibbling will be required, do the trimming as you go along. Just work in smaller sections and keep in mind the open time of the adhesive, so you have enough time to work before it sets up.

To set the tiles, hold them in position just above the adhesive, then place them gently with a very slight twisting motion to align them. Do not set and slide them into place, because this will drive adhesive into the joints. If a tile is out of alignment, wiggle it gently with slight pressure until it lines up. Set trim tiles like caps or bullnoses the same way, as called for in the design. If the tiles do not have spacing lugs, you can drive 6d nails part-way into the sheathing, two to a side at each tile to maintain the proper spacing. Or, you can set lengths of thick string between the tiles, or work with thin wood shims. If you have a good eye for alignment, you can work freehand; many do.

After a section of tile has been set, beat the tiles in by sliding a flat piece of scrap wood across them and tappng the wood with a hammer or mallet. Use just enough force to squish the adhesive flat and align the tile faces. Then inspect the section for proper align-ment, wipe any adhesive smears off the tile faces, and pick bubbles out of the joints with a toothpick.

Allow plenty of time for the adhesive to cure, then grout the joints. The process is the same as described for "Ceramic Tile Flooring" in Chapter 2. There are several kinds of grout that might be used, depending upon the kind of tile and adhesive or mortar you have used, and the decorative effect you want. Your supplier can recommend the best product. After the grout has had plenty of time to cure, go over the whole surface with a recom-mended tile and grout sealer.

Note: You can set ceramic tile on counter, vanity, and table tops using the same meth-ods and materials.

WAINSCOTING

Wainscoting originally was an architectural feature consisting of wooden linings applied to the lower part of an interior wall surface. Over the decades, however, the term

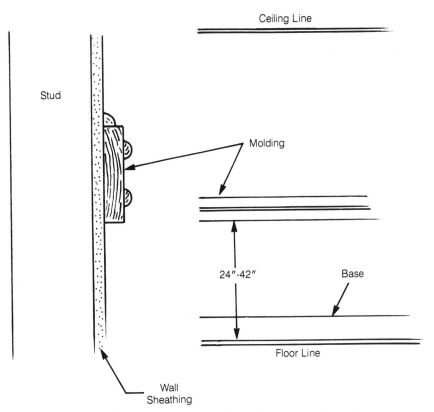

Fig. 3-68. *A false wainscot consists of a molding, either single piece or a combination of two or more different patterns, applied directly to the wall surface.*

has come to mean the lower 3 or 4 feet of any inside wall that is finished or decorated differently than the rest of the wall. Wainscoting can have great decorative impact, and can be carried out on all the walls of a room, or just one, or whatever arrangement appeals.

There are two kinds of wainscoting: true and false. the false wainscoting leaves the entire existing wall surface in the same plane (Fig. 3-68). With a true wainscoting arrangement, the wainscot is a separate and usually different material that is applied to the existing wall surface and extends outward from it 1/4 inch to 2 inches or more (Fig. 3-69).

Installing False Wainscot

The false wainscoting is the easiest to arrange, because all you have to do is install a strip of molding across the wall. Then decorate above and below the molding with paint/ paint, paint/paper, paper/paper, or paper/paint. The molding itself can be painted or clear-coated wood of any appropriate species. Preparation depends upon your decorating plans. If the wall is now painted and you are going to repaint or hang wallpaper, just wash the walls and repair any defects. If the wall is papered and you will repaper or paint, strip

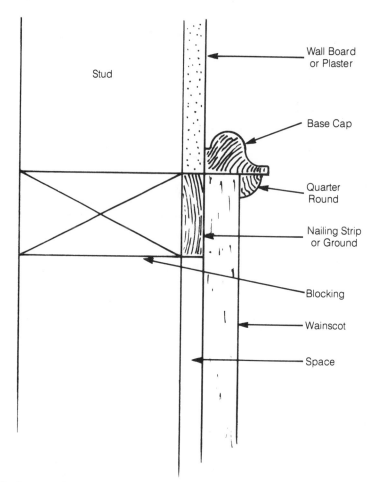

Stud

Wall Board or Plaster

Base Cap

Quarter Round

Nailing Strip or Ground

Blocking

Wainscot

Space

Fig. 3-69. A true wainscot consists of a separate wall material, usually paneling or planking, applied over the existing wall and capped with molding to form a ledge of variable size.

the old paper off, wash off all the old glue, and repair defects. Or, you could leave the old paper or paint above or below the molding, and strip and/or clean just the surface that will be re-covered.

Once the prep work is done, select a transition point at a suitable height above the floor. Traditionally, this point was high enough so that the backs of chairs placed near the walls would rub against the paneled wainscoting and not against the more delicate painted or papered wall. However, for modern wainscoting any height that looks good to you (usually from about 24 to 42 inches) can be selected. Snap a level chalkline across the wall at this point. Select a fairly sizable flat-faced molding; a tiny one has little decorative impact and gets lost on the wall. You can use plain 1-×-4 S4S stock, chair-rail molding, mullion casing molding or apron, or any of a number of stop or hook strip moldings. You can also make up a design of your choice from raw stock or by combining two or more standard moldings of different kinds. Pine is the common choice for a painted molding, and pine or other species for natural finished or stained molding.

If the wall is to be redecorated both above and below the molding, nail the molding to the wall studs with 8d finish nails, aligned along your chalkline. For a painted finish, prime the molding first, then fill the nailholes and joint cracks with vinyl paste spackle, sand smooth, and apply the finish coatings. For a natural finish, apply the finish first, sanding lightly between coats, then fill the nail holes with a matching wax putty stick. Then go ahead with redoing the wall surfaces.

If the upper or lower part of the wall will be left as is, it is easier to prefinish the molding strips. Do all the painting and sanding to the uncut molding strips. Cut the strips to fit with a miter saw and nail the pieces up with 8d finish nails. For painted moldings, fill the holes and joint cracks with glazing putty, then smooth the surface of the plugs true and flush with the molding surface using your finger.

Allow the putty to harden for several days, then touch up the putty patches with paint and an art brush. For a natural finish, just fill as usual with a putty stick. If a stain was used under a clear coating, you might have to apply a tiny dab of the same stain at cut edges where wood fibers have torn away and exposed specks of unstained wood. An alternative method is to paint/paper the upper/lower sections of the wall first, then install the prefinished molding to cover the break point between the wall sections.

Building True Wainscot

To build a true wainscot, first remove base moldings, electrical outlets, heat registers, and other objects that might interfere. Determine the height of the wainscoting. If the upper part of the wall will be redecorated, do whatever paper stripping and cleaning is necessary, down to just below the wainscot line. If the wainscoting will be glued in place, strip and/or clean that portion of the surface as well. If the proposed wainscot will be of any great thickness when completed, pipes might have to be extended slightly or wires repositioned to suit. Constructing a built-out wainscot on furring strips, by the way, is a good way to cover the installation of new wiring or piping without breaking into the existing wall.

Construction details of the wainscoting depend upon the material you have selected. Thin hardboard or plywood is usually glued directly to the old wall surface with panel or construction adhesive. Plywood and planking in the $3/4$-inch-thick range is best fastened with nails. Unless there happens to be some usable continuous blocking within the wall, horizontal furring strips to which the wainscoting can be attached should be nailed through the sheathing into the wall studs (Fig. 3-70). Ceramic tiles can be set with organic mastic on the sheathing or on backer board secured to the wall, or set in mortar. Resilient floor tile or vinyl sheet flooring can be glued right to the wall surface with an adhesive recommended for that particular product. Vertical surface plastic laminate should be glued to a plywood underlayment with contact cement; use the nonflammable variety.

In general, whatever material you select is applied with the procedures and methods usually associated with that material. Leave a small gap at the floor line—usually $1/4$ inch

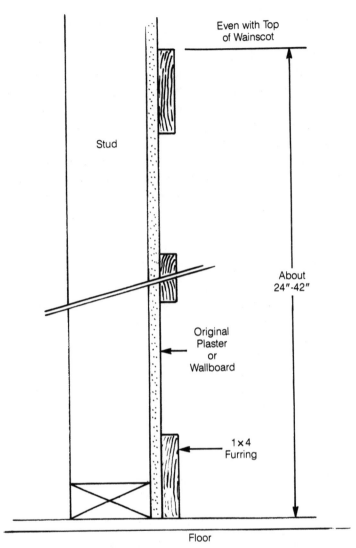

Fig. 3-70. *A wainscot can be applied to furring strips retrofitted over any existing wall sheathing.*

is ample—and install the wainscoting to a level line at your chosen height. Replace the base moldings, trimming to fit as necessary. The last step in construction is to apply a cap or top molding (Fig. 3-71). Depending upon the nature of the wainscoting, two or more stock moldings might have to be combined in order to skirt effectively trim out the raw edge. Or, you might prefer to design a custom one-piece molding that is made for you (or by you in your own shop).

For thin wainscoting, a single chair rail or similar molding with an appropriate rabbet cut in the back might do the job. To embellish it further, a center strip molding of smaller material, perhaps a narrow panel or screen molding, could be applied along the cen-

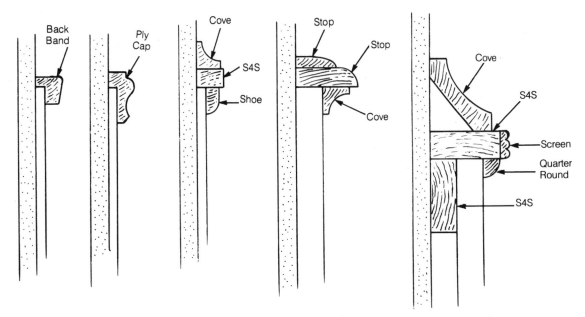

Fig. 3-71. *These are just a few of the many ways that a wainscot can be capped.*

terline. A very thick construction might be fitted with a flat ledge molding that has a shoe on the top and a skirt beneath. The skirt effectively hides the nails holding the top of the wainscoting in place.

Finishing the wainscoting also depends upon the materials used. If the wainscoting is prefinished and requires no further attention, such as paneling or ceramic tile, you might find it easier to prefinish the moldings and touch them up as necessary after mounting them. If the entire installation is to be painted or varnished, you can do the whole job at once.

WALL TRIMWORK

There are two kinds of wall trimwork to consider: decorative field molding and edge or perimeter molding. Decorative field molding can be applied to any walls in any kind of rooms in a limitless number of ways. If edge or perimeter molding is chosen, there are two further categories. One is base molding, which is applied to the bottom of the walls to conceal the wall/floor joint. This type was discussed in the section "Floor Trimwork" in Chapter 2. The second is ceiling or cornice molding, which is seldom used today, especially in newer, small, low-cost houses. This trimwork is applied to the tops of the walls to conceal the wall/ceiling joint, and will be covered in Chapter 5.

Perhaps the simplest example of trimwork is a plain chair-rail molding attached to the surface of the walls to form a narrow belt around the room at approximately waist height. This creates a false wainscot, and the upper and lower portions of the walls can be decorated differently, if desired. The molding itself may be painted to match or contrast the wall finish or the other wood trimwork in the room, or both.

While a chair rail is usually installed on all the walls of a room, a plate rail (Fig. 3-72) is more often mounted on only one of the principal walls. Used to display antique or collector plates, the rail is usually mounted at about eye level or a bit higher and is easy to construct. The shelf portion is a narrow strip of S4S, and the support can be made from crown, bed, or wide cove molding (Fig. 3-73). The edge restraint typically is small quarter round, but could be any other small molding of suitable pattern. The shelf rail (Fig. 3-74) is a variation on this theme. Used to display knickknacks or small collectibles, it lacks the restraint molding, and the shelf portion is wider than the plate rail shelf.

Other possibilities lie in applying vertical, horizontal, or diagonal strip moldings to the wall surface in any of a variety of patterns (Fig. 3-75). Plain lattice, which is available in several stock widths that can also be ripped to provide other, nonstock widths, are excellent for this purpose. Screen molding, mullion, shelf mold, astragal, half round, panel strips, or batten might also be used.

Another popular treatment is to mount frames on the wall surface. The frames can be of any desired size and shape, and finished to match or contrast the wall finish. The frames are commonly plain squares or rectangles set either horizontally or vertically, but they may be made in more complex shapes as well. They may extend almost from floor to ceiling, or be mounted only on the upper or lower portions of the walls (Fig. 3-76). Many arrangements are installed in conjunction with a chair rail, with the frames mounted both above and below the rail, or only above (Fig. 3-77).

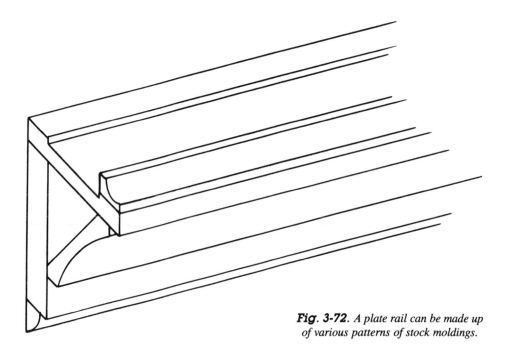

Fig. 3-72. *A plate rail can be made up of various patterns of stock moldings.*

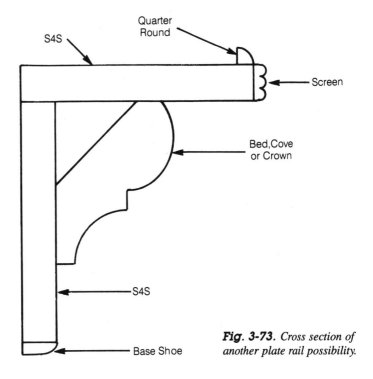

Fig. 3-73. *Cross section of another plate rail possibility.*

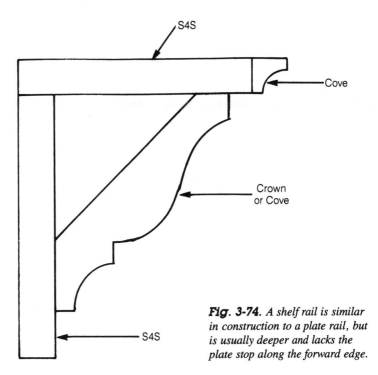

Fig. 3-74. *A shelf rail is similar in construction to a plate rail, but is usually deeper and lacks the plate stop along the forward edge.*

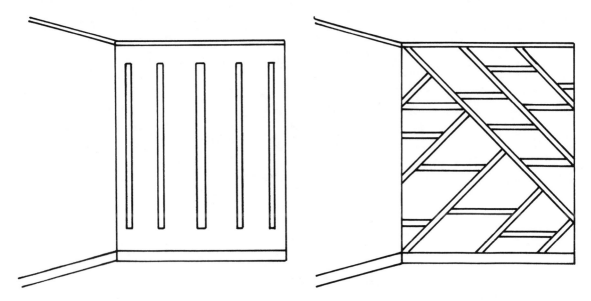

Fig. 3-75. *Walls can be decorated in a great number of ways with simple strip moldings of several patterns.*

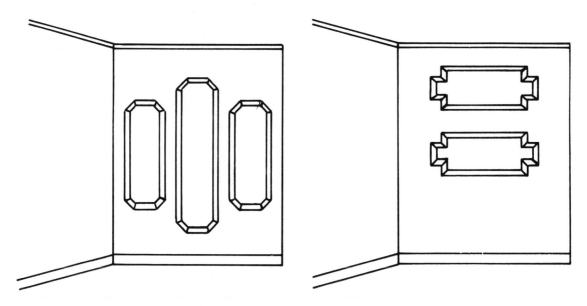

Fig. 3-76. *More complex frame patterns made up of moldings can be applied to wall surfaces as decoration.*

Sometimes the wall area contained within the frames receives a decorative treatment emphatically different from the remainder of the wall surface, such as a figured or printed wallpaper, an exotic fabric, or perhaps just a contrasting color of paint. The moldings most commonly used for the purpose are small cove, base cap, half round, stop, lattice,

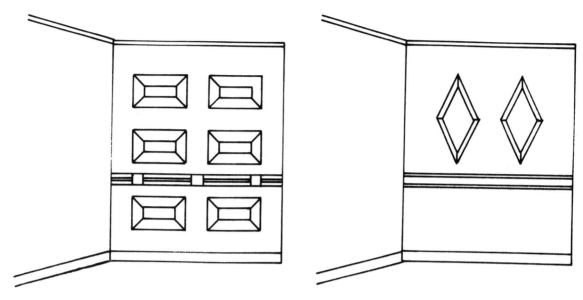

Fig. 3-77. *Block, frame, or strip patterns made from various types of stock or custom moldings can be used in conjunction with either true or false wainscots as wall decorations.*

screen, astragal, and panel patterns. For large, wide-band frames, consider batten, casing, wide stop or astragal, panel strip, mullion, and shingle moldings.

Decorative field moldings are usually painted, and pine is the common choice of wood. For a natural finish many species are suitable, but only pine, oak, and mahogany are likely to be available from ordinary lumberyard stocks. Other varieties can be obtained from custom woodworking shops, and milled to whatever pattern you choose. Installation and finishing of wall moldings follows the same procedures as for decorative ceiling moldings (although the work is easier). Refer to "Ceiling Trimwork" in Chapter 5 for further details.

4
CHAPTER

Windows and Doors

THE WINDOWS IN YOUR HOUSE ARE NOT JUST YOUR VIEWPORTS TO THE OUTSIDE WORLD, they are also part of the exterior architecture and the interior design. Your windows can be redecorated, remodeled, replaced, even removed or added to, in order to change the appearance and/or the liveability of the rooms. You can make simple but attractive and effective changes for just a few dollars and a little time, or undertake extensive (and usually expensive) window replacement, or anything in between.

As for doors, some houses have too many; others not enough. Sometimes doors are in the wrong places, or swing the wrong way. You might have doorways with no doors, or perhaps the doors are rickety and ill-fitting. Maybe you have lights where you don't want them, or no lights where it would be nice to have them. Possibly the door style or the trimwork is out of touch with the rest of the decor, or will not fit in with projected redecorating schemes. The doors might be cheap, with cracks in the wood, peeling paint, skimpy trimwork, and crummy hardware. Some doors are so featherweight that you can't even get a decent slam out of them when you're mad at something.

Whatever the problems, you can cure them without a great deal of cost or effort and coordinate window and door arrangements exactly to your liking. There are numerous options you can consider as you lay your remodeling plans, and with a few noted exceptions, these are good do-it-yourself projects. This chapter will explore some of the possibilities.

WINDOW TYPES

You have several distinctly different types of windows to choose from as you develop your window remodeling plans. Their appearances differ, as do their operating and venti-

Fig. 4-1. *A typical double-hung window.*

lating characteristics. Some houses are fitted with only one type throughout, but most installations call for two or more types to fulfill specific ventilation, decor, and space requirements. Those applicable to residential installations are as follows:

- *Fixed* windows are stationary in the frame. They can be specially made in any size and shape, but commercial fixed units are constructed to match in size and appearance each operable unit made by the same manufacturer.
- *Double-hung* windows (Fig. 4-1) are sometimes square but usually vertically rectangular, with two sections of glass divided horizontally. In some older styles one section, usually the upper, is taller than the other, but in modern styles the two sections are equal. There might be one or several panes of glass in each section; the lower section slides fully up, and the upper slides fully down. Maximum ventilation is 50 percent of the window area. Older styles are fitted with counterbalance (sash) weights in wall pockets beside the window. Modern units are fitted with integral springs. Operation is by manual sliding.
- *Single-hung* windows look the same and are built the same way as double-hung, but the upper glass section is fixed. Other details are the same as double-hung units.

- *Casement* windows (Fig. 4-2) are sometimes approximately square but usually vertically rectangular. The entire glass unit is hinged at one side or the other and swings outward either to the left or right, depending upon hardware placement. Maximum ventilation is usually 100 percent of the window area. Operation may be by manual pushing and locking in place, or by crank.
- *Awning* windows (Fig. 4-3) are horizontally rectangular. The entire glass unit is hinged at the top and swings outward and upward. Some models are fitted with sliding friction hinges, so that as the bottom of the glass swings outward, the top slides down. The glass usually swings out to 45 degrees or less. Maximum ventilation is generally considered to be 100 percent of the glass area, but in fact they do not ventilate as well as full opening types. Operation may be by manual pushing and locking in place, or by crank.
- *Hopper* windows (Fig. 4-4) are horizontally rectangular. The entire glass unit is hinged at the bottom and swings inward and downward. Other characteristics are the same as awning windows.
- *Horizontal sliding* windows (Fig. 4-5) are usually either approximately square or horizontally rectangular. There are two or more glass units. If two, one slides to left or right to cover a fixed section, or both may slide. If three, the center unit is fixed and the outer two slide to the center. In most designs, maximum ventilation is 50 percent of the glass area, but it can be somewhat less. Operation is by manual pushing.

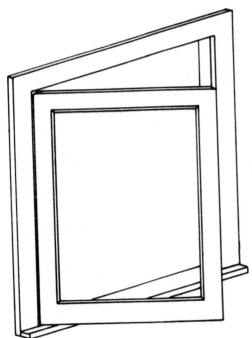

Fig. 4-2. A typical casement window.

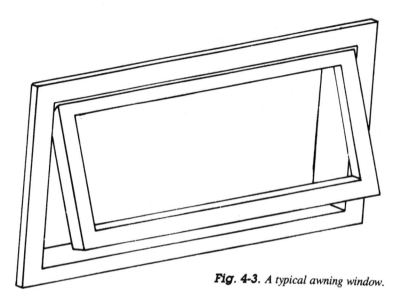

Fig. 4-3. *A typical awning window.*

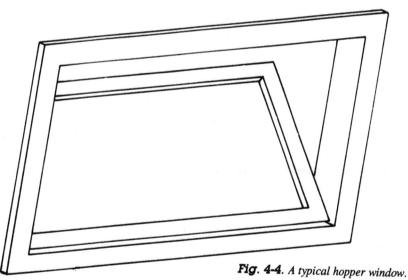

Fig. 4-4. *A typical hopper window.*

- *Jalousie* windows (Fig. 4-6) also called *louver* windows, are vertically rectangular and composed of a series of slightly overlapping horizontal glass louvers. The louvers operate with a common mechanism to simultaneously swing outward and upward, awning-fashion, usually to a full horizontal position. Unlike other window types, they are prone to air and dust infiltration, and are not totally weatherproof. Also, they are suitable for use only in temperate climates or in enclosed areas where heat loss/gain is of no consequence, like a vestibule or porch. Maximum ventilation is 100 percent of the glass area. Operation is by crank.

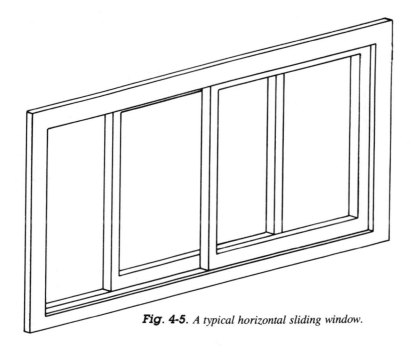

Fig. 4-5. *A typical horizontal sliding window.*

Fig. 4-6. *A typical jalousie window.*

Fig. 4-7. *A composite window is made up of two or more window styles in one frame or frame set—here, two double-hungs and one fixed light.*

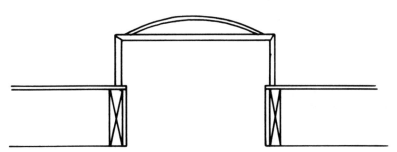

Fig. 4-8. *A typical skylight in cross section.*

- *Combination* or *composite* windows (Fig. 4-7) are large units composed of several smaller ones stacked vertically, joined side by side, or both. Two or more types of windows may be used. Typical combinations: One or more stacks of awning or hopper windows, or combinations of the two; a large, fixed central section flanked by one or two stacks of hopper or awning windows; a central fixed section with one or more hopper units below; the same with one or more hopper units above; or a two-way horizontal sliding unit flanked by narrow fixed sections. *Bow* and *angle bay* windows are also combination units, but of different configurations.
- *Skylights* (Fig. 4-8) are roof-mounted units primarily used to admit added light. Some models are called *roof windows*. These windows are sometimes installed for solar heat gain purposes, and a few have ventilating capability. The glazing may be plastic or glass, either domed or flat. Single through quadruple glazing is available in a wide range of sizes, both square and rectangular.

WINDOW ANATOMY

Researching manufacturers' literature, window planning, discussing windows with your supplier, and working with or building your own windows is easier if you understand at the outset how windows are put together and the terminology involved.

To begin with, the generic term for the transparent or translucent part of all windows, which may be either glass or plastic, is *glazing*. A cut piece of glass for installation in a window is a *pane*, the plastic counterpart is a *sheet*. A single pane or sheet in a window or a door is a *light*. A window may have one or several lights, and windows are often referred to in terms of the number of lights in each sash. For example, a fixed window with four panes is a four-light window, while a colonial-style double-hung window might be called a six-over-six or a twelve-over-eight.

In a single-light, openable window, the glazing is held in a surround called a *frame*. The frame is composed of a *stile* at each side, a *top rail*, and a *bottom rail*. This assembly is the *sash*. If there is more than one light in a sash, the individual panes are separated within the frame by slender members called *muntin bars*, or just muntins. Some windows actually only have one light but are fitted with false muntin bars that snap into place over the glass to create the illusion of a multiple-light window. The glazing is bedded in *glazing compound* or *glazing putty* applied to the muntins and/or frame. It can be held in place by several small bits of metal called *glazier's points*, plus a beveled bead of glazing compound on the exterior side. This is the traditional method. Alternatively, a wood stop can be attached to the frame, or a grooved frame assembled right around the glass. Figure 4-9 shows a typical sash assembly.

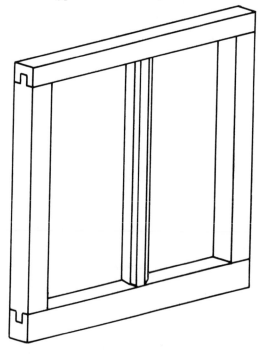

Fig. 4-9. *A typical simple sash assembly.*

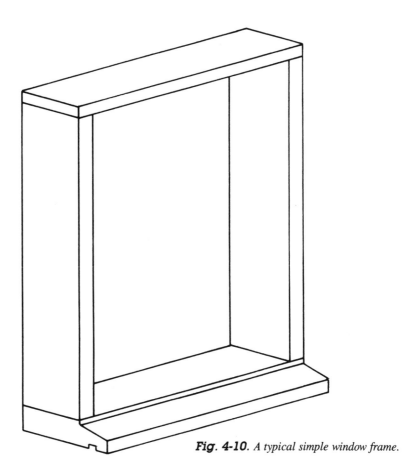

Fig. 4-10. *A typical simple window frame.*

Some windows have only one sash, such as a single casement unit. A double-hung has two, a combination unit may have a dozen or more. Where two sashes are located one above the other and meet, as in a double-hung, the top rail of the lower sash and the bottom rail of the upper become the *meeting rails*. In a horizontal sliding window where two sashes meet vertically in the middle, the two abutting stiles are called *meeting stiles*.

The glazing units of a window must be set in a *window frame*. If it is an openable window, the sash is mounted in the frame. A fixed window with either one or multiple lights might also have a sash assembly, but the glass or plastic might also be mounted directly into the window frame. This frame is made up of four basic parts: vertical *side jamb* at each side, a *sill* at the bottom, and a *head jamb* at the top (Fig. 4-10). A simple frame assembly is often built up of heavy dimension stock (or formed of steel) and is called a *buck*. Bucks are typically installed in poured concrete, log, or concrete block walls, and sometimes in other wall constructions as well. Usually a buck is employed as a strong subframe, and a complete window assembly with its own frame is installed in it. Occasionally, however, a sash or even a single light is directly installed.

In most cases, a window frame is a bit more complex, having both a sill and a *subsill* (Fig. 4-11). In a single- or double-hung window, a narrow *parting stop* is mortised into the side and head jambs to separate the two sashes. To keep the sash in place within the frame, *blind stops*, also called *face stops*, are installed on the side jambs, outside of the sash (Fig. 4-12). In other designs, such as the awning type, there is no blind stop on the outside of the frame (which would prevent the sash from opening), but there is a *back stop* on the inside. This is reversed for hopper type windows. Where two or more complete window units are installed side by side, they are separated and supported by a vertical member called a *mullion* (Fig. 4-13). In some cases the adjacent side jambs are eliminated and the mullion also serves as a double-sided side jamb for both units.

The window frame with its sash installed is mounted within a framed opening in the wall structure called a *rough opening* (or r.o.). This opening is larger than the window unit, so the unit must be installed, plumbed and leveled, and held in place with *shims*. The raw edges of the frame are exposed and there are gaps between the unit and the rough opening framing; these must be covered with trimwork. On the interior side, this consists of three parts. The *casing* is applied to the side and head jambs, the *stool* or *stool cap*—often incorrectly called the windowsill—covers the inner edge of the sill, and the *apron* is

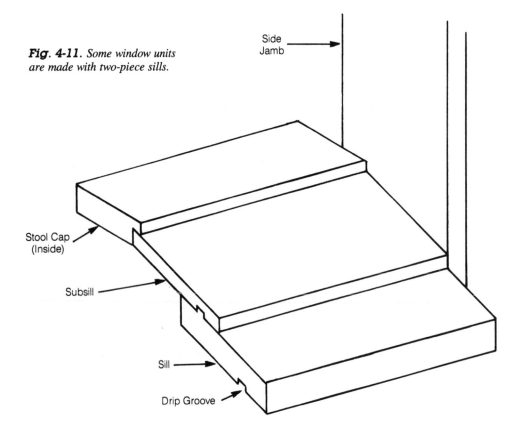

Fig. 4-11. *Some window units are made with two-piece sills.*

Side Jamb

Stool Cap (Inside)

Subsill

Sill

Drip Groove

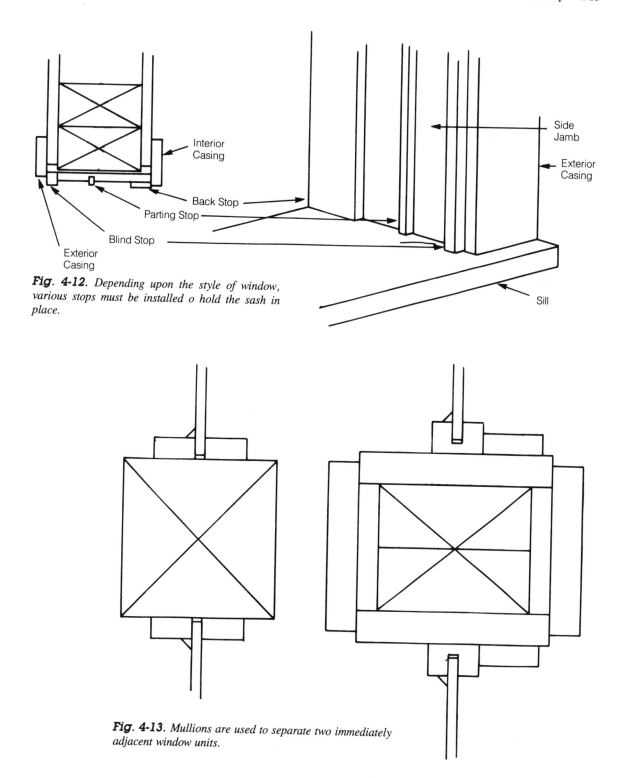

Fig. 4-12. *Depending upon the style of window, various stops must be installed o hold the sash in place.*

Interior Casing

Side Jamb

Exterior Casing

Back Stop

Parting Stop

Blind Stop

Exterior Casing

Sill

Fig. 4-13. *Mullions are used to separate two immediately adjacent window units.*

mounted immediately beneath the stool (Fig. 4-14). The casing may be small, plain, and simple. It may also be wide and ornate in configuration, with any of various molding patterns applied to it, or perhaps embellished with a cornice molding across the top or decorative blocks or medallions at the upper corners. Depending upon the window design, a back stop might be needed along the inner face of the side and head jambs of the frame, to restrain the sash and help to weatherseal it. A stool can serve this function across the bottom of the sash.

On the exterior side of the window frame, the sill is already in place as part of the frame. An apron might be installed beneath the sill, but usually the exterior siding is simply butted up to it. In many window designs, side and head casings are attached to the frame; the exterior siding butts up to them (Fig. 4-15). In some designs there is no casing, but rather a mounting flange that wraps over the frame's outer edges and lies between the exterior siding and the sheathing. Casing is sometimes separately attached to the wall sheathing around the window for effect. At the top, most windows are fitted with a *drip cap* or metal flashing, or both (Fig. 4-16).

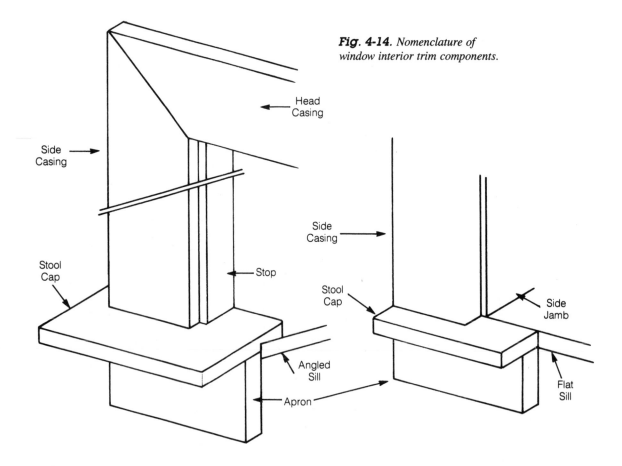

Fig. 4-14. *Nomenclature of window interior trim components.*

Head Casing

Side Casing

Side Casing

Stool Cap

Stop

Stool Cap

Side Jamb

Angled Sill

Apron

Flat Sill

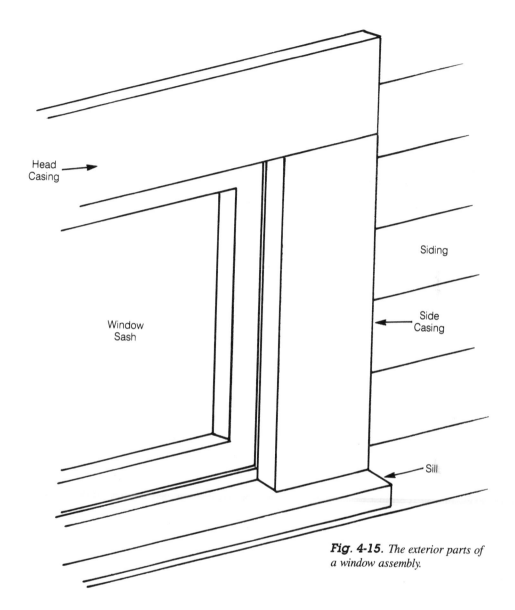

Fig. 4-15. The exterior parts of a window assembly.

Head Casing

Siding

Window Sash

Side Casing

Sill

Modern commercial window units are also equipped in various ways with different kinds of weatherstripping and sealing materials to reduce air infiltration/exfiltration.

WINDOW GLAZING

A few decades ago, residential window units were glazed with panes of ordinary window glass, often quite thin and full of bubbles, waviness, and distortion. Today there are many kinds of high-quality glazing tailored to meet specific needs and conditions, and new products are constantly appearing in the marketplace. Making the right selection for

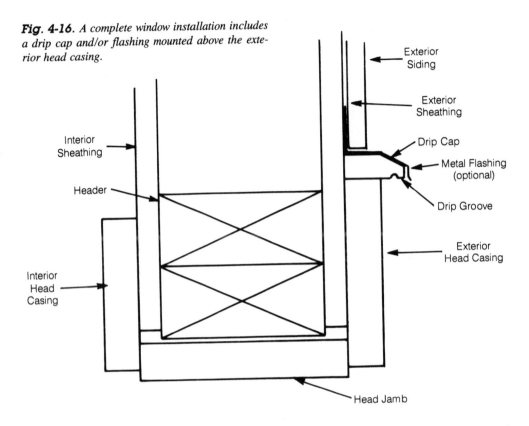

Fig. 4-16. *A complete window installation includes a drip cap and/or flashing mounted above the exterior head casing.*

the specific needs in the various parts of your house now requires some study and an investigation of manufacturers' most current literature.

Windows can be glazed with either glass or plastic. In residential applications, however, plastic is almost never used because it is easily scratched and can become cloudy from repeated cleaning and aging, and it quickly loses clarity. Plastic glazing is very tough, though, and so is useful in areas such as basement windows or doors, or garage door windows.

Glass is by far the preferred glazing material. Of the many different products available, several are popular specifically for residential installations. As you make your glazing selections in accordance with the requirements for each unit, compare and contrast these factors:

- Weight
- Thickness
- Clarity
- Color of transmitted light
- Color and appearance of the glass (from both inside and outside)
- The percentage of visible light transmitted through the glass from outside

- The percentage of solar energy transmitted from the outside, and conversely the percentage excluded
- The percentage of radiant energy blocked from the inside
- Does the window location indicate a tempered glass?
- Can the glass be cut to size or purchased in specific sizes?

Types

Window glass is still sometimes used in inexpensive residential glazing operations, particularly for replacing lights in old windows. It seldom appears in new commercial units. Window glass lacks strength, and the surfaces are neither wholly parallel nor distortion-free. However, it is the least costly type, and adequate for many unexacting situations. Window glass comes in two thicknesses: 3/32 inch (single-strength, designated SS) and 1/8 inch (double-strength, designated DS). There are three quality levels: AA, A, and B. The AA is specially selected top-grade material, A for quality glazing, and B for general purposes. Few if any suppliers stock all three grades, and some might not stock any. SS window glass weighs 1.22 pounds per square foot, and DS weighs 1.63 pounds per square foot. Both can be cut. Light transmission through both is 91 percent. Window glass is clear and the color of the transmitted light is natural. Solar or radiant energy transmission is not considered in window glass applications.

Heavy sheet glass is in the same family as window glass. The thicknesses are 3/16 and 7/32 inch, with weights of 2.5 and 2.85 pounds per square foot. Light transmission is 90 percent and 89 percent respectively. Otherwise, the characteristics are the same as for window glass. Heavy sheet is not used in modern factory made residential window units, nor are you likely to find it at a glass supply house. However, it has been used in some older residential glazing applications, and might turn up as second-hand salvaged material. This is a good and inexpensive source for glass of all kinds, by the way; it can come from an old store front or a glass table top, or similar sources.

Regular plate or *float* glass is today's choice for ordinary clear glazing purposes. Plate glass is uniformly flat with polished surfaces and has been the traditional top-quality glazing material for years. Float glass is almost indistinguishable from plate and has the same characteristics as plate, but is made by a newer and different manufacturing process. Both have excellent clarity with virtually no distortion. There are three quality levels: silvering, mirror glazing, and glazing. Glazing is used in residential glazing applications. The two thicknesses are 1/8 and 1/4 inch, weighing 1.64 and 3.28 pounds per square foot, with a light transmission percentage of 90 and 89 respectively. The glass is clear and the light color transmitted is natural. Solar or radiant energy transmission is not considered. The material can be cut.

Heavy plate or *float* glass is similar to the regular variety. Although seemingly thick and heavy, most weights have bee used in special residential applications, primarily in relatively large fixed windows. The available thicknesses and weights and their respective light transmission percentages are shown in Table 4-1. Other characteristics match those of regular plate or float glass.

Thickness Inches	Weight Lbs./Sq. Ft.	% Light Transmission
5/16	4.10	87
3/8	4.92	86
1/2	6.54	84
5/8	8.17	82
3/4	9.18	81

Table 4-1. Typical Characteristics of Heavy Plate or Float Glass.

Tinted glass is available in sheet, plate, and float versions. Admixtures are included in the glass to provide heat-absorbing and glare-reducing properties, which are dependent upon the tint and thickness of the glass. The tint colors are bronze, gray, and green or blue-green. Unless the glass is very thick, or more than one tint is employed in multiple glazing, there is little distraction from the coloring after a short period of adjustment, and the thinner versions do not limit vision. Incoming light through the thinner sheets appears approximately natural and interior colors are not changed appreciably. Typical properties are shown in Table 4-2. The glass can be cut, and can be combined with clear glass in multiple glazing. Installation must be carefully and correctly done to avoid later breakage from thermal stress.

Tempered glass can be made from any of the preceding kinds of glass that is 1/8 inch or more in thickness. The primary residential application is as safety glass, typically in shower or tub enclosure doors and patio doors. It is also used in large lights installed close to or at floor level, and in door sidelights. It can, in fact, be installed in any kind of window, either alone, as insulating glass, or in combination with other glazing. Tempered glass is much stronger than untempered glass. When broken, rather than shattering into shards, it crumbles into a pile of relatively harmless pebbles that look like oversized rock salt. Except for the tempering aspect, it has the same characteristics as the glass from

Table 4-2. Representative Characteristics of Tinted Plate or Float Glass.

Color	Thickness Inches	Weight Lbs./Sq. Ft.	% Light Transmission
Bronze	1/8	1.64	68
	3/16	2.45	58
	1/4	3.27	50
	3/8	4.90	37
	1/2	6.54	28
Green	3/16	2.45	78
	1/4	3.27	74
Gray	1/8	1.64	62
	3/16	2.45	51
	1/4	3.27	42
	3/8	4.90	28
	1/2	6.54	20

which it was processed. Tempered glass cannot be cut, but must be installed in whatever sizes are commercially available, or custom manufactured to specific sizes.

Laminated glass consists of two or more layers of glass with one or more layers of thin plastic sandwiched in between. The most familiar product is automotive window glass, but there are numerous residential applications as well. Among these are skylights, windows susceptible to accidental damage (such as basement or garage door lights or large picture windows fronting a golf course), windows exposed to storm damage (as on a seacoast), or as a burglar-resistant security measure in ground floor windows. Laminated glass can be cut, although it is more easily used in stock sizes. The characteristics of products most applicable to residential installations are shown in Table 4-3.

Insulating glass is another composite type of glazing. It is manufactured in certain stock sizes and types, and cannot be cut. A wide variety of glass types and thicknesses can be made up into insulating units. Metal-edged units consists of two or more lights separated by a metal or rubber edge space, sealed airtight, and usually set into a metal frame (Fig. 4-17). Several thicknesses of glass are available, and spacing between the glazing layers can range from $1/4$ to $1/2$ inch. Glass-edged units consist of two lights fused together around the edges, usually with a $3/16$-inch space that can be filled with a gas or dry air (Fig. 4-18). The available size range of this type is somewhat limited. In modern construction, insulating glass is almost a must in either hot or cold locations.

Reflective coated glass is plate or float glass that has been coated with a very thin, transparent metal or metal oxide film. Although primarily employed for commercial glazing, they have residential applications as well. The purpose of the coating is to reduce solar gain and glare, and also to act as an ultraviolet ray screen. From the exterior the glazing is mirror-like and cannot be seen through under normal daytime light contrast conditions. Visibility from the inside out, however, is good—depending upon the nature and thickness of the film. There is a wide range of coated glass characteristics, differing somewhat from one manufacturer to another. Typical colors are blue and blue-green, copper, gold, bronze, and silver. The colors can be used for architectural emphasis. Coated glass is usually used along with clear or other glazings, to protect the delicate film and also to introduce different glazing properties not inherent in the coated glass itself. These glazings are particularly useful in hot country where cooling is the major consideration.

Low-E (low emissivity) glass refers to a type of glass coated with metallic film. It has recently become popular and is available for residential glazing purposes. The glass is

Table 4-3. Some Types of Laminated Glass Used for Residential Purposes.

Glass Type	Thickness Inches	Weight Lbs./Sq. Ft.
SS-SS	$13/64$	2.45
$7/64$-$7/64$	$15/64$	2.90
DS-DS	$1/4$	3.30
$1/8$-$1/8$	$1/4$	3.30
$3/16$-$3/16$	$3/8$	4.80
$1/4$-$1/4$	$1/2$	6.35

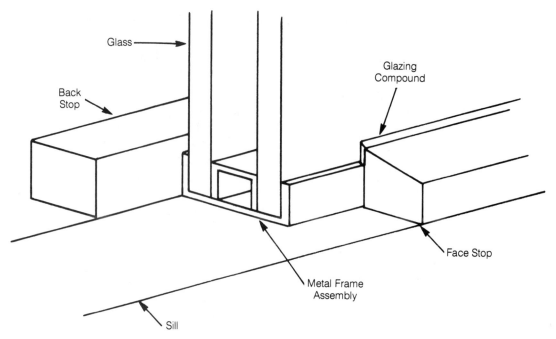

Glass

Glazing
Compound

Back
Stop

Face Stop

Metal Frame
Assembly

Sill

Fig. 4-17. *Metal-edged thermal insulating glass consists of two separate sheets glazing sealed into a special metal frame.*

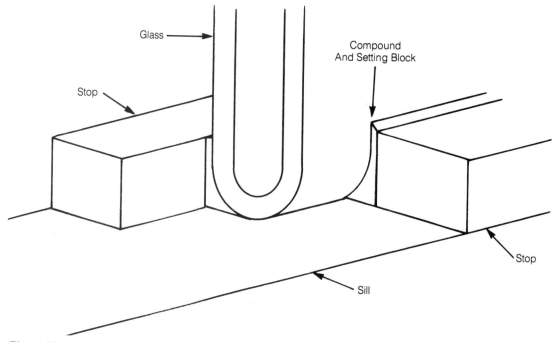

Glass

Compound
And Setting Block

Stop

Stop

Sill

Fig. 4-18. *Glass-edged thermal insulating glass is formed by fusing the sheets of glass together in rounded edges.*

fully transparent and is used together with clear, plain glass in a thermal insulating glazing. Its function is to reduce heat loss in winter by reflecting a large portion of the interior radiant energy back into the house, and to block out solar heat gain in summer by reflecting sunlight off the outside of the window. This glazing is not much more expensive than plain glazing, and is useful in all climates. Consult with a supplier or refer to manufacturer's specifications for full details; the product line is changeable.

Low-I (low iron) glass is another type of glass that has recently become available for residential glazing applications. Most glass has a high iron content and absorbs a substantial amount of the solar energy reaching it. Low-I glass has a very low iron content, and therefore passes into the house a large amount of the solar energy reaching it. This is a useful glazing in cold climates for southern exposure windows or for solar rooms, sunspaces, or greenhouses because of the added solar heat gain that can be realized. Again, consult with suppliers and/or manufacturers to see what is currently available.

Patterned glass is translucent, and made in a wide variety of patterns such as ribbed, floral, stippled, and hammered. The pattern may be embossed on one or both sides. Usual thicknesses are $1/8$ and $7/32$ inch; it can be cut to suit. The degree of obscurity and light transmission varies widely with the pattern. These are sometimes further lessened by etching or sandblasting to create a frosted look. Most patterned glass is untempered, but some sizes are available tempered. For residential applications, this glass is used for decorative impact in interior windows or wall screens, for privacy (as in a bathroom window), or for tub or shower enclosures in the tempered variety.

Glass block is not a glazing material in the strict sense of the term. It deserves mention here, however, as a fixed, light-transmitting architectural feature that might be installed in lieu of a window, with an entirely different appearance. For decades this was almost a forgotten material, but it has made a comeback in recent years and is again being manufactured in the United States. Glass block is installed much like concrete block, and ranges in clarity from distorted clear to obscure. It may be plain, patterned, or coated. Factors such as light transmission, solar heat gain or blocking characteristics, thermal insulating value, sound reduction properties, and light-directing attributes vary, and the product lines change from time to time. There are many potential residential applications. Contact a manufacturer for the latest information.

STOCK WINDOWS

There are many circumstances where window units must be custom-made to fulfill certain requirements; stained glass is one example. Also, in some residential designs fixed lights are either fabricated as sash in a local shop or the lights themselves are fitted and installed on the job site as part of the construction project. Most residential installations, however, are made with factory-produced, complete window units supplied by local window dealers. There are numerous manufacturers who offer an amazing array of products to cover nearly any kind of window requirement. All of the styles and most of the glazing options discussed here are available somewhere, from someone, in stock window units.

Finding the combinations, sizes, and styles that you want is a matter of investigation. Such investigation will invariably lead to still more choices, because stock windows are made in a variety of frame options and not all window styles, glazing choices, or sizes are available in all frames.

Types

Solid aluminum window frames have been around for decades, but have fallen into disfavor and are produced in limited quantities. Plenty of second-hand and some new units are available, but they are not recommended for normal residential applications. They conduct heat (or cold) readily and are prone to serious condensation problems.

Thermal break aluminum frames are widely available. These frames are made with a thermally nonconducting material between the inner and outer frame halves, which alleviates the heat conduction and condensation problems. Some are unfinished, most are coated with a baked-on enamel finish. Advantages: The frames are rust- and rot-proof, lightweight, maintenance-free, strong, and can be readily made in custom sizes. Disadvantages: They have a low insulating value, susceptibility to corrosion in seaside or high pollution areas, and limited color selection. The frames cannot be repainted, they are prone to denting, and they are not recommended for winter climates.

All-vinyl frames (Fig. 4-19) have recently become popular. They have a modern, non-traditional appearance. Advantages: They are rust-, rot-, and corrosion-proof, require no maintenance, the color does not fade, the insulation value is good, and they can be readily made in custom shapes and sizes. Disadvantages: The frames might crack on impact, especially in extreme cold; they are susceptible to sunlight and are not recommended for prolonged sun exposure—especially at high altitudes; they can warp, shrink, or deform slightly under temperature extremes and consequently jam or stick; color selection is typically confined to white or brown; and they cannot be painted.

All-wood windows are the perennial favorite, and have a traditional appearance. Advantages: There is a wide range of models available, the frames are compatible with any house style or decor, and they are readily repairable (for the most part) if damaged or worn. The insulation value is high, they can be painted or stained any color, temperature extremes or strong sunlight do not bother them, condensation on the frame is very rare, and they are very weatherable if properly maintained. They are also easy to install, trim and retrim, and fit out with a variety of hardware. Disadvantages: Maintenance is high, they can stick or warp, the wood can crack if not properly protected, they are susceptible to rot, they lack the strength of some other types, and condensation on the glazing can cause problems with rot or finish deterioration.

Aluminum-clad wood frames (Fig. 4-20) have recently become popular. Advantages: There is a good range of models available, the traditional wood appearance remains on the inside because the cladding is only on the exterior, exterior maintenance is very low, insulation value is high, and the frame is not bothered by weather or temperate extremes. Disadvantages: Cladding color selection is limited, denting and scratching is possible, the cladding cannot be repainted, the interior wood surface must be periodically refinished,

Fig. 4-19. *Typical all-vinyl window frame construction. (Courtesy Fiberlux, Inc.)*

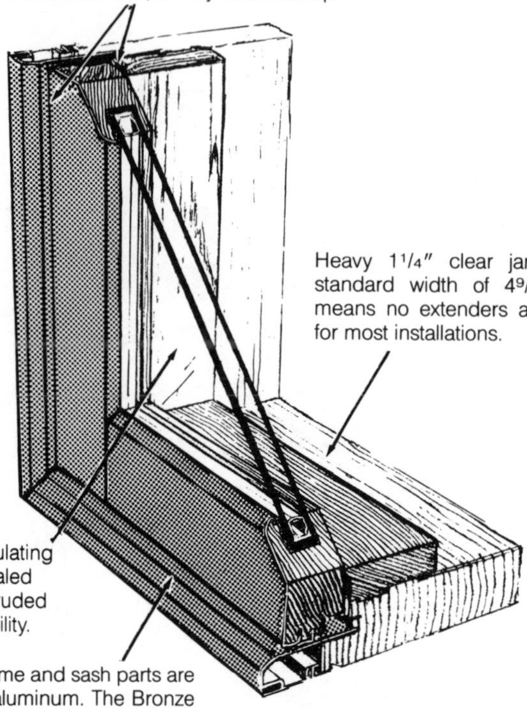

A dual durometer bulb weatherstrip compresses lightly between the sash and frame. Also applied to all four sides of the sash is a leaf vinyl weatherstrip.

Heavy 1¹/₄″ clear jamb has a standard width of 4⁹/₁₆″ which means no extenders are needed for most installations.

All units are glazed with ⁵/₈″ insulating glass which is hermetically sealed and bedded with a durable extruded butyl for maximum weatherability.

All exposed exterior frame and sash parts are clad with long lasting aluminum. The Bronze or White finish is baked on to give years of virtually maintenance free service.

Fig. 4-20. *Typical aluminum-clad wood window frame construction. (Courtesy of BiltBest Windows.)*

condensation on the glazing can affect the interior wood, and the wood within the cladding can rot if leaks develop.

Vinyl-clad wood frames (Fig. 4-21) are also very popular, and many models are available. They have the same attributes and almost the same appearance as aluminum-clad frames. Vinyl, however, is susceptible to sun damage under prolonged exposure, especially at high altitudes. Advantage: The cladding typically includes a perimeter mounting flange, which makes for easy installation and a weathertight seal.

Fiberglass frames are new in the marketplace. Because of this there is little experience history, but they may well prove to be the all-around best frame. Advantages: Fiberglass can be formed in any style, so is easily custom made in odd shapes or sizes. They are very strong, weather-resistant, and not susceptible to degradation from sunlight or temperature extremes. Insulation value is good and they are rot-, rust-, and corrosion-proof. Fiberglass is maintenance-free if not painted, although these frames can be painted if desired. Condensation on the glazing creates no problems, and the frames are light-

Fig. 4-21. *Typical vinyl-clad wood window frame construction. (Courtesy Andersen Windows.)*

weight. Drawbacks: High cost, limited availability, limited number of models being produced, and no performance history.

Frame Characteristics

Frame characteristics are an important part of a stock window selection process, but there are other considerations as well. One of them is the glazing. Apart from various glass options discussed earlier, the number of layers of glass and how they are put together plays a part. Although there are slight variations, the insulating value for a single thickness of clear glass is R-1. This arrangement is known as single-glazing. Conventional double-glazing has a value of R-2, triple-glazing or the addition of storm sash to double-glazing results in R-3, quadruple-glazing has a value of R-4.

In some windows the air has been evacuated from between the panes, in others the space is filled with an inert gas. The result is improved energy efficiency. The introduction of special types of glass, such as low-E system, changes the insulation value markedly. In this case, direct comparisons of manufacturers' specifications for particular window units must be made. In terms of the R-value of the most commonly available window units, single-glazing, then triple-glazing. Low-E double-glazing tops the list. The greater the R-value, the higher the window cost, so an assessment must be made as to which level of thermal efficiency is most cost-effective for any given installation. As to the quality of multiple glazing, check the manufacturer's warranty on the seal—it should be 15 to 20 years. Also, look for the initials SIGMA, which means that the unit meets the standards of the Sealed Insulation Glazing Manufacturer's Association.

Weatherstripping

Another important item on openable window units is weatherstripping. Each unit must be fully weather-stripped around the perimeter of the movable sash. Double stripping is good, triple is better. The material should be smoothly fitted and snug with no humps or gaps or rounded corners. Numerous materials have been used in window weatherstripping: felt, foam plastic or rubber, spring bronze strips, and multifiliment pile. None of these are overly effective. Vinyl (in the form of polyvinyl chloride or PVC) and polypropylene in various forms are widely used, but both tend to harden in cold and soften in heat, and so fail to do their job well when most needed and finally fail altogether. Many of these weatherstrippings can be replaced or augmented with new material as needed, but that means extra maintenance. The best weatherstripping materials to date are polyurethane and EPDM (ethylene propylene diene monomer) rubber, and new developments will probably appear as a result of continuing research.

Bear these two additional points in mind with respect to weatherstripping: Unless the best materials are installed with the best design and craftsmanship, a window that is airtight when new might become leaky after only a few hundred opening/closing cycles. And, window units with the fewest and simplest joints are the most airtight and the likeliest to remain so. Other than fixed windows, the least susceptible to air infiltration are casement, hopper, and awning styles. Horizontal sliding and double-hung are the most susceptible. Most large, reputable window manufacturers can provide you with air infiltration ratings for each model they produce. These ratings are established under standard testing conditions and are the same for each manufacturer, so they can be readily used to compare all rated window units. The ratings are given in terms of the number of cubic feet of air infiltration per minute per foot of window joint crack length under a 25 mile per hour wind, the lower the number, the tighter the window. The present minimum standard for the industry is 0.37 cfm/ft (cubic feet per minute per foot).

RETRIMMING WINDOWS

Sometimes it is more advantageous to replace rather than refinish interior window trimwork, especially if the wood or finish has become badly scarred and scratched or if numerous coats of thick enamel have been applied over the years. Retrimming is also a good way to help change the appearance and decor of a room: Install larger/smaller or plainer/fancier woodwork, and then finish it differently. In most cases you can change just the casing, or the casing and apron—or both—along with the stool.

Removing Old Trim

The first step is to remove the old trim. Do this carefully in order not to damage the sash, frame, or working parts of the window. Start with the casing. There might be a back stop installed to keep the sash restrained and aligned, and it might overlap the joints between the edge of the frame and the casing. If so, remove the stop first. It might be nailed or screwed to the faces of the side and head jambs of the frame, and possibly nailed into the casing as well. If nailed, gently pry the stop outward, working a bit at a time at several locations along the length of each stop piece until it comes free. If screwed, back the screws out, then break the stop free of the surrounding paint. This will also allow the sash to come out of the frame, so have a few small nails handy to drive partway into the side jambs to keep the sash temporarily in place.

There will be a joint of some sort, probably either miter or butt, at each top corner of the casing. If they are hidden by corner blocks, a cornice, or other applied molding, pry these away. If the side casing butts the bottom of the head casing, it will probably be easiest to remove the head casing first. Do this by tapping a wide chisel blade pry bar under the end of the piece. Do not pry directly against the wall; place a wide piece of scrap wood under the heal of the bar to avoid damage to the wall sheathing. If the side casing butts against the end of the head casing or there is a miter joint, it will probably be easiest to start with one of the side casings.

Begin prying about 2 or 3 inches down from the top. With a butt joint, the side casing will probably come away cleanly from the head casing. Continue prying the side casing away from the wall in steps at various points along its length. If a miter joint, the head casing might be nailed to the side casing, down through the joint. Both will come away from the wall, and you might have to pry alternately at the top corners to work the whole casing set outward until you can knock the head casing off, then free up the sides all the way down to the sill. If the bottom ends of the side casings are held by nails driven up through the sill, work the sides back and forth while pulling upward until they come free. If the window frames are wood, the casing will be nailed to the edge of the frame as well as to the surrounding wall studding. If not, the casing will be secured only to the studding.

The apron is usually a single piece. If nailed only to the studding, it should come away easily be prying gently first from one end, then the other. Often, however, the stool is nailed to the top edge of the apron. If it is nailed only to the apron, both will come off together. If it is also nailed to the window frame or toenailed clear into the studding or a nailing block, you will have to pry a little bit at a time, here and there, until you can see where the nails are and the best way to pull everything apart without causing damage. Sometimes it is helpful to pry the pieces out far enough to expose the nail shanks, then clip them off with wire cutters.

Choosing Replacement Trimwork

A wide range of choices are available for replacement trimwork. For single-piece casing, you can select any of a number of stock molding patterns (Fig. 4-22), such as several different casing shapes, base molding, batten, panel or shingle molding, mullion, or even chair rail. Plain S4S (surfaced four sides) stock such as 1 × 4 or 1 × 6 is another possibility. Any wood species can be used. For a painted finish, the common choice is kiln-dried clear pine of the highest available grade, No. 2 at a minimum. Or, you can have any mold-

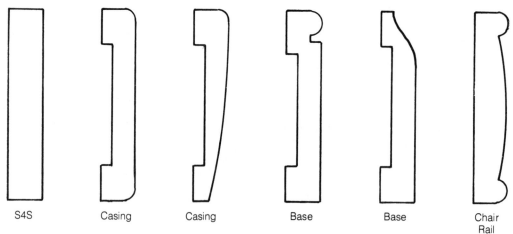

S4S   Casing   Casing   Base   Base   Chair
                                                                         Rail

Fig. 4-22. These are just a few of the common molding patterns that can be employed as window casing material.

ing pattern of your choice custom milled. For the apron, material the same as the casings is often installed, but a different configuration might look better to you. For the stool, stock stool or stool cap molding is often fitted directly against a standard angled sill. However, this molding forms a fairly narrow stool. Many modern window frames are not fitted with the traditional thick, downward-sloping sill but rather a narrow, horizontal framing member. In either case, it is easier to fashion a stool of the width desired from nominal 1-inch pine or other stock. The inner, squared edge butts the window frame and the outer edge can be left squared, rounded, or otherwise molded with a router or shaper. Or, you can attach a small molding of identical thickness and any desired shape, such as half-round, quarter-round, or base shoe.

Replacing the Trimwork

Before starting to replace the trimwork, check the window frame for squareness by placing a framing square at each corner. If the frame is square or just a tiny bit out of square, you can start retrimming. If it is out of square by so much that the sash works only reluctantly, the frame should be squared up, if possible, and reweatherstripped. This probably will mean taking the window out (see "Window Removal"). If the sash will not operate at all, or is badly worn or damaged, the window should be replaced (see "Installing Window Units"). If the sash is square within reason, check next to see if the window unit sits plumb and level in the wall opening. In rare instances, windows are installed cock-eyed. Hang a plumb bob alongside both side jambs, and set a spirit level across the sill and head jamb. If all is in order, go ahead with the retrimming. If not, you can either take the window unit out and reset it, or live with it as is.

edge lies close to or flush with the interior wall surface, and you are using a narrow stool fashioned from S4S stock—install the stool first. If the edge of the sill is flush with the inside wall surface, the stool is a straight-sided piece. If the sill is indented, "ears" must be cut at the stool ends so that the inner edge butts against the sill and the inner edges of the ears butt against the wall surface (Fig. 4-23). The minimum length of the stool must be exactly equal to the distance between the outside edges of the side casings, but the stool ends can extend beyond the side casings to any extent you wish. Sometimes the stool is continuous across two or more windows that are separated by several inches of wall space.

In any case, fit and trim the stool piece as necessary. If it is to be edge-molded, do that first, but if it will be fitted with separate edge molding, leave that until later. Nail the stool to the walls at the ends—either through the leading edge or by toenailing at the ends—with finish or casing nails. Drill pilot holes if necessary. In some cases the stool might be nailed

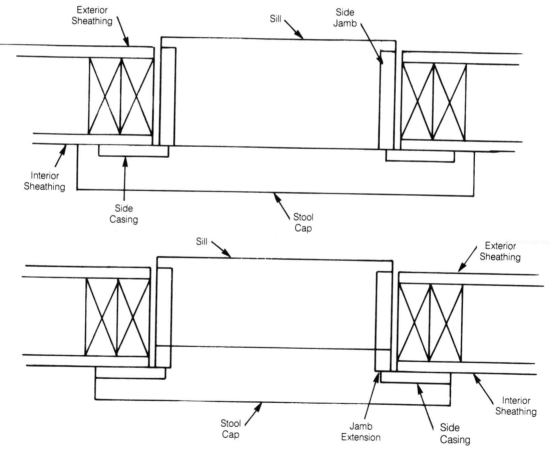

Fig. 4-23. *Above, the window sill inner edge lies flush with the surface of the interior wall, so the stool cap is a straight length of stock. Below, the stool cap must be "eared" to project into the window opening and meet the inset edge of the window sill.*

to the sill as well, but be careful not to damage the sash or crack the glass from the impact of the hammer.

Another choice involves trimming without a stool. This can be done with any window frame that has a square-edged, flat-bottomed sill. The apron is moved up slightly, which results in a uniform band around the window edge like a picture frame. Install the apron first, with its top edge flush with the sill upper surface (Fig. 4-24). If the sill outer edge is not flush with the interior wall surface and leaves a gap between the sill and the apron, rip a suitable filler strip. Nail the strip to the sill edge, then install the apron. If the apron is the same size and configuration as the casing stock, miter the end joints. However, if the stock is all S4S you can also use butt joints. If the apron is S4S but a different width than the casing, you can use simple or double rabbets or offset miter joints (Fig. 4-25). Center the apron and nail it to the sill edge and the rough opening framing with finish or casement nails.

Yet another method is to start with the side casings. You can set the casings flush with the sides of the side jambs, or set them back $^1/_8$ to $^1/_4$ inch (Fig. 4-26). The side casings must be plumb and the head casing level for tight joints at the corners. If the window frame is a bit out of square, set the casing edges back from the faces of the side jambs and

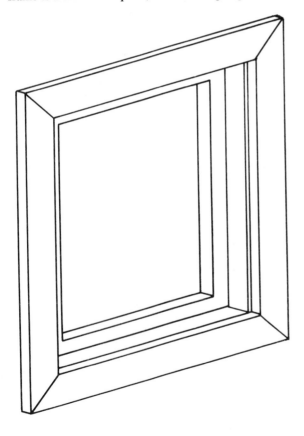

Fig. 4-24. *If no stool is desired, the apron can be moved up and joined with the side casings to frame the window, somewhat like framing a picture or a mirror.*

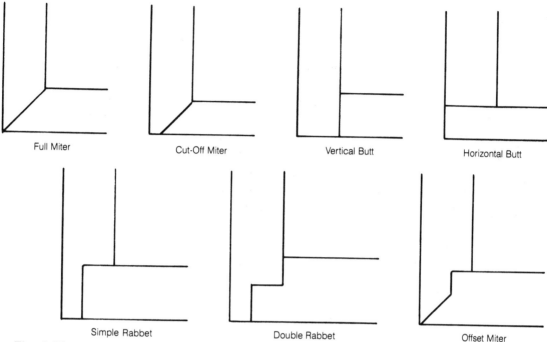

Fig. 4-25. *Any of these joints can be used to join the apron and side casings, depending upon the width and configuration of the stock and the appearance desired.*

Full Miter

Cut-Off Miter

Vertical Butt

Horizontal Butt

Simple Rabbet

Double Rabbet

Offset Miter

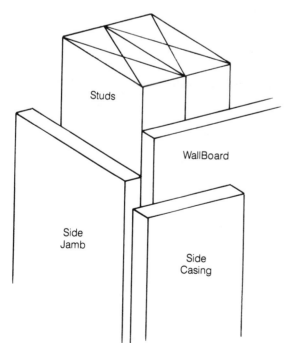

Studs

WallBoard

Side Jamb

Side Casing

Fig. 4-26. *Casing edges can be set flush with jamb faces, or set back a fraction as shown here. This arrangement usually poses fewer problems in finishing.*

adjust them plumb. The setback distance will vary from top to bottom, but this will be barely noticeable. The bottom end of each side casing should lie exactly at stool-top level. At the top, you can butt the ends against the bottom edges of the head casing, or cut a miter joint. Measure the length of the side casings to suit, including the setback for the head casing if needed. This should be approximately the same as for the side casings. Nail each side casing to the studding with one partly driven nail at top and bottom.

Cut and fit the head casing and tack-nail it in place. Adjust the three pieces as necessary so that they are fully aligned and the joints fit tightly, then complete the nailing. Usually 8d finish or casing nails spaced about 16 inches apart work well, but use your judgement. If the window frame is wood, face-nail through the casing into the frame edges with a few 6d finish or casing nails. If the casing is set back, these nails will have to be angled slightly into the frame. Make sure that the points do not come through the jamb face.

If you have already installed the stool, drive one or two 6d nails (depending upon the size and shape of the casing material) up through the stool and into the ends of the side casings. If you have not, trim and fit the stool as explained earlier. Hold it in position against the bottom ends of the side casings and nail it in place. Nail through the ears and into the studding, or up into the side casing ends, or both (Fig. 4-27).

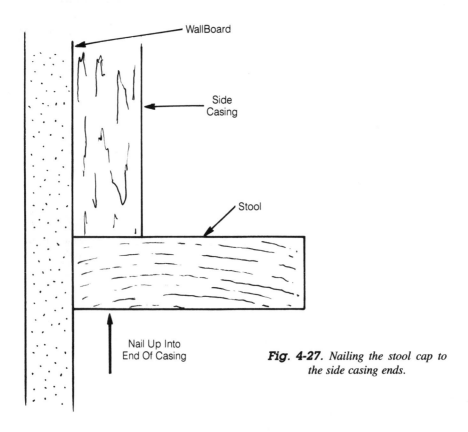

WallBoard

Side
Casing

Stool

Nail Up Into
End Of Casing

Fig. 4-27. *Nailing the stool cap to the side casing ends.*

Now trim and fit the apron. This is sometimes exactly the same length as the stool, sometimes a bit shorter. Position it beneath the stool and nail it to the studding with 8d finish or casing nails. Place two nails at each end and space others along its length. Drive a few 6d finish or casing nails down through the stool, into the upper edge of the apron (Fig. 4-28). If required, install edge molding around the outer edges of the stool. If an existing stop molding was removed, that must be replaced. It need not be the same kind, however: any form of molding that will fit properly on the faces of the side and head jambs and has a flat inner edge to bear against the sash will do the job.

Multiple piece or built-up window trim begins with the installation of the basic casing, the stool if required, and the apron. Select either stock molding of such shape and size that other moldings can be added to it, or use S4S plain stock in the desired width (usually anywhere from about 3 to 5½ inches). Then attach the additional moldings to the casing and apron. There are dozens of possibilities for these casings; a few are shown in Fig. 4-29. Corner blocks can also be added at the top corners, to serve both as decoration and cover the corner joints. Though the head casing is generally treated the same as the side casings, sometimes a cornice is built up instead.

Most stools are relatively narrow but they can also be made wide and sturdy. A shelf-like stool is perfect for house plants (Fig. 4-30). You must, however, provide adequate support beneath the stool; an ordinary apron is not enough. The simplest method is to install a wide apron and attach a pair of metal brackets to it and the underside of the stool.

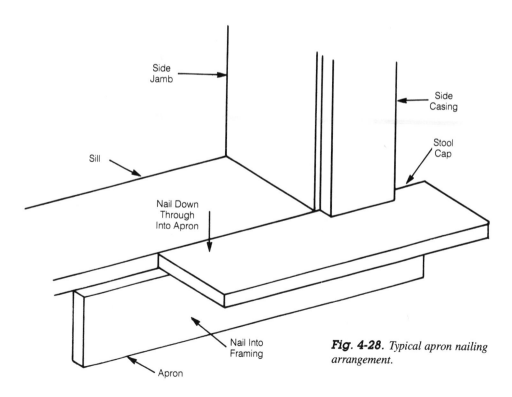

Fig. 4-28. *Typical apron nailing arrangement.*

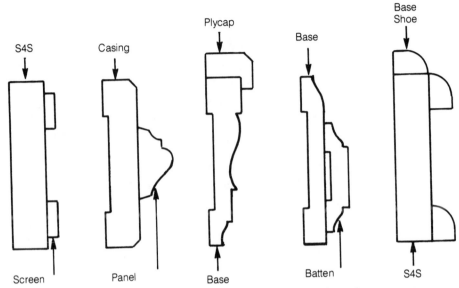

Fig. 4-29. *These are a few possibilities for built-up window casings; there are many more.*

Fig. 4-30. *A window can easily be fitted with a wide piece of S4S stock in place of ordinary stool cap, thus becoming a usable shelf.*

Alternatively, fashion wood brackets and either nail or screw them in place. This looks fine with plain trimwork, but not with more ornate styles. A workable solution in this case is to construct what amounts to a cornice at the bottom of the window. The stool is the top of the assembly, and support is provided by a wide crown or bed molding that runs full length of the apron and stool (Fig. 4-31). If a sufficiently wide stock molding is unavailable, substitute a length of edge-beveled S4S to suit. That, in turn, can be embellished with small stock moldings such as base cap, panel, shelf edge, or screen moldings, or astragal. In any case, the open ends of the assembly are blocked with simple triangular gussets.

After all the trimwork has been installed, it should be prepped and finished as soon as possible to seal and protect the wood. For a paint finish, first apply a coat of appropriate primer, then plug all nail holes and defects with vinyl paste spackle or wood filler. If there is any visible end grain, wipe a thin coat of spackle across those areas as well. Sand the patches smooth, and scuff-sand all the surfaces with fine sandpaper to remove dust nibs

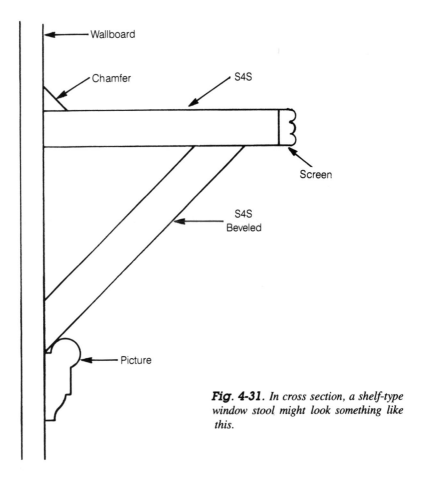

Fig. 4-31. *In cross section, a shelf-type window stool might look something like this.*

and stray wood fibers. Then apply one or two coats, as needed, of either oil-based or emulsion semi-gloss or high-gloss paint. For a clear finish, apply a coat of sealer first. Scuff-sand all the surfaces, and apply two topcoats of semi-gloss or high-gloss varnish or similar finish. Then plug all the nail holes with matching color putty sticks.

WINDOW REMOVAL

The recommended first step in window removal is to take the sash out of the frame, especially if the window unit is to be salvaged for reuse or sale. This lessens the possibility of damage, or injury from broken glass, and also greatly reduces the weight that has to be wrestled out of the wall.

Sliding sash is removed by pushing the sash upward in the track and pulling the bottom outward, toward the interior. The upper track may be adjustable up and down by means of screws; tightening them pulls the track upward and allows the sash to be removed.

Removal of single- and double-hung sash depends upon how the window is constructed. In older types, taking the interior stops off the side jambs will release the lower sash. To get the upper sash out you might have to remove either the exterior stops or the parting stops between the sashes. Removing sash cords and weights or spring balances might also be necessary. In some newer units both sashes come out of the frame together, held by a metal sash guide on each side. This requires removing the outer blind stops, or in some cases the exterior casing trim, and taking the sash out from outside. In other types, the sash can be pushed sideways against the metal sash guide to release the opposite side from the guide, then pulled back into the room.

Awning and hopper types are usually removed by taking off part of the operating hardware (which might be held in place with clips) and removing the hinges from the frame. Casement sash is removed in much the same way—by taking off the pivot hinges at the top and bottom corners and disconnecting the operators.

Removing the sash from fixed windows sometimes is not worth the trouble, especially if the unit is small, and could cause irreparable damage. Nonetheless, it is best to take them out if that can be reasonably done. Usually the best approach is to remove the exterior stops all the way around the frame, then gently push the sash outward. It might be sealed in with caulk, which can be slit apart with a utility knife. In cases where fixed lights are installed directly into the frame without benefit of a separate sash frame, exercise plenty of caution in prying the lights loose.

Fixed lights in combination units, especially in clad wood frames (a large center light with a horizontal slider at each side, for example) are often held in a channel in the frame with small plastic cleats at the exterior corners. Remove the screws and take the cleats off, then push the light out from the inside. It might be stuck in a sealant or caulk and require a little gentle persuasion; be sure to have someone positioned outside to catch the light as it comes free.

With the sash out of harm's way, the next step is to remove the interior trim, casing, stool, and apron. If there is a vapor barrier stapled across the crack between the stud framing and the edge of the window frame, cut it away if it will not be reused. If a new same-size window will be put back, pull the staples out, fold the flap of material back, and tape it to the wall for reuse later. If insulation has been stuffed into the cracks around the window, dig it out with a screwdriver. This might allow you to see any exposed screws or nails that have been driven through the window frame, into the stud frame of the rough opening. You will also be able to see the shimming, if there is any.

Detective work comes next—you have to figure out how the window was anchored to the framing. Metal units will most likely be held by screws through the side jambs, perhaps through shims and perhaps not. Remove the screws and pull the frame out from the outside.

With all-wood frame windows there are a few possibilities. For many years, factory units have been supplied with the exterior casing attached to the frame. These were usually installed by nailing along the outer edges of the side casings, and sometimes the head casing as well. Locate these nails and pry the heads up with a cat's claw, then pull them. Or pry the exterior casing loose from the sheathing and rough opening studding. Then pull the window frame out from the outside. Note that nails might also have been driven through the side jambs of the frame. If you can see them in the cracks around the frame, clip them off with wire cutters, saw them off with a chunk of hacksaw blade, or pry them out with a cat's claw. If they are not visible, chop the shims apart with a hammer and an old screwdriver to see if they hide some nails. Cut or pull them, then pull the frame out.

Some wood windows, particularly those several decades old and double-hung, were partially built on the job site. There might be nails driven everywhere—through casings, side jambs, head jamb, even the sill. All you can do is keep chasing nails until you find all of them and the various parts come free.

The newer vinyl- or aluminum-clad units with factory-fitted mounting flanges around the outside of the frame are a different matter. The flange is fairly wide, and it lies beneath the exterior siding. Usually the siding butts directly against the face of the frame jambs. In some cases, however, a wood molding is installed around the window and over the flange to give the window a dressier look, and the siding butts against the molding. In that case, pry away the wood molding first, without disturbing the siding. This might expose all or most of the flange. If not, or if there is no molding, you have no recourse but to remove some of the siding.

If the siding is clapboard or bevel siding; cedar shingles; channel or a similar type of lapped board; vinyl, steel, or aluminum siding; or any kind that is installed starting from the bottom and working up, begin taking the siding off from the top, above the window. Pry off only as many pieces as necessary to expose the window flange. If the siding is plywood, locate the vertical joints closest to the window and start there. Either up the wood or, better yet, remove the nails with a cat's claw. For vertical board siding, break loose one of the boards either above or below the window and pry the others up, working in both directions to clear the window. Usually it is not possible to salvage all the old siding, so first make sure that you can match the old with new replacement stock, if necessary.

Once the window flange is exposed, removal is easy. The flange will be nailed to the framing with broad-headed roofing barbs. Pry the heads up and pull the nails. These can often be started with electrician's diagonal cutting pliers and then pulled with a claw hammer, without damaging the flange. Pull the window frame out from the outside.

FRAMING AND REFRAMING WINDOW OPENINGS

When framing a new rough opening for a window, you should either have the window unit on hand for its measurements, or at least know exactly which model you will install, along with the specified rough opening size for that window. The minimum opening size is 1 inch wider and 1 inch higher than the outside dimensions of the window frame, to allow a $1/2$-inch adjustment gap all around. There is no specific maximum, but once the gap becomes wider than $3/4$ inch, shimming becomes more difficult. Also, if the window has a narrow exterior casing and no mounting flange, properly securing the unit might prove troublesome. **Note:** If the new window is more than 5 or 6 feet wide and you plan to install it in a load-bearing exterior wall, get professional advice or help.

Preliminaries

The wall studs in most houses are placed on 16-inch centers, but there are often a few that do not coincide with this module. In some houses the studding might be on 24-inch centers, and in very old houses there might be no particular spacing. The first step, then, is to find all the wall studs in the vicinity of the proposed window location, and mark their locations.

The space between 16-inch on-center (OC) studs is approximately $14^1/2$ inches; between 24-inch OC studs it is about $22^1/2$ inches. If your window is narrow enough to fit between two studs, position it there, if possible, to save yourself extra work.

Next, check to see if there are pipes or wires within the wall area that you plan to cut out. Usually pipes are not placed in outside walls, but sometimes they are. Wires can often be moved or rerouted, but this is much more difficult with pipes. Look in the attic or basement to see if either go up through the floor or down through the ceiling at the proposed window location. If there is a convenience outlet in the wall or baseboard below the window location, turn the circuit off and remove the cover plate, pull the outlet out of the box, and try to determine if the wires go up through the wall. If it is impossible to avoid pipes or wires by moving the window location, do so. If not, you will have to make whatever changes are necessary. Also, check outside the house to make sure that you are not placing the window right in front of a downspout or some other obstruction.

If the coast is clear, mark the position of the window opening on the interior wall, preferably alongside a wall stud. Start a hole around the middle of the window location and remove the wall sheathing chunk by chunk. Trim the sheathing cleanly along one stud face, floor to ceiling, removing any wall/ceiling joint molding that might be in the way. Then remove the remaining wall sheathing back to the nearest wall stud beyond the window location. Remove the insulation at the same time. If there is blocking between the studs, knock that out and clip off or pull any old nails that are left.

Installing Window Framing

A typical window framing arrangement as installed in new construction is shown in Fig. 4-32. Your chore is to replicate this arrangement as closely as possible within an existing wall. There are a number of approaches to doing so, depending upon the nature of the existing construction; the size, shape, and type of window; its placement in the wall; and the kind of exterior siding and sheathing involved. No one way is easier or necessarily better than another. Following is one workable method, which can be modified to suit existing conditions.

For convenience, assume that the wall stud against which the window will be placed is on the right when faced from the inside, and that the stud is plumb. Measure along the sole plate from the face of that stud a distance equal to the width of the rough opening, plus 4½ inches, and mark the sole plate. Do the same at the top and mark the top plate.

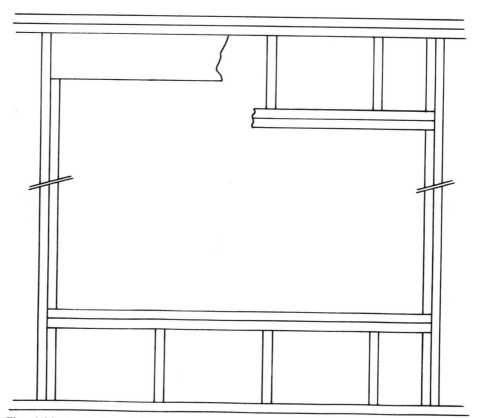

Fig. 4-32. *A typical rough framing arrangement for a window opening, using either a full header (top left) or a simple header and cripple stud installation (top right).*

Cut and fit a full-height stud and toenail it in place with its outer (away from the window) face on the marks. Cut and fit another full-height stud against the leftmost existing wall stud, beyond the new one, and nail it to the existing stud with 10d common or box nails. This will be used only as a nailing base when the wall sheathing is reinstalled (Fig. 4-33).

On the existing studs that interfere with the rough opening, measure up from the floor a distance equal to the point the bottom of the rough opening will be, minus 3 inches. Mark the studs across the edge and both faces with a try square for a square cutoff line. Cut the studs through, flat and square. You can do this entirely with a small handsaw, or make the first cut with a portable circular saw and finish with a handsaw, or use a reciprocating saw fitted with an offset woodcutting blade. Do not cut into the exterior sheathing.

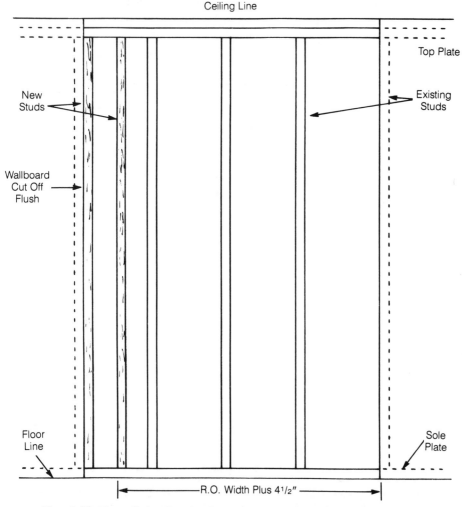

Fig. 4-33. *The wall sheathing has been cut away and a nailing stud and a side support stud installed.*

Leave the lower stud pieces in place, and tap the upper ones back and forth with a hammer until they begin to come loose from the exterior sheathing/siding nails and the nails at the top plate. Then carefully pry the pieces sway. Save these pieces; they might come in handy when framing the new opening, especially if they are the old type and larger than new stud stock (Fig. 4-34).

Next, cut two trimmer studs to match the length of the cut-off stud ends remaining in place (these are now called cripple studs). Nail one to the bottom inner face of the right existing stud, the other to the bottom of the left new stud, with 10d nails. Cut and fit a sill piece across the tops of the cripples from full stud to full stud. Fasten it flat to the cripple

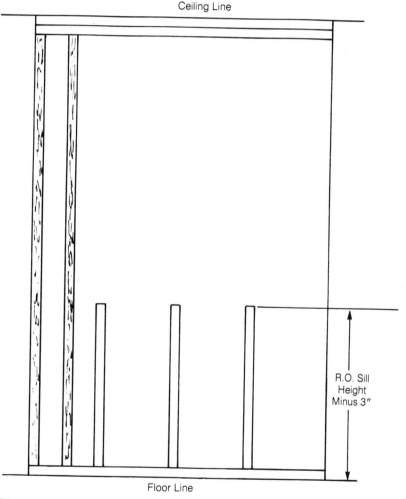

Fig. 4-34. *The existing studs in the way of the new opening have been cut off at an appropriate level and the upper portions removed.*

ends and the full studs with 16d nails. Cut a second sill piece and nail it to the first with 10d nails, and toenail the ends to the full studs to form the framework shown in Fig. 4-35.

Measure up the inner (toward the window opening) face of the left- and right-hand studs, from the sill top, a distance equal to the height of the rough opening. Make squared lines, which will mark the bottom of the header. The way the header is built up depends upon the width of the opening and the distance from the lines to the bottom surface of the top plates. Usually the bottom of a window header is calculated to be about 6 feet and 10 or 11 inches from the floor in order to match the door tops. However, this rule of thumb is not always followed, nor need it be.

The usual minimum header sizes for various spans (rough opening widths) are shown in Table 4-4. These can be made larger, however, and often are when close to the ceiling, because this eliminates the need for small, hard-to-install cripple studs between the header

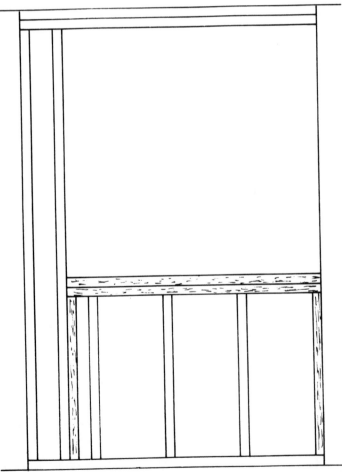

Fig. 4-35. *Here the doubled sill and the side cripple studs have been added.*

Table 4-4. Header Specifications Considered Amply Safe in Residential Frame Construction.

Max. Span (Feet)	Header Size (Inches)
3.0	Two 2- × -4
4.0	Two 2- × -6
6.0	Two 2- × -8
7.5	Two 2- × -10
9.0	Two 2- × -12

and the top plate. Headers are usually made by standing the dimension pieces on edge with a shim between them, so that the combined width is equal to that of the other framing members (Fig. 4-36). The shims can be plywood, hardboard, or whatever is handy. Headers can also be made box fashion (Fig. 4-37). Also, shim stock can be inserted between the top of the header and the bottom of the top plate to close a small gap and make a solid construction.

Make up the header to fit snugly between the two outer studs. Secure it in place with 16d nails driven through the outer faces of the studs into the header ends; toenail when the stud faces are inaccessible. If the header does not extend up to the bottom of the top plate,

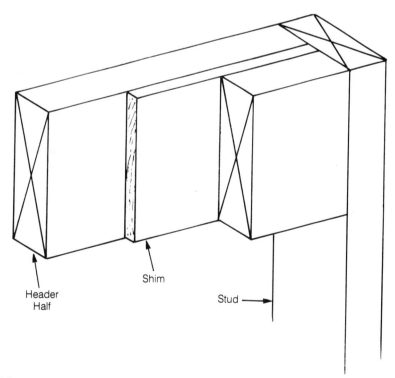

Header
Half

Shim

Stud

Fig. 4-36. A common method of building up a header is to insert shim stock between two lengths of dimension stock stood on edge, to make the assembly match the face width of the wall studs.

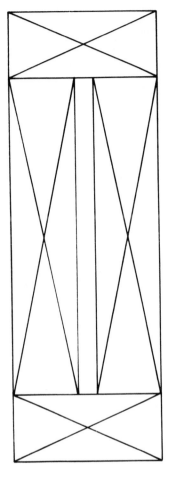

Fig. 4-37. *A box header can be constructed by nailing 2 × 4s to the top and bottom edges of wider stock.*

cut and fit cripple studs and toenail them in place (Fig. 4-38). Spacing distance is not crucial, but they should be on no more than 16-inch centers and placed to make insulation installation easier. If only one cripple is needed, it is usually centered above the opening.

As the last framing step, cut a pair of trimmer studs to fit between the header and sill at each side, to stiffen the full-length studs (as in Fig. 4-38). Fasten these with a series of 10d nails along their faces, and toenail to the header and sill at the ends.

Now, from the inside, drill a small hole out through the exterior siding and sheathing at each corner of the rough opening. On the outside, use a straightedge to draw perimeter lines connecting the holes. If the siding is flat surfaced and has no channels, grooves, or bevels—as is the case with some plywood and plank sidings—the exterior casing of the window can lie flat against it. First, drive a series of nails of appropriate kind and length, depending upon the siding type and thickness, through the siding and sheathing into the framing. Drive these about 3/4 inch outside the opening line all the way around. Set the shoe of a portable circular saw so that the blade will just pass through the siding and sheathing, and cut them both along the outline of the opening. Finish the cuts at the cor-

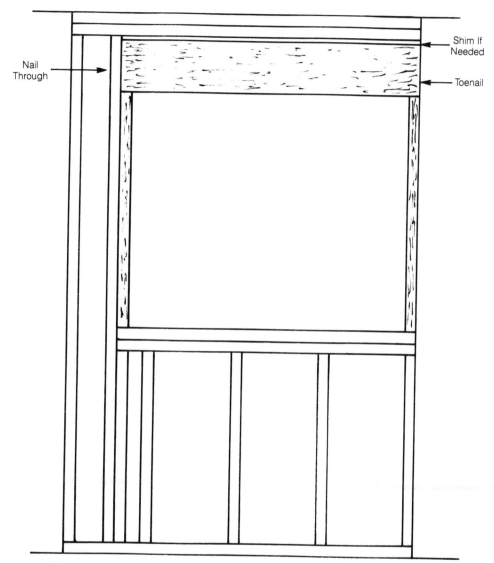

Fig. 4-38. *A full header and a shim have been installed just below the top plate, and the trimmer studs installed.*

ners with a handsaw. Or, bore large starter holes at the corners, and make the cuts with a handsaw, jigsaw, or reciprocating saw from either inside or outside, whichever is more convenient.

Any siding other than flat surfaced must be removed around the window area. If plywood, find the nearest vertical joint and pry the sheets away. Bevel siding, clapboards, shingles, and channel or similar lap sidings are installed from the bottom up. Start from the top, over the window location, and remove it piece by piece in whatever sequence is logical, down to a few inches below the rough opening. Work carefully to avoid damaging the pieces because they will have to be trimmed and refitted after the window is installed. Make a new rough opening outline by connecting the drilled holes in the sheathing. Nail the edges of the sheathing to the rough opening framework about $3/4$ inch to the outside of the line. Nail roofing barbs about 5 inches apart for fiberboard, nail 6d nails for $3/8$-inch plywood or 8d for $1/2$-inch plywood about 7 inches apart. Cut the sheathing out and trim the cut edges as necessary so they are flush with the edges of the rough opening framing. The proceed with the window installation as explained in the next section.

Reframing Existing Window Opening

Reframing an existing window rough opening to accept a window unit that is either wider or taller than the old one usually involves the same procedures as cutting in a new opening. If the window is taller but narrower, you can leave the original header in place, remove the interior sheathing below the opening, knock out the old sill and the cripples beneath it, and install shorter cripples and a lower sill. To narrow the opening, put in as many trimmers and shims as needed to reduce the width. This procedure depends upon how the opening was originally framed up (Fig. 4-39).

If the new window unit is smaller than the original opening, the job is simple. Leave all the framing members in place. Block down from the header if necessary, or install a false header—which can be a single dimension stock member set flat. Build up from the sill with blocking or with a false sill on cripples attached to the old sill. Make the opening narrower with trimmers and shims as necessary (Fig. 4-40).

A window rough opening can be framed into an existing door rough opening in much the same way. The header is already in place. Install bottom trimmers at the sides with a cripple in the middle, and set a double sill across them. Adjust the opening to coincide with the existing header, or install a false header at some other height. The set trimmers and shims to the sides, blocked out as necessary to position the opening, and adjust it to the correct width (Fig. 4-41).

With all such rough openings, the keys are to provide a proper header to support the structure above; a rigid sill to strengthen the opening; and stiff, sturdy side members. Blocking and shimming can be introduced in any way at any point, as necessary to make a strong, solid framework. Exactly what it looks like when finished is really immaterial. When the rough opening is set, replace the interior wall sheathing, trimmed flush with the edges of the rough opening. If an existing opening has been reduced in size, you will have to patch in the exterior sheathing. This might require attaching a few wood strips to some of the faces of the old opening so there is something to nail the sheathing to. After the window has been installed, cut and fit the exterior siding to butt tightly against the exterior casing of the window.

INSTALLING WINDOW UNITS

Installing window units—whether commercially manufactured, locally custom built, or made in the home shop—is not a difficult job. It does need to be done with care, and

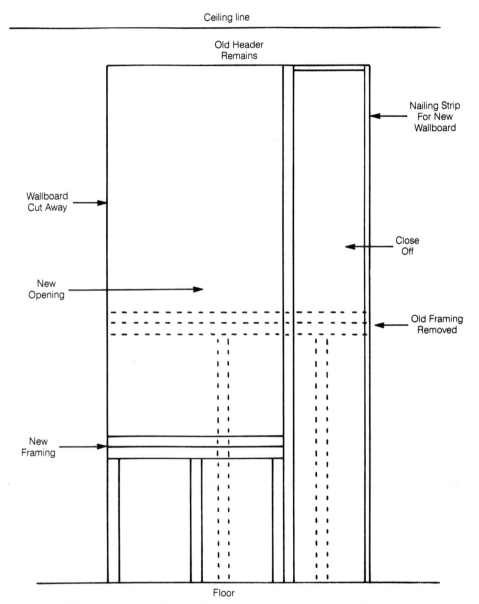

Ceiling line

Old Header
Remains

Nailing Strip
For New
Wallboard

Wallboard
Cut Away

Close
Off

New
Opening

Old Framing
Removed

New
Framing

Floor

Fig. 4-39. *An existing window rough opening reframed to accept a taller, narrower unit.*

usually a helper is needed to boost the unit in place and hold it there while the installer
makes the final adjustment before securing it.

Preliminaries

Commercial units generally are factory fitted with one or more temporary cross
braces to hold the assembly in square during shipping and installation. Shop-built win-

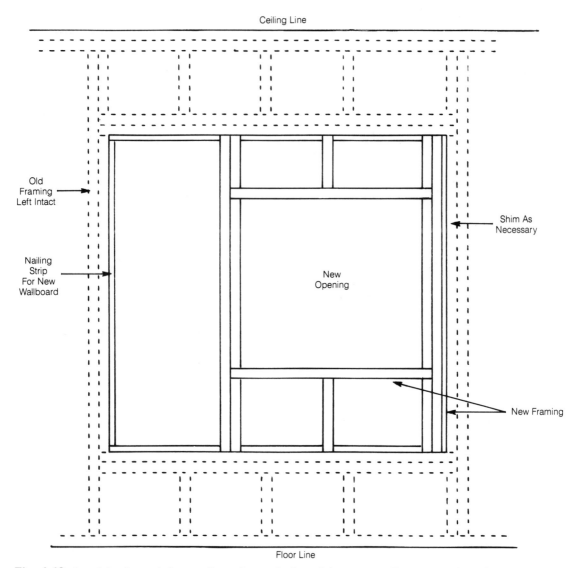

Fig. 4-40. *An existing large window rough opening can be framed down to a smaller opening along these general lines.*

dows should also be braced in square immediately after assembly. Braces, however, are not foolproof. Just before installation, always check all four corners of the unit using a framing square, or by measuring the diagonals (they should be equal). If the unit is out of square, release one end of the braces, set the unit upright on a flat surface, and square it up by pushing laterally on one top corner or the other to rack it slightly. When the frame squares up, reattach the bracing.

The next step is to install an air infiltration barrier. This is often ignored, but should be done if you want a tight installation, especially if the exterior sheathing is boards. Cut

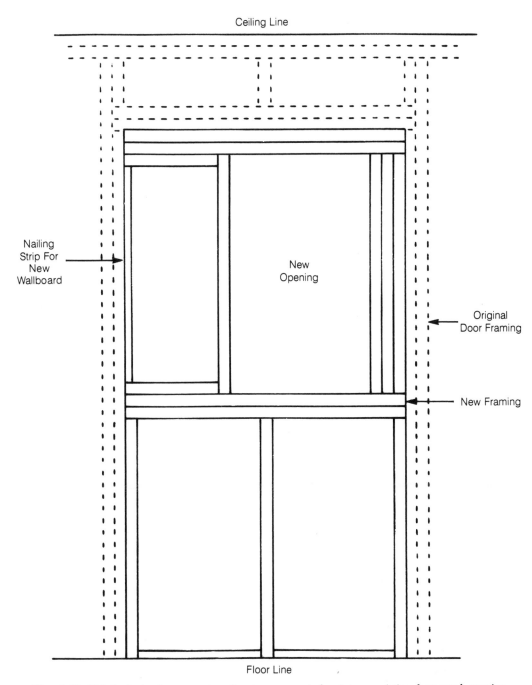

Ceiling Line

Nailing
Strip For
New
Wallboard

New
Opening

Original
Door Framing

New Framing

Floor Line

Fig. 4-41. *This is the basic approach to framing a new window into an existing door rough opening.*

strips of 6-mill polyethylene plastic sheeting (known as "construction plastic") wide enough to wrap around the rough opening framing and extend about 6 or 8 inches outward over the exterior siding, and 3 or 4 inches over the interior wall sheathing. Staple the outer edges of the strips to the exterior sheathing, fold the plastic through the opening and over the framing, and tape or temporarily staple the inner edges to the wall—just enough to keep them out of the way. Make sure there are overlaps and no gaps at the corners.

Procedure

Most windows are put in from the outside. Set the unit on the rough sill and have a helper hold it there. From the inside, check the height of the frame header. If necessary, raise the header to the correct height by shimming with small wood blocks or chunks of shim shingle at the bottom. At the same time, center the window frame in the rough opening. Use a spirit level to make sure the unit is level; if not, adjust the bottom shims as necessary until it is.

If the unit will be attached with nails through an exterior wood casing, or if it is fitted with a mounting flange, go outside while your helper holds the unit firmly against the outside wall and drive one nail in near each bottom corner. Use galvanized nails that are long enough to drive at least 1 inch into the framing. Use casing or finish nails for wood frames, roofing barbs for mounting flanges. For wood frames, drive the nails only partway; for flanges, drive the nails almost all the way but position them at the top of the oversized, factory-punched holes. If the unit is designed to be secured with screws through holes provided in the side jambs, drive a pair of screws part way home at the lowest attachment points.

While your helper continues to hold the unit in place, especially toward the top, go inside and check the side jambs with a level or a plumb bob and line. Make sure they are perfectly vertical; adjust as necessary with shims. Drive two more nails or screws at the top corners of the unit. If the window is more than about $2^1/2$ feet wide, it is a good idea to place wedge blocks 1 foot apart beneath the sill so that it will not sag as it ages. The shims should be just snug; don't drive the sill upward.

While keeping the unit pressed firmly in place, remove the braces and operate the sash to make sure it opens and closes smoothly and evenly. If it does not, readjust the shimming until it does. The finish securing the unit. Nail spacing for wood casing is 10 to 12 inches on sides and head, and on other frame types it is predetermined by the holes provided. If the window is tall and has relatively thin and limber wood side jambs, some installers also insert shim shingles at intermediate points along the side jambs, and drive a nail through the jamb and shims at each point to keep them rigid and aligned.

If the window was supplied with a wood drip cap or metal drip flashing, or both, install by nailing to (or through) the exterior sheathing—but never into the window frame or casing. Run a bead of caulking all around the joint between the window frame and the exterior sheathing, but make it small and narrow so that it does not interfere with snug fitting of the finish siding. When the siding is reinstalled, the butt ends or edges of it should also be caulked, preferably in the joints as the siding is put up.

Finally, stuff the crack all around the window frame full of fiberglass or mineral wool insulation. Pack it well. If the exposed inner edge of the window frame is wood, fold the plastic infiltration barrier loosely over the edge, staple it at just a few points, trim the material off so it will not protrude when the interior casing is installed, and drive the staples flush to the wood with a hammer. Cut narrow strips of plastic and staple them over the crack, to the frame edges and the wall, all around the window. This vapor barrier will be hidden beneath the interior casing and apron. If the window frame is not wood, you might be able to tape the infiltration barrier edges to the frame edges or sides. If not, leave it against the wall, trimmed to disappear under the casing. If you can also install a vapor barrier strip so that the interior trim will hold it in place, perhaps with the aid of some tape, do so. However accomplished, the object is to double-seal the unit (Fig. 4-42) so that neither air nor water vapor can pass through the joints.

SKYLIGHTS

Over the past few years the installation of skylights—some of which are also known as "roof windows"—has become an increasingly popular type of home improvement. There

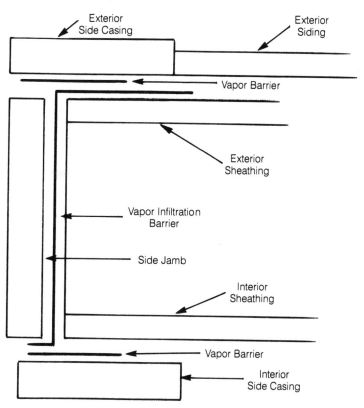

Fig. 4-42. *Vapor barrier strips should be installed all around a window opening and interleaved with the framing, trim, and window frame components so that air and moisture cannot pass through.*

are good reasons for this. For one thing, commercial skylight products are now much improved, widely available, and reasonably priced. Also, skylighting brightens up the interior of any house with even, diffuse daylight; opens up the interior to the outside world for a more spacious atmosphere; and in some instances can be used for solar heating gain. Certain units also have ventilating capabilities.

Choosing the Skylight

Commercial skylight units are available in a wide range of sizes, from approximately 14 inches square to as large as 10 by 12 feet. Some are glazed with a dome of plastic, others with flat lights of either plastic or glass. Although there are several models that are only single-glazed, most are offered with at least double-glazing, and some with triple- and quadruple-glazing. Fixed units remain the most common and are the less expensive, but operable units that allow increased room ventilation and that act as very efficient heat exhausters, are becoming more popular. Various accessories such as screens, sunshades, blinds, remote operators, and electrical motorization are also available.

Quality is a crucial consideration when selecting skylights. The unit must be well constructed and effectively sealed. If operable, it must work smoothly and be properly weatherstripped. A good unit has an integral flashing system; the best incorporate a built-in curb as well, fitted with copper flashing rather than aluminum. The glazing must be tightly sealed with long-lived materials.

When choosing glazing, bear in mind that plastic is susceptible to easy scratching and might degrade somewhat in strong sunlight, especially at high altitudes. If the plastic is tinted this is less of a problem, but it can be very noticeable if the plastic is clear. Also, a certain amount of visual distortion is inherent in dome type plastic glazing. If clarity of vision is important to you, plastic is not the best choice. However, plastic glazing is tough and long-lived. If the glazing is glass—which affords the best clarity, and light, and solar heat transmission—it must be some type of safety glass. Those most used are tempered, wired, and laminated glass. Note, however, that some building codes disallow tempered glass; check on this before buying or building your units.

If you live in a hot country, multiple glazing is a must. In winter climates, triple- or quadruple-glazing should be considered. In any case, skylights should have added protection: shades for solar units and movable insulation for all units (especially solar) where heat losses/gains can be expected. Glass multiple glazing is made up of factory sealed insulating glass and should not be susceptible to internal condensation and clouding or convective currents. Look for a substantial warranty, as with ordinary windows. Plastic multiple glazing is done differently. All the layers are sealed into the frame, but there are tiny vents to allow air circulation between the layers and equalize inside and outside humidity. This system should be checked carefully, because in some inexpensive units, air from within the house can also escape through the vents.

Installation Considerations

Skylights can be installed in roofs of any pitch, and practically at any location. However, placement should not be haphazard. Consider the appearance the skylight opening will make in the room, which will be partly determined by where you prefer the light to fall. Consider the compass orientation, too. Depending upon roof pitch, north-facing skylights will admit no sunlight and the daylight will be of the coolest tones. South-facing will admit maximum sunlight and maximum warm tone daylight. In-between compass directions will admit various combinations of sun and daylight, depending upon the season of the year and the time of day.

Most skylights are installed as added light sources, which means placing them to best advantage to illuminate the desired areas without overdoing it. In many cases, two or more skylights do a better job of light distribution in a room than one big one. Although the overall cost is a bit higher, the installation is easier and the results are better. If solar heat gain is your primary consideration, proper placement is crucial for best results. There are many variables involved here, and each installation must be evaluated and engineered for specific conditions and requirements. If you are knowledgeable about passive solar heating and its ramifications, this is something you can do yourself. If not, consult a solar architect.

Most factory made skylights come complete with installation instructions. Details vary from product to product, so be sure to follow the directions faithfully. The general procedures, however, are the same for all units. The first requirement is a hole in the roof. The rough opening should be framed exactly to the manufacturer's size specifications. The easiest installations involve skylights that are small and placed between two adjacent rafters in a roof/ceiling system. The most difficult involve cutting one or more rafters and framing the opening into a plain roof system over an attic, then building a lightwell to join the roof opening to a finished opening in the ceiling below. No great technical problems arise if the roof system is made up with conventional rafter framing. However, if the roof is supported by trusses, get professional engineering advice before you make final plans or purchase skylight units. You might not be able to cut into the trusses without creating major structural problems.

Procedure

To cut in and frame up the rough opening, first locate the skylight position from inside, under the roof. If a lightwell is required, locate that first on the ceiling, then adjust the skylight position to it. The ceiling opening is framed in the same way as the roof opening, although the size will be different in order to accomodate the material that will form the sides of the lightwell. Outline the opening, cut away interior sheathing, and remove thermal insulation as necessary. Then, form the inside, drill one or more locater holes up through the roof sheathing and weather surface. Using these holes as a reference for the skylight position, remove a section of the roof weather surface and underlayment and cut away the roof sheathing. All of this material will have to be cut back far enough to allow you to install the rough opening framing; it will all have to be pieced back in after the framing is done and as the skylight is installed.

To frame a small opening between two adjacent rafters (Fig. 4-43), install a pair of headers between them of the same size stock as the rafters. If the space between the rafters has to be narrowed just a bit, secure shim stock in the necessary thickness to one or both rafter faces between the headers. If the gap must be narrowed substantially, install a trimmer rafter on one or both sides between the headers. Sometimes the headers are doubled; this is most easily done after trimmers are installed.

If the opening spans more than one rafter space, one or more rafters must be cut. If more than one rafter is cut in an area where roof wind or snow loads are heavy, the two side rafters to which the rough opening framing will be attached should be doubled from wall plate to roof ridge. In areas of light roof loading, as many as three rafters can be cut without doubling the side rafters. Note, though, that much depends upon rafter size and spacing, roof pitch and construction, and local building code requirements.

The general procedure, as shown in Fig. 4-44, is to cut the interfering rafters away 6 inches more than the height of the rough opening. Then install upper and lower doubled

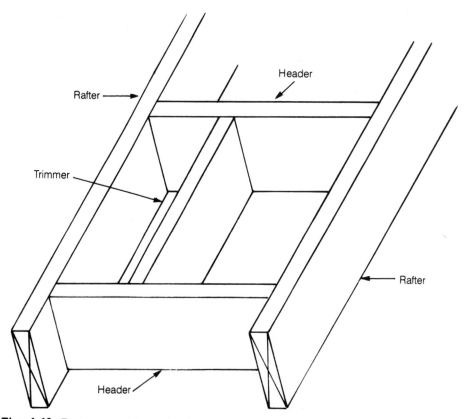

Fig. 4-43. *Framing a rough opening for a small skylight between two adjacent rafters involves installing single headers and trimmers to box it out.*

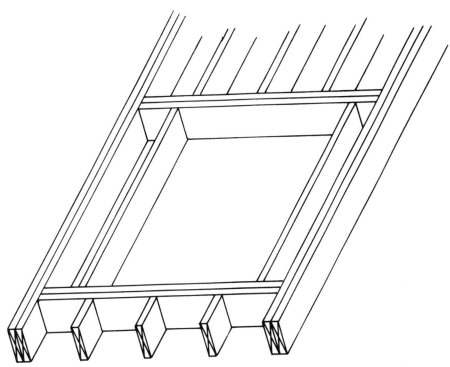

Fig. 4-44. *Framing large skylight openings involves structural reframing with double headers, doubled side support rafters, and single or double trimmers as required.*

headers, attached at the ends to the side rafters and also to the ends of the cut rafters, which have now become tail rafters. Install single trimmers between the headers and toe-nail to them to form the rough opening (or they can be doubled if you wish). Replace the roof sheathing, cut to fit around the rough opening and flush with its edges. Cover the sheathing with an underlayment such as 15- or 30-pound roofing felt, lapped at least 1 foot over the existing underlayment at the bottom and 1 foot under at the tops and sides.

Most lightwells flare out from the skylight opening in the roof to a substantially larger opening in the ceiling below. This is almost a must if the lightwell is more than about 2 feet long and the skylight itself is small. This prevents the lightwell from looking like an unattractive air shaft, and also allows a well-disbursed light spill into the room. Lightwell framing consists of a series of 2 × 3s or 2 × 4s running from the skylight framing to the framing of the ceiling opening, with the edges facing into the well (Fig. 4-45). Their placement must be calculated so that the sheathing mates properly with both the bottom of the skylight and the bottom edge of the ceiling opening, and also allows for the installation of trimwork if desired. The inside corners must be elled (Fig. 4-46) to provide both rigidity and a solid nailing surface for the sheathing edges. Gypsum wallboard is commonly used for sheathing, with the joints taped in the usual way. However, you could also use plywood and cover the corner and edge joints with wood molding. The typical finish for lightwells is gloss white enamel for high reflection.

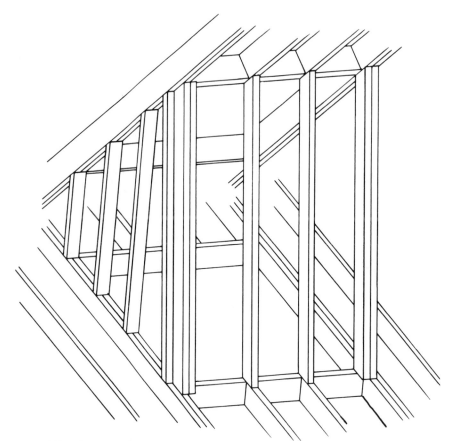

Fig. 4-45. *Typical framing arrangement for a lightwell with an angled back.*
(The front is vertical, but to avoid confusion is not shown here.)

Installing a factory made skylight in the rough opening is a matter of following the directions and using common sense. The crucial element is the flashing. This may be one piece, several pieces, flashing and counter-flashing, integral with the unit, or set in proper position as the installation proceeds, depending upon the product. In any case, the flashing at the top of the unit extends upward on the roof, and the flashing at the sides extends outward for a substantial distance to lie between the underlayment and the weather surface (shingles, etc.). The flashing at the bottom lies on top of the weather surface, and if a separate piece, also lies beneath the side flashing. All lap seams or joints in both roofing and flashing must be arranged so that water flows down over and past them, not into them (Fig. 4-47). This sounds elementary, but it is amazing how often that simple rule is ignored. In addition, lapped flashing joints are usually sealed with roofing compound or soldered, depending upon the metal. Many installers also seal the roofing to the head and side flashing, and the bottom flashing to the roofing.

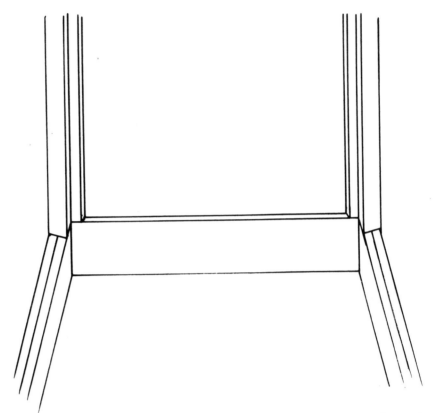

Fig. 4-46. *The corners of the lightwell frame, seen here straight on from inside the opening, must be elled to provide nailing points for the sheathing.*

DOOR TYPES

There are two general categories of doors: exterior and interior. Both can be further classified by the way they operate: swinging, sliding, or folding. Exterior doors include standard entryway doors, which are the swinging type, and patio or atrium doors, which may be either swinging or sliding. The most common interior doors are the familiar solid wood types. Other interior specialty doors used less often are accordian folding, bifolding or multifolding, batwing or cafe, and exterior patio or atrium doors installed for interior use.

Many newer houses are fitted with exterior doors made of steel sheet over a core of thermal insulation, which incorporate a thermal break between the inside and outside faces. These are occasionally retrofitted. Fiberglass entry doors, relatively new to the marketplace, are made in much the same way as steel doors, but are much easier to retrofit because they can be trimmed and fitted to an existing finish frame. Both of these designs simulate popular wood door styles. Most exterior doors, however—especially those bought for replacement purposes—are made of wood in a wide range of styles, with and

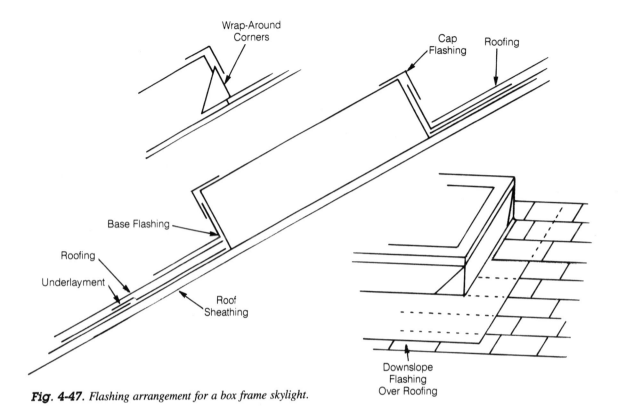

Fig. 4-47. *Flashing arrangement for a box frame skylight.*

without lights. All exterior entryway doors, unless custom made, are manufactured in standard heights of 6 feet 8 inches and 7 feet, and in widths of 2 feet 6 inches, 2 feet 8 inches, and 3 feet. Standard thickness is $1^3/_4$ inches.

With the exception of some accordian folding doors that are made partly or entirely of various plastics, interior doors are constructed of wood. The standard thickness is $1^3/_8$ inches, and the standard heights are 6 feet 8 inches and 7 feet, though a few are available in a height of 6 feet 6 inches. Because these doors are used in so many different ways, standard widths range from 1 foot to 3 feet 6 inches, variable with the door style and manufacturer. For passageway use, widths of 2 feet 8 inches and 3 feet are common; for closets, widths 2 feet and 2 feet 6 inches are often selected. Bifolding and multifolding units usually come in sets for 4-, 6-, and 8-foot (finished size) openings, although other sizes are available. Accordian folding doors are in a class by themselves, because there are no set standards. They may be from 5 to 10 feet high and of almost any length.

Wood doors, whether interior or exterior, are classed by the way in which they are made. *Flush* doors are constructed with faces of wood veneer, plastic, or hardboard bonded to a core. The faces are smooth and uninterrupted, unless by a light or a louver or grill insert. There are two varieties: Hollow- and solid-core. A *hollow-core* door (Fig. 4-48) has the faces bonded to either a honeycomb core or a series of closely spaced wood

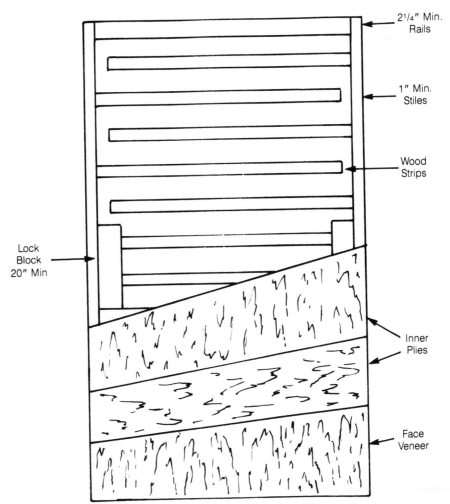

2¹/₄″ Min.
Rails

1″ Min.
Stiles

Wood
Strips

Lock
Block
20″ Min

Inner
Plies

Face
Veneer

Fig. 4-48. *One common hollow-core flush door construction method.*

strips. These doors are characteristically lightweight and flimsy, can be easily damaged, are susceptible to warping and delaminating, and are not weatherable. A *solid-core* door (Fig. 4-49), on the other hand, has a core of solid wood blocks butted together and arranged in various ways, or solid composition board. They are much heavier and more rugged than the hollow-core type, will stand up to hard use, and are reasonably weatherable.

Panel doors (Fig. 4-50) are the traditional choice for both interior and exterior use. They are made up of an assembly of horizontal and vertical main members that frame and support one or more thinner, inset panels. A wide range of patterns is available. Though the arrangement and size of the panels can vary, these doors are often designated by the number of panels. *Sash* doors (Fig. 4-51) are made in the same way, the difference being

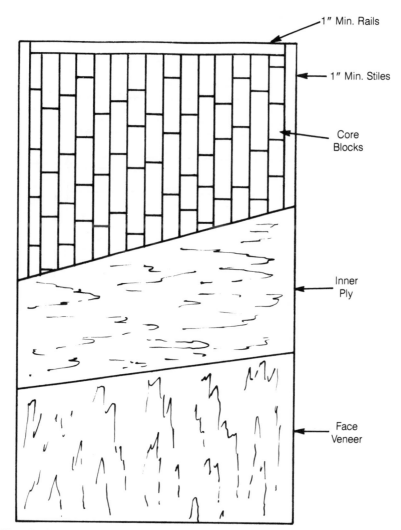

1" Min. Rails

1" Min. Stiles

Core
Blocks

Inner
Ply

Face
Veneer

*Fig. 4-49. Solid-core flush doors are cored in several different ways—this is one—
but contain no hollow spaces.*

that they incorporate one or more lights in place of panels. These are used mostly in exterior applications. If all of the panels are replaced by lights, the door becomes a *casement* or *French* door (Fig. 4-52). Further, if there is only a single, full-sized light in the frame, it is a *rim* casement door; if there are multiple lights, it is a *divided-light* casement door.

Unlike other wood doors, *louver* doors, (Fig. 4-53) are not available in 1³/₄ inch thickness because they are not for exterior use, 1³/₈-inch is standard. However, they do come in a 1¹/₈-inch thickness for closet and utility applications. These are a bit cheaper, but also a bit flimsier. Louver doors consist of an outer frame into which slightly overlapping thin wood slats are mortised. There are two slat configurations (Fig. 4-54). *Standard* slats are

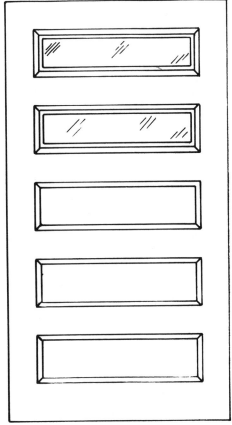

Fig. 4-50. *A typical panel door; there are many designs or patterns.*

Fig. 4-51. *A sash door contains one or more lights along with panels.*

individual pieces set at a steep angle and separated by a space of about $1/4$ inch. Viewed from one side of the door the louvers slant downward, and from the other they slant upward. *Chevron* slats, on the other hand, are inverted V's stacked one atop another. They may be set solidly together or spaced slightly apart, and the slat faces slant downward on both sides of the door. Almost the entire height of the door can be louvered, or a cross rail in the middle, or the bottom third of the door fitted with a panel or two instead of louvers. These doors are used primarily as visual screens and the open louver types will also allow a certain amount of ventilation. Except perhaps for the solid chevron style, these doors should not be installed where sound privacy or noise control is a concern.

Accordian folding doors (Fig. 4-55) are produced in four typical types. In the first, narrow, solid, thin wood slats are set vertically and attached to one another with metal, nylon fabric or vinyl hinges. Second, the wood slats may be even narrower than the first and connected with thin cords interwoven through them. Third, very thin strips about $1/2$ inch wide, commonly in a basketweave pattern, are interwoven with vinyl or nylon rein-

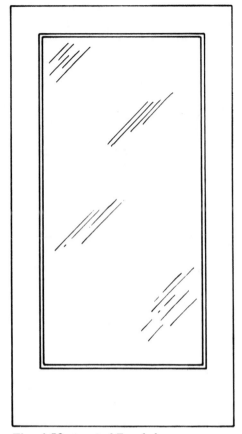

Fig. 4-52. *A typical French door.*

Fig. 4-53. *A typical louver door. It can also be completely louvered, or have a panel in the bottom section.*

forced vinyl tape. Fourth, flexible vinyl sheeting or vinyl webbing strips are interwoven to make an unbroken panel. The panels fold back upon themselves in loops as they close, like a drapery. These doors are primarily intended as visual screens between areas, or as closures for closets or storage areas. They do allow a certain amount of ventilation, but are unsuitable where either sound isolation or privacy are needed.

Patio or atrium doors (Fig. 4-56) capture the best of both door and window. They are made as factory assembled units, although they might be knocked down for shipping, to be reassembled at the job site. The units are typically comprised of either one opening door and one fixed light of the same size, or two fixed light sections with a center door opening either left or right. The door portion of the unit is a single framed light, which can be bypass sliding or single swinging. Approximate sizes range from 5 to 12 feet wide, with the standard height being 6 feet 8 inches. Several types of safety glazing can be used, including insulation glass. The frames may be all-wood, vinyl-clad wood, or aluminum. Although designed for exterior use, they can just as well be installed to separate interior

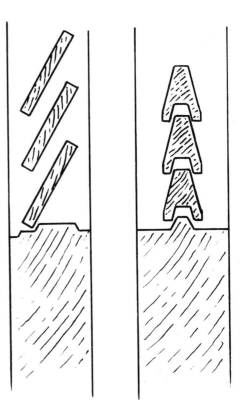

Fig. 4-54. Two different louver styles are used in louver doors: slat (left) and chevron (right).

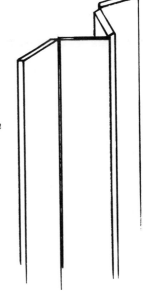

Fig. 4-55. A typical accordian folding door.

Fig. 4-56. *A typical patio door with one sliding door and one fixed light.*

parts of a house. To gain the added flexibility and ventilating capability of two or more doors, and to create one version of a window wall, multiple units can be installed side by side.

Solid wood doors, intended for exterior applications, are uncommon but available. They are largely one-of-a-kind doors, produced by small, custom woodworking shops. They are made in a variety of sizes, but mostly adhere to the standard dimensions—except for thickness. A $1^3/4$-inch-thick door is considered thin; 2 to 4 inches is more common. Most of these doors are artistically made in a myriad of designs, often including hand carving on one or both faces. Rather than the usual pine or fir, heavy planking of various domestic and exotic hardwoods are generally employed. The wood is shaped and glued up as necessary to gain the desired thickness and patterns. Special hardware and framing are often required to install these doors. As you might expect, though handsome, these doors are expensive.

DOOR ASSEMBLIES

All interior an exterior door assemblies have much the same makeup. There are three main elements: the door, the door frame, and the hardware complement. The pocket door requires a fourth element: the pocket.

Door Frame

The door frame consists of the structural framing hidden within the wall to form the rough opening for the doors and the finish frame (Fig. 4-57). The basic parts of the finish frame are the two side jambs, the head jamb, and the casing. The casing in turn is comprised of two side casings and a head casing mounted on each side of the doorway. In addition, exterior door frames are fitted with a sill at the bottom. Interior door frames do not have a sill, but might have a threshold (Fig. 4-58).

If a single-swinging door—which is hinged at one side or the other and swings only in one direction—is to be hung, the door frame is also fitted with a stop. This stop can be just a small piece of wood or metal mounted on a jamb face opposite the door latch assembly. However, for the sake of appearance and proper sealing of exterior doors, the stop nearly always consists of a full-length molding applied to the side and head jambs.

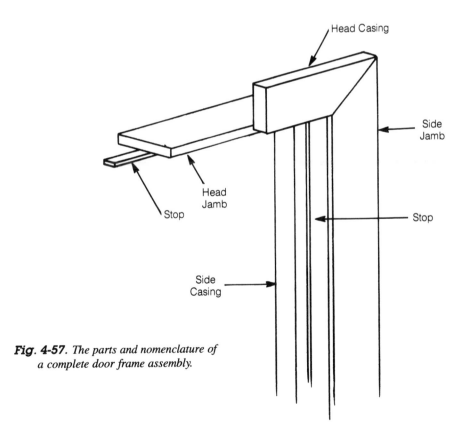

Fig. 4-57. *The parts and nomenclature of a complete door frame assembly.*

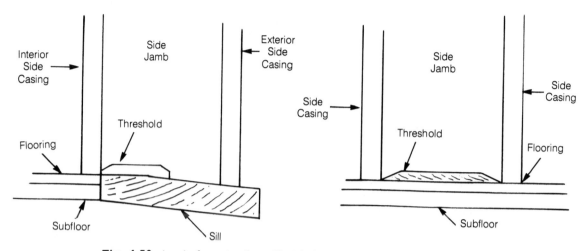

Fig. 4-58. *A typical exterior door sill with threshold, often weather stripped (left), and a typical interior threshold in cross section.*

Fig. 4-59. *Typical exterior door embellishments include transoms or side lights (left), pediments (center), and pilasters (right).*

Smaller trim moldings or corner blocks of various sorts may be applied to the casings as embellishments. Exterior door assemblies can include one or two sidelights and/or a transom, side or overhead panels, or assorted ornate trimmings on the outside such as a cornice, pediment, or pilasters (Fig. 4-59). They are also typically fitted with weather-stripping at the stops and the threshold, and sometimes the door bottom as well.

There is a vast array of door hardware available. Selection depends primarily on the type of door being installed and the way it will operate, and secondarily upon the decorative appearance of visible components. Accordian folding, patio, bifolding, and multifolding doors when bought new are usually supplied with the proper operating hardware, such as tracks and hinges. Knobs, pulls, latches, or catches are left to the purchaser to provide, except in the case of sliding patio doors. Individual wood doors are not supplied with any hardware at all, because that must be selected for the purpose at hand. The exception to this is prehung interior passageway and exterior entryway doors. These doors have been factory fitted to an assembled door frame. The frame is built, usually without a threshold for interior use, and temporarily braced for shipping. The door is installed in the frame complete with hinges and stops. In some cases a latch assembly might be included, and the unit can be ordered with or without the trim casing materials, which are bundled but not attached.

Single-Swinging Doors

Single-swinging doors fit snugly within their frames and are restrained from backswinging by a stop (Fig. 4-60). They are attached to one side jamb with loose-pin hinges (the hinge pin can be removed from the barrel), sized to fit the door thickness and weight. Two are typically used, but if the door is large and heavy, a third hinge can be mounted at

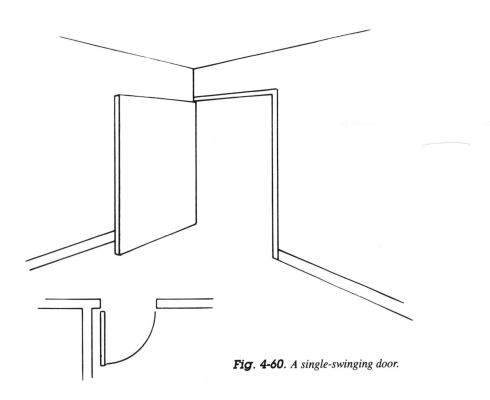

Fig. 4-60. A single-swinging door.

midpoint. Exterior doors are sometimes fitted with ball bearing hinges for longer life and smoother operation.

Interior single-swinging passageway doors are fitted with tubular locksets or the slightly more expensive cylindrical locksets. There are several variations: Knob latches have a knob on each side to retract the latch bolt. Privacy locks have a pushbutton or push knob lock on one side and a means to unlock the mechanism through the other knob. Exit locks have a push-locking knob on one side and no means of unlocking on the other. Outside locks have an automatic deadlock, pushbutton, or push-knob lock on the inside and a keyed pin tumbler lock on the outside. Matching dummy knobs are available in some styles, and latch bolt extension links can be obtained to allow the lockset to be installed in the center of the door instead of at one side.

Although the standard outside lock type of lockset is often installed in exterior doors, there is a wide variety of sturdier, larger, and more ornate locksets available for such purposes in knob, lever, and thumblatch styles. These often include a separate deadbolt and various security features, and may come with matching door knockers. Exterior doors are also sometimes fitted with relatively complex, mortised box locks. When selecting locksets, be sure to investigate manufacturers' literature. Low-cost locksets commonly stocked by hardware suppliers do not represent all that is available. You can find better quality, more attractive models, and greater selection if you consult the catalogs and go the special order route.

Double-Swinging Doors

A double-swinging interior passageway door is fitted into its frame with a substantial amount of clearance (Fig. 4-61). Because it must swing fully open in both directions, no stop is installed on the jambs. A full-sized door swings on stout pivot or pin hinges that are usually mortised into the top and bottom door edges close to one side, and into corresponding locations in the floor and the head jamb. Smaller doors, such as the cafe or batwing type, swing on lighter-weight pivots that are surface-mounted to the upper and lower door corners and the side jamb of the frame. Most of these hinges include an automatic stop position that holds the door open at slightly past 90 degrees. The full-sized doors are usually installed as a single unit in residential applications, while the cafe type is installed in center meeting pairs within a single opening. No latches or locksets are used. Full-sized doors are usually fitted with a metal or plastic push plate at about chest height, close to the nonhinged side, on both sides of the door. A kick plate might also be fitted along the bottom on both sides. Cafe doors require no additional hardware, although small push plates can be added.

Bifolding Doors

Bifolding doors may be installed as a single two-door set, opening from one side and folding toward the other. However, they are more often installed in two sets of two doors each that meet at center and fold back to each side (Fig. 4-62). Multifolding doors are installed as sets of three or more doors, hinged together and folding from one side to the

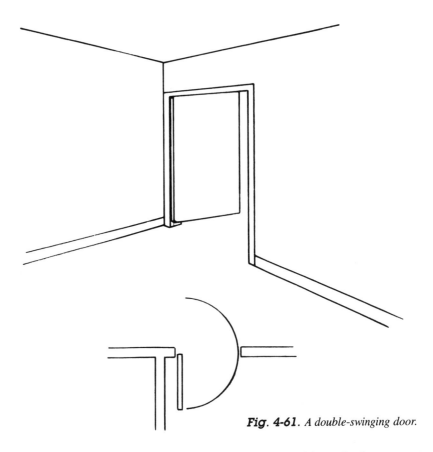

Fig. 4-61. A double-swinging door.

other (Fig. 4-63). Both types are set in a plain door frame with ample clearance on each side. No stops are applied to the door jambs, but moldings are often mounted around the outer perimeter of the frame to hide the relatively wide closure gaps at the sides and the track at the top.

Both of these folding types are supplied with all the necessary operating hardware: Special surface-mounted or mortise fixed-pin hinges for the door panels, pivot-pin assemblies that mount at the top and bottom corners on one side of each set, track guides or rollers that mount in the door top edges, and a track assembly that mounts on the face of the head jamb of the door frame. Installation and adjustment instructions are also included. The selection of door knobs or pulls is left to the purchaser.

Accordian Folding Doors

Accordian folding doors are likewise mounted in a plain door frame (Fig. 4-64) without stops or other added moldings (unless installed for decorative purposes). All hardware is included: Mounting screws, a track that mounts to the face of the door frame and head jamb, and a latch or catch assembly. Installation consists of securing the track to the head jamb, setting the door into the track and anchoring one side of it to the face of a door

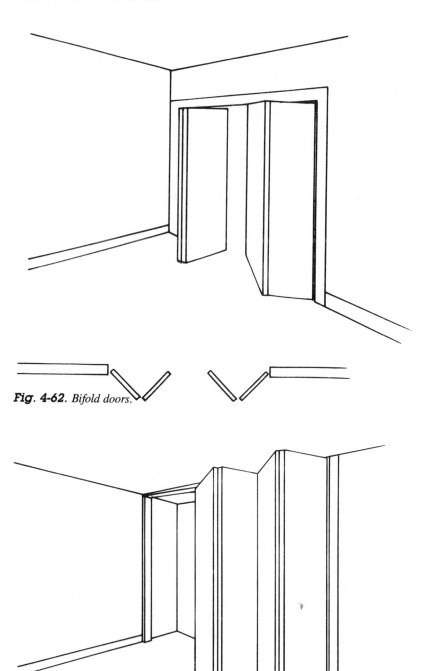

Fig. 4-62. *Bifold doors.*

Fig. 4-63. *Multifold doors.*

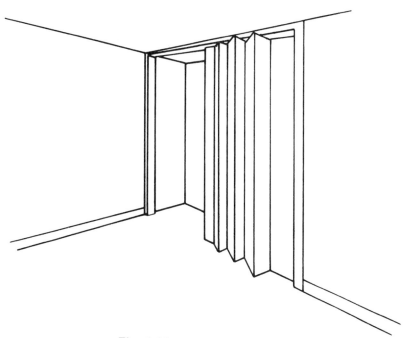

Fig. 4-64. *Accordian folding door.*

frame side jamb, and attaching the latch assembly to the other side jamb. Large units might include a second track to guide the bottom of the door. This track can be surface-mounted or mortised into the floor, depending upon the nature of the finish flooring.

Surface Sliding Doors

A surface sliding door is not set within an ordinary door frame, but instead covers over a usually frameless but finished opening in the wall (Fig. 4-65). The door, which can be of any rigid type, is suspended by adjustable hangers that run in a track mounted to the wall above the opening. The track and hangers are hidden by a canopy or valance which may be metal or plastic, or custom made of wood in any style or finish. The bottom of the door is restrained by any of several types of guides or tracks. Usual practice is for the purchaser to select the door, then match the required track hardware to it. The only other item needed is a suitable knob or pull; no latches are used. This type of door is usually installed as a single unit, but double center meeting doors might be installed over a large opening.

Bypass Sliding Doors

Bypass sliding doors are installed in pairs within a plain door frame (Fig. 4-66). The doors can be of any rigid type, but flush doors are the most common choice. No stops are needed, but moldings may be added to conceal the closure joints at the side jambs and the overhead track and hanger assembly on the head jamb. Alternatively, the door edges a the

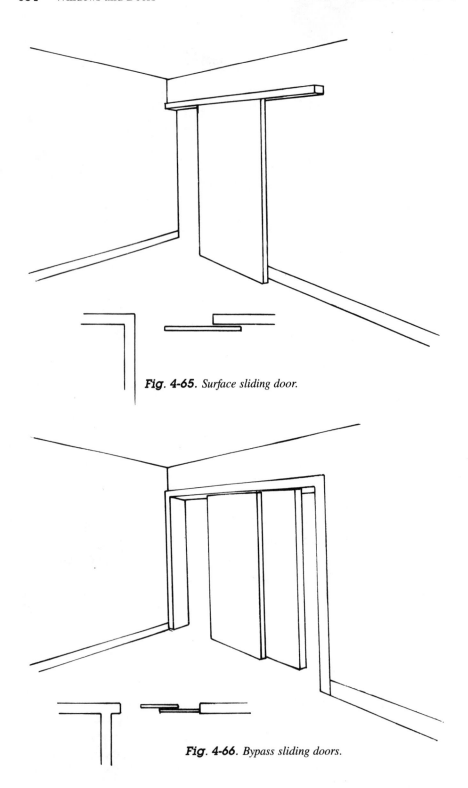

Fig. 4-65. Surface sliding door.

Fig. 4-66. Bypass sliding doors.

closing sides may slip into shallow, full-height mortises fashioned in the faces of the door frame side jambs. The doors are suspended on adjustable hangers from a double track attached to the face of the head jamb. The bottoms are restrained by guides or a track that is either surface-mounted or mortised into the flooring. Usually the doors, track hardware, and knobs or pulls are all bought separately.

Pocket Sliding Doors

A pocket sliding door is the most difficult to install and requires a special structural and door framing (Fig. 4-67). Because the door, when open, is fully enclosed within the wall, the wall must be thicker than normal and double-framed to form the door pocket. This involves building two individual but interconnected stud frames, separated by the thickness of the door plus a clearance allowance, then building the door frame with a slotted side jamb at one side and a full-width opposite side jamb. The head jamb is usually constructed to conceal the track and hanger assembly, and the unslotted side jamb may contain a shallow mortise into which the closing edge of the door slides. This operation hides the closure joint. The casing trimwork can be done in whatever manner is desired, following conventional practice. For the sake of appearance, the entire wall that contains the door must be of the same thickness. For small openings a single door is installed, but for larger ones, a center meeting pair sliding into pockets on each side of the opening is a

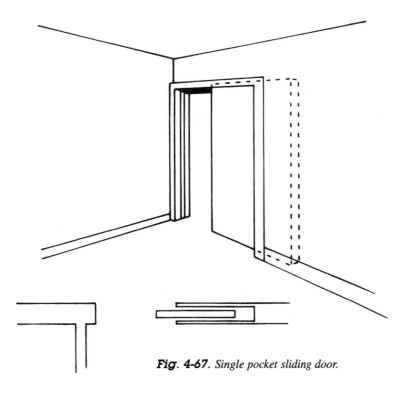

Fig. 4-67. *Single pocket sliding door.*

more satisfactory arrangement. The advantage of the pocket door is that it opens 100 percent of a doorway area, yet interferes with no part of the living area.

Any rigid type of door can be used for this purpose. A popular choice is a 1³/₄-inch-thick, 7-foot exterior panel door. The hardware consists of a single track and hanger assembly, plus a track or guides for the door bottom, and the door pulls. Surface-mounted knobs or pulls cannot be used; they must be the recessed type. If the door is designed to slide back entirely into the pocket, a special folding lever can be mortised into the leading edge. This allows the door to be pulled out far enough to grip the pull recessed in the face of the leading stile. Special locking mechanisms are also available to lock a single door into a side jamb slot, or to lock together a pair of center meeting doors. All of the materials must be matched up and bought separately.

Patio Doors

A patio door (Fig. 4-68) is supplied as a complete assembly (although it might be partly knocked down for shipping) within a full frame, including sill. Aluminum and vinyl-clad wood models do not include any casing trim, while all-wood models might or might not. Installation consists of inserting the unit into a framed rough opening of specified size, adjusting for level and plumb, and securing it—similar to window installation. The exterior casing for aluminum units is applied to overlap the frame of the unit, after it has been properly sealed into the opening. The interior casing is installed in the same fashion, but might require two or three built-up moldings for full coverage and a complete trim-out. With vinyl-clad wood units, the exterior siding can abut the door frame, or wood

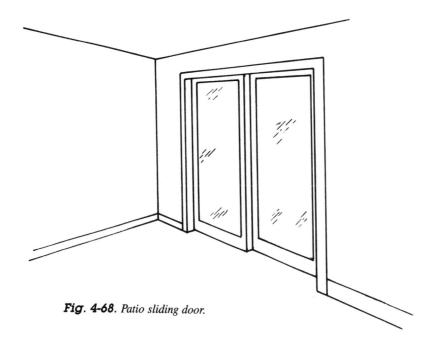

Fig. 4-68. *Patio sliding door.*

moldings can be added alongside the frame for extra emphasis, with the siding butted to the molding. The interior casing is applied directly to the inner edge of the door frame, just as with any other door frame. All-wood units can be cased in the usual fashion.

RETRIMMING DOORS

There are several approaches to changing the appearance of a door or doorway: Add moldings to an existing casing, replace an existing casing with another of a different size or configuration, install a finished door frame and casing where none previously existed. If no door was installed, as is often the case when a doorway has no wood finish frame, you could install one. This procedure is covered later in the chapter. You can also cover an old panel door with a veneer to create a flush door, or add moldings in a pattern to a flush door to change its look entirely. And lastly, you might replace an old door with a new one of a different style.

Adding New Moldings to Existing Casing

Adding new moldings to an existing casing requires that the original be in decent or repairable condition—tight and well aligned, and not badly gouged or otherwise damaged.

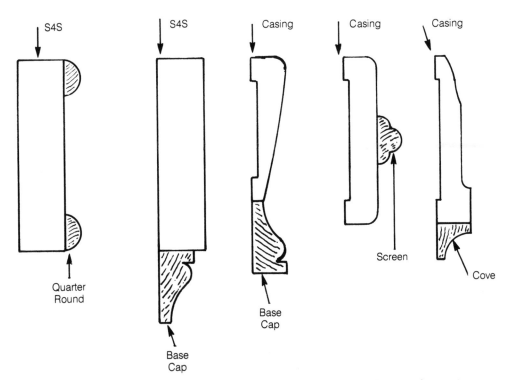

Fig. 4-69. *A few of the numerous possibilities for changing the appearance of existing door casings by adding one or more stock moldings.*

If the casing is made up of flat-faced S4S stock, the job is easy. A molded pattern, especially if irregular, makes the job more difficult, but good results are still possible. Figure 4-69 shows some possible combinations. Check with your local supplier to see what molding patterns are readily available, then work out an arrangement that appeals to you.

Regardless of the pattern of the existing casing, the first step is to patch the holes, cracks, or gouges with wood putty or vinyl paste spackle, then sand the patches smooth. If the old finish is glossy or rough and full of dust nibs, sand the surface with fine sandpaper. Prime or seal the new molding stock, and touch-sand the surfaces smooth afterward. Then trim and fit the pieces, and secure them to the casing with finish nails of appropriate length. Don't skimp on nails; space them every few inches to ensure that the moldings will lie flat, tight, and straight. Set all the nail heads and fill the holes with vinyl paste spackle, sand smooth, and apply a finish coating to the whole affair.

Replacing the Casing

Often it is easier and more effective to replace the casing. First check the casing/wall joints and also the casing/door jamb joints. If a layer of paint solidly covers those joints so that the casing is stuck down, slit the joints open with a razor blade or utility knife. To remove the casing, first check the joints at the top corners. If they are butt joints, the head and side casings will not be nailed together. Pry the head casing off, working first from one end and then the other, until it comes free. Use a broad, chisel-bladed prybar with a wide piece of scrap wood under the heel to protect the wall. Then pry the side casing loose, starting from the top and working your way down.

If the top corner joints are mitered, the head and side casings might be nailed together through the edges, close to the corners. In that case, you will have to pry alternately at the head casing ends and the tops of the side casings until you can separate them. The casing will be nailed to the edges of the door frame and side jambs, as well as into the wall rough framing. Be careful not to split away or damage the jambs as you pry, or those will have to be replaced, too. Sometimes it is possible to pry the casing partway off, then tap it back into place, leaving the nail heads sticking out. You can then pull the nails with a claw hammer. The bottom ends of the side casings might run below the level of the finish flooring. With careful prying and pulling you should be able to break them free without causing any damage. If there is a shoe molding along the baseboard that extends across the casing bottom ends, that will have to be removed first.

There probably will be a ridge of old paint left along the edges of the door jambs where the casing edge was. If so, scrape it flat and smooth with a knife or wood chisel blade. Cut and fit the new casing material next. If the walls are not to be refinished, the casing stock must be wider than the old, to completely cover the original casing location and extend out onto the existing wall finish. Fasten the new casing with finish nails long enough to extend about 3/4 to 1 inch into the door jambs, and at least 1 inch into the rough framing. You can use either miter or butt joints at the corners, depending upon the molding shape. Set all the nails, then prime or seal all raw wood. Fill the nail holes with vinyl paste spackle, and at the same time make any necessary repairs to the door jambs and stops. Sand the whole assembly smooth. If the old finish is glossy, degloss it by sanding with fine paper or by applying a liquid deglosser. Finally, apply the finish coatings.

Veneering and Laminating Door Faces

You can rejuvenate an old panel door, cover over a louver door, or rehabilitate a damaged flush door by veneering one or both faces. If a flush door surface is perfectly smooth, or can be repaired to be so, you can cover it with a thin wood veneer (which vary from $1/28$ to $1/64$ inch), or a sheet of plastic laminate. A wide variety of laminates are available, and many different domestic and exotic wood veneer sheets can be obtained from specialty wood dealers. The finished results can be both practical and handsome. A thicker overlay is needed for panel or louver doors, such as 5-millimeter or $1/4$-inch plywood, or $1/8$- or $1/4$-inch hardboard.

The first step is to remove the door (see the next section). For the best and most permanent results, strip the paint from the stiles and rails of a panel door with a remover, or by sanding or scraping. If the door is a flush type, strip a band about 6 inches wide around the perimeter, plus two or three intermediate stripes across the door. Or strip the whole door if you prefer. Exception: If veneer or laminate will be applied, strip the door faces completely and sand them smooth and clean, leaving as few traces of old finish as possible.

Cut one panel of the covering material to size. For best results, make it $1/16$ to $1/8$ inch oversize all around, to be trimmed in place later. For materials other than thin veneer or laminates, lay the door on a flat, solid surface and apply a thin layer of aliphatic resin (carpenter's) glue to the stripped area of the door and to a like area on the underside of the panel. Lay the panel in place and align it carefully, then secure it with short brads around the perimeter and across the rails and stiles, and set the nail heads. If the door is a flush hollow-core, however, nail only around the edges—there is nothing to nail into elsewhere. Then lay a clean, smooth sheet of thick plywood, a series of 2 × 4s, or some other load-spreading layer over the door. Place heavy weights on top distributed as evenly as possible. Allow a day for curing.

If the panel is oversized, trim the edges flush with the door edges using a hand or power plane, or better yet, a router fitted with a straight edge-trimming bit. Follow up by beveling or rounding the sharp corners slightly to a $1/16$- to $1/8$-inch radius. If required, flip the door over and repeat the process on the other side. Finally, fill the nail holes and any gaps along the door/panel edge joint with vinyl paste spackle or wood filler, then apply a finish (Fig. 4-70).

Note that this job can also be accomplished without stripping off the old door finish and without glueing. Just cut and fit the panel, nail it on, trim and round the edges as necessary, fill the nail holes and sand, and apply a finish. However, the results of this procedure are seldom as satisfactory. Gaps around the edges are likely to appear, and because of the thinness of the panels, they are susceptible to warping or buckling, which creates a wavy effect that cannot be cured.

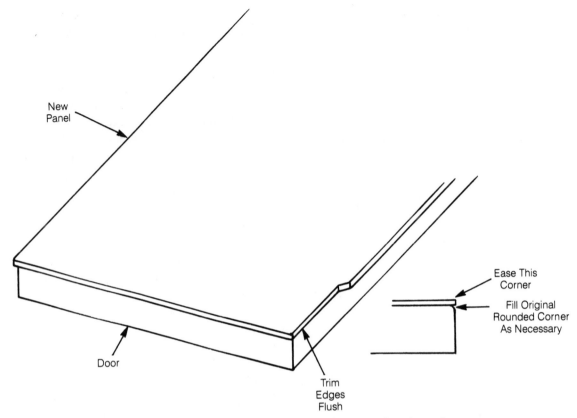

Fig. 4-70. *An old door can be covered with a new facing panel to change its appearance.*

If you are applying a plastic laminate, follow the usual installation procedures for that material. Apply contact cement to the mating surfaces, following the manufacturer's instructions. After the cement cures, lay the panel over a row of sticks or a layer of waxed paper on the door and position it. Maintain the position and slowly remove the sticks or paper until the laminate lies flat. Then roll or "beat" the surface, as is done with tile flooring, using a fair amount of pressure to form the final, complete adhesive bond. Trim the edges with a laminate trimmer router fitted with a carbide bit.

If you are applying a veneer, follow the usual techniques for that process. It is a rather specialized operation that requires some study, practice, and preferably a little experience. Several sheets of veneer might have to be fitted together to cover an entire door face, and you will need some special tools. However, you also have the opportunity to create some stunning natural finish patterns in matched or contrasting woods, as well as some fancy inlay work.

Regardless of the kind of door or panel material, the finished product will be thicker than the original, which means that it will no longer fit properly in the door frame. To solve this problem, first pry loose the original stop molding from the door jambs. Scrape and sand down the paint ridges that remain on the jambs. Replace the stops. Set them forward on the jambs a distance equal to the amount of thickness you have added to the door, plus a clearance allowance of about $1/16$ inch.

Cut the hinge mortises in the door edge back to the inner face of the door, and fill the narrow gap left in the mortise on the opposite side of the hinge leaves. If the difference between the new and old hinge positions is slight, however, you may not be able to drive the hinge into new holes. They will drift into the old holes and the hinges will not seat properly. In that case, move each hinge position up or down enough to allow you to drill new holes well away from the old ones. Fill the mortise gaps that remain with thin pieces of wood glued and nailed in place and sanded flush. Recut the mortises in the frame side jamb to corresponding positions as necessary, attach the hinges, and hang the door. The door face should be flush with the outer edges of the door jambs, with about $3/32$ inch of clearance between the door and the jambs, and $1/16$ inch between the inner door face and the stop.

Adding Trimwork

Changing the appearance of a flush door by adding trimwork is easy, and fun (Fig. 4-71). All you need is a few lengths of either stock or custom-milled molding. You can use single pieces, or combine molding for a different effect, and place them in whatever arrangement your imagination suggests. No alterations need be made to the door hardware, but for best results the door should be taken down, the hardware removed, and the old finish stripped off.

Work out the pattern or paper first, complete with dimensions, then transfer it full-sized to the door face. Cut the pieces of molding accordingly and lay them out on the door face to make sure they fit up well and the cuts are correct. With most layouts there is a logical progression in fitting the pieces together; set them aside in order. If the door is a solid-core type, fasten the first piece with a thin coat of a aliphatic resin glue and set nails every few inches. If the door is a hollow-core, you will have to rely mostly on the holding power of the glue. Drive an occasional brad primarily to hold the pieces in position, or use a few short, small flat-head wood screws and countersink them. Work out from the first piece to complete the pattern. Fill the nail or screw holes and sand smooth. Repeat the process on the other door face if desired, then apply a new finish.

If you want to use a different finish on the moldings than on the door, or prefer not to strip the original finish from the door, the procedure is a little different. If the new finish is the same color as the old and will be applied over it, first make any necessary repairs to the door faces and sand or degloss the surfaces as necessary. Apply a primer or sealer to the lengths of molding, then cut and attach them to the door as just described. Coat the whole assembly with a new finish. If the molding will be a different finish, prime or seal the molding lengths, then apply one or two coats of finish as necessary. Cut and attach the pieces to the door faces, and fill the fastener holes with glazing compound. Smooth the compound off flush with your fingertip or an artist's palette knife. Let it harden for about a week, then touch up the patches and joints with a small art brush. Alternatively, you can

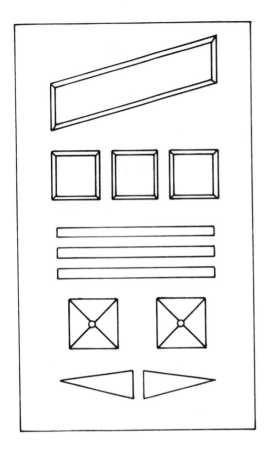

Fig. 4-71. *An existing plain flush door can be perked up by trimming it with any of several different kinds of stock moldings; many patterns are possible.*

reprime the molding, cut and install the pieces, fill with vinyl paste spackle and sand smooth, then apply masking tape all around the edges of the molding and finish it with a small brush. Finally, remount the hardware and rehang the door.

DOOR AND/OR FRAME REMOVAL

Single-Swinging Doors

Single-swinging doors are usually mounted on mortised butt hinges, sometimes on antique style (Colonial or Early American) surface-mounted L, H, or strap hinges. To remove a door hung on surface-mounted hinges, close the door and slide scraps of wood or other wedges under the bottom edge until they are barely snug and in a position to take the full weight of the door. Remove the screws from the portion of the bottom hinge that is attached to the casing. Then take out the corresponding screws at the top hinge, holding the door in place against the stop as you do. Release the latch and, using the knob as a handle, open the door slightly. Pull or tilt it toward you until you can get your fingers around the hinged edged and lift it free.

If the door has butt hinges, check to see if they are the loose-pin type; most are. There will be a small hole in the bottom of each hinge barrel. Close the door and set wedges beneath the bottom edge. Tap a 6d or 8d nail up through the hole in the bottom hinge, forcing the pin upward until you can pull it free. Keep a knee or shoulder against the door to hold it, and tap the top pin out. Then release the latch, open the door a bit, and pull it outward until the hinge knuckles separate, and you can lift the door free.

If the hinges are fixed-pin, open the door as far as it will go and set wedges snugly beneath the bottom edge. Remove all the screws from the leaf of the bottom hinge that is attached to the door jamb. Then take out the top screws, starting with the bottom one in the leaf and working up. As you break the top screw free, grasp the top edge of the door with one hand and steady it. When the last screw is out, lift the door away from the jamb.

Folding and Sliding Doors

To remove an accordian folding door, locate the screws on the inside of the door that secure one edge to the door casing, and remove them. Lift the door and hangers off the overhead track, then remove the track and latch hardware.

Most surface and bypass sliding doors suspend from hangers that roll along an overhead track. Depending upon the hardware used, it might be first necessary to remove a bottom guiding system, or it might not—there might not even be one. If possible, release the door by grasping each side and lifting it a half inch or so, meanwhile tilting the top of the door inward slightly until the hangers or rollers slip out of the track. Or, you might have to release the mechanism first from the inside on a by-pass door, or remove the track valance from a surface door.

Removing a bifolding or multifolding door set can be tricky, especially if the panels are large and heavy. They tend to flop around and you loose your grip on them. It is advisable to have a helper, who can hang onto the doors while you release them. At a bottom corner next to the door frame there is an adjustable pivot pin. This pin screws up or down, and its base fits into a shallow locking socket. At the top corner there is a spring-loaded straight pin extending up into an adjustable socket that locks into the overhead track. At the opposite top corner of the door set there is either a spring-loaded roller that travels in the track, or another spring-loaded pin that fits into a socketed slide in the track.

If the door is a small, lightweight set—a pair of 1-foot-wide louvers, for example—open the door about third of the way. Grasp the side edge of the open panel with one hand and top edge of the pivoted panel with the other. Lift straight up, and at the same time, ease the bottom of the set out toward you until the adjusting screw comes out of its socket. Then bring the door bottom upward and downward until the top pins or rollers come free of the track. If the door set will not lift high enough to release the bottom adjusting screw, slide a length of 1 × 2—or something similar that will not mar the finish—under the bottom edge of the door, next to the adjusting screw. Pry upward with the 1 × 2 just enough to take the weight off the screw, and run the screw down to gain more clearance at the top. Then lift the door out.

The approach for larger, heavier sets is a bit different. First, open the door set partway and push the track roller or pin at the upper corner of the open panel straight downward. If it is a pin, you might have trouble getting hold of it to do so. If the pin or roller will clear the bottom edge of the track, hold it down and swing the panel outward to free it, leaving the whole set swinging on its corner pivots. If the pin or roller will not clear, slide a 1 × 2 under the bottom edge of the door, next to the bottom corner adjusting screw, and pry the door upward just a bit. With the pressure off the adjusting screw, run it down to gain more top clearance until you can free the pin or roller from the track. Fold the sections almost together, and have a helper steady the set by hanging onto the outside edges near the top. Then pry upward at the bottom corner until the adjusting screw releases from its socket. Pull the bottom corner outward a bit and lower it toward the floor. As you do, the top pivot will release from its socket and the whole unit will come free. Be prepared to hang onto it, and watch your fingers.

These are the easy removals. Sometimes doors are fitted with clearance allowances so slight that the installer can barely cram them into place. Sometimes they are fitted and hung, then a skirt molding is installed around the jamb to hide the closure cracks and the overhead track. Getting the door out again without causing any damage can be very difficult. If the methods previously discussed will not work, try this: First, back the adjusting screw down as far as it will go, to gain maximum top clearance. Put your helper inside the closet, preferably with a light, and close the doors. Slide several props of scrap wood under the door edge to minimize, as much as you can, the distance the door can drop. This will also give your fingers clearance. Insert a 1 × 2 pry stick under the middle of the pivoted panel, and get a firm grip on the other panel bottom with the other hand, as far away from the pry as you can comfortably manage. Boost the set straight up, compressing the top pins or rollers into the track, pull the panels outward at the bottom, and have your helper restrain and then grab the top of the set is it tilts inward and drops free. Open the set slightly (that is, fold the panels a bit) to clear the jambs at the sides, and slide the whole set outward at an angle on the wood scraps until it is clear of the head jamb, then stand it up.

To make matters worse, you might have only a tiny bit of clearance at the door bottom. In that case you will have to do the prying with a pair of screwdrivers, chisel-bladed prybars, or whatever will fit. The process is the same, just a little trickier and more difficult. In any case, the thing to watch out for is the doors suddenly folding up or flailing around, or unexpectedly falling off the track.

Door Finish Frame

To remove a door finish frame, first strip off any remaining hardware, then take off the casing. If there is a stop molding attached to the jambs, pry that off. The next steps depend upon how the frame was put together and installed. Salvaging it intact is unlikely; the main object is to get it out without causing any damage to surrounding materials. If there is a threshhold across the bottom with ends butting against the faces of the side jambs, remove that first. If the threshold ends go under the jamb ends, leave it for now. If the ends of the side jambs rest upon the finish flooring, carefully pry the side jambs outward. Start at the bottom and work upward until you can free the side jambs enough to pull and twist the top ends free from the head jamb. Then pry the head jamb away from the rough framing . However, if the jamb bottom ends extend below the finish floor level, this maneuver won't be possible.

Check the joints at the top corners. If the side jamb ends butt up against the head jamb, you can pry them outward at this point. The corner nails will split out of the side jambs, given enough pressure. Work downward until you can pull them free. Then pry the head jamb off. If the ends of the head jamb butt against the side jambs, or the two are joined in a rabbet or dado joint, this will not work either.

The final recourse is to saw through the side jambs at any handy point. Pry the bottom pieces off the rough framing and pull or cut whatever nails are necessary to allow you to pull the pieces out of their sockets in the flooring. Then do the same with the upper pieces and wrench them free of the top corner joints. You might have to work the head jamb loose at the same time, pulling or cutting off nails as necessary. If there is a fitted sill or threshold, pry that up last. It should come up readily. If it does not, examine the top surface closely. Some thresholds are installed by driving screws into counterbored holes and then covering the heads with matching wood plugs. Here you have three choices: Locate the plugs, chisel the wood away, and remove the screws; split the wood into chunks with a chisel; or keep prying mightily until something gives.

DOOR AND/OR FRAME INSTALLATION

Replacing a Single-Swing Door

To install a wood single-swinging door in place of an old one, first measure the width of the opening from jamb face to jamb face, and the height from the floor to the face of the head jamb. Select a new stock door that is no less than $1/16$ inch smaller in both dimensions at any point. An exact fit is unlikely, and you will have to trim the new door. Most wood doors can be trimmed at least 1 inch on each side and the top, and as much as 5 or 6 inches at the bottom. Be sure to check the manufacturer's trim allowance figures before buying.

Next, inspect the old hinges. If they are loose-pin mortise butt or surface-mounted hinges in good condition, and you wish to reuse them, that presents no problem. If they are worn, paint-covered, or the fixed-pin variety, replacement is advisable. If possible, select new hinges of the same leaf size as the old ones, which will save some work. If the old lockset is also in good condition, you might wish to use that as well. To do so, you need to have sufficient carpentry skills to install the old lockset in the new door, using the old door as a guide. If you select a new lockset, it will come with complete installation instructions and a dimensioning template, making the job a good deal easier.

The first step is to size the door. Check all four corners of the opening with a framing square. If the opening is in square or just a little bit out, you can make your initial trim cuts square. If the opening is badly out of square, as is sometimes the case in an older house, make your initial trim cuts correspondingly out of square. Use an adjustable angle measuring device to measure each corner, and transfer the angles to the door. When the installation is finished, you should have a clearance gap of $^1/_{32}$ to $^1/_{16}$ inch on the hinged side, $^1/_{16}$ to $^1/_8$ inch on the latch side, and at least $^1/_{16}$ inch at the top of the door (Fig. 4-72). The minimum clearance at the bottom is usually about $^1/_4$ inch over the finish floor for an interior door when it is closed, but this is variable. It can be set to whatever you wish, and more clearance will be needed if the floor rises or humps along the opening path of the door.

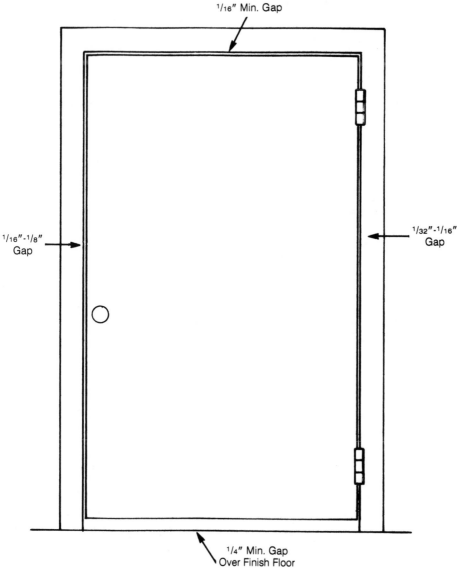

Fig. 4-72. *These are the usual clearances for single-swinging hinged doors.*

Lay the new door flat on a bench or sawhorses and use a straightedge to mark the trim lines. When trimming, remove an approximately equal amount from both side edges and reduce the height by trimming the whole amount from the bottom, if you can. Trim the top edge only slightly when necessary to compensate for an out-of-square head jamb. If more than about $3/8$ to $1/2$ inch of material must be removed at any point, it is easiest to make the first trim cuts with a power saw, cutting to the outside of the trim line. Otherwise, use a hand or power plane. Do the final trimming in several steps, standing the door in the frame between steps to check the fit. Remove the material gradually, compensating for slight irregularities in the frame as you do so (the jambs may be slightly bowed, for instance). Once the door is fitted, ease the sharp edges slightly with sandpaper or a block plane to a $1/16$- to $1/8$-inch radius.

Next, stand the door on its side with the hinge edge upward and lay out the hinge outlines. In most installations the top of the upper hinge is located 5 to 8 inches down from the top edge of the door, and the bottom edge of the lower hinge is located 10 to 11 inches up from the bottom edge of the door. If there are other doors in the room, however, locate the hinges to match those. If a third hinge is needed, center it between the top and bottom hinges.

Remove the pins from the hinges to separate the leaves, and use one to trace around at each location. In most cases the straight edge of the leaf will lie aligned flush with the door edge. With a router or a hammer and a sharp $1/2$- or $3/4$-inch wood chisel, cut a mortise to a depth equal to the thickness of the leaf. Set a leaf with two barrel knuckles in the mortise and mark the centers of the screwholes. Drill a pilot hole at each point, then drive the screws home. Take great care to run the screws in straight and centered, otherwise the leaf will be cocked and you have a terrible time hanging the door. The centerline of the knuckles (and the leaf edges) must be perfectly aligned with the door edge.

Repeat the process with the other hinge(s). Then lay out the leaf locations on the door jamb by tracing around one of them. The leaves must be set so that when the door is closed its outer face is flush with the outer edges of the jambs. Your measurements must be exact, and bear in mind the top clearance allowance. If your top door leaf is 7 inches down from the door top, the top of the upper jamb leaf (which has three knuckles) must be 7 inches plus a minimum $1/16$ inch down from the face of the head jamb. The distance from the bottom of the upper jamb leaf to the top of the lower jamb leaf must be exactly the same as between those of the door leaves. Cut these mortises into the door jamb and attach the jamb leaves in the same way as the door leaves. Figure 4-73 shows a properly installed hinge.

Next, install the lockset. Typically, these are centered 36 to 38 inches above the finished floor level, but this is variable. If the door has a single cross rail, usual practice is to center the lockset on the rail. If there are other doors in the room, set the lockset to the prevailing height, or close enough to it that the difference is unnoticeable. Installation details vary; follow the manufacturer's instructions, or duplicate the old installation.

Now comes the moment of truth. Put the hinge pins where you can reach them easily, stand the door up, and work it into the opening. Grasp the door by the edges and fit the

Fig. 4-73. *A correctly installed door hinge.*

hinge knuckles together so that they all line up. If your measurements and mortising were accurate, they will mesh nicely. Slip the top pin into the barrel, then the bottom one, and tap them home with a hammer if necessary. If the knuckles will not mesh, set the door aside and loosen the screws in the jamb leaves until the leaves are just a bit loose in the mortises, and try again. If they go together, install the pin, open the door carefully, then tighten the screws. If they do not, check your measurements and reset the hinges as necessary. With the door hung, close it and check the clearances all around. You might have to take the door down and do more edge planing in places to achieve a proper fit.

The next consideration is the door stop. If you are installing the door in an old frame with no stop or in a new frame, you will have put one in. If there is an existing stop, it might turn out to be too snug or too loose for the new door. To install a new stop, close the door so that the face opposite the stop location is just flush with the jamb edges. Mark the stop location on the hinge jamb face, allowing about $1/32$ inch of clearance. Some workers prefer to wedge the door shut in its proper position with shim shingles, then install the stop molding using the closed door face as a guide—which is an easy course to follow. Just cut and fit the side and head stop pieces and nail them in place, following the door face with a bit of clearance. In the case of an improperly positioned existing stop, remove the molding (you probably won't be able to save it) and scrape and sand down the old paint ridges. Then install a new molding in the proper position.

Now locate the striker plate and box—if your lockset has a striker box, not all do—and install them according to the instructions. The spring latch should fit snugly against the striker plate edge and the door edge snugly against the stop so that the door doesn't rattle back and forth. The fit should not be too snug, however—remember that two or three coats of paint will take up some space later. Finally, take the door down, remove all the hardware, and sand and finish it. Allow plenty of time for the finish to cure before putting the hardware back on and rehanging it, to avoid damage from hand pressure on the fresh, soft coatings.

Installing a New Finish Door Frame

There are two situations that call for the installation of a new finish door frame. One is when a structurally framed rough opening, either interior or exterior, for a door or doorway must be finished out. The other is when a finished but unframed interior doorway is to be fitted with a door that requires a frame, or when that opening is to be fitted with a conventional finished frame and trimwork as a matter of style. In the latter circumstance, sometimes doorways are made by simply wrapping the plaster or plasterboard around the rough opening framing as a continuation of the wall, with no conventional wood jambs, casing, or other trimwork. Many find the appearance undesirable, and subsequent door installation may be difficult or impossible, depending upon the kind of door installed.

A complete interior door frame consists of nine parts: two side jambs and a head jamb, four side casings, and two head casings. Various kinds of molding or other trimwork may be added to these basics. An interior door frame can have a threshold added after installation, over the finish flooring. An exterior door frame is usually fitted with a sill that rests upon the outside floor joist and butts flush against the finish flooring, and is capped by a threshold installed afterward. All of these part are usually sold separately, with selection up to the installer. Sometimes the parts are sold in jamb or jamb-and-casing bundles, for assembly at the job site. The exception is prehung doors, which are factory assembled units ready to plug into the rough openings.

To assemble an interior door frame, cut the side jambs to the proper height to fit in the door opening. Cut the head jamb to fit between them so that the width of the opening equals the width of the door. You can include the door clearance allowances in the width if you wish, but many installers prefer to make the frame a bit tight and trim the door to fit the installed frame. The theory is that you can easily trim a door a bit, but if you miscalculate and the frame opening is slightly wide, you can't very easily add to the door width.

Lay the side jambs on the floor on edge and set the head jamb between them, and nail and glue the top corners together. Aliphatic resin (carpenter's) glue and 6d finish nails will do the job. Sometimes the corner joints are simple butts, but rabbet or dado joints are preferred (Fig. 4-74). Stand the frame in the rough opening and center it, wedging it in place just below the top corners with shim shingles run in over one another from opposite sides. Drive a pair of 8 finish nails partway in—through each side jamb, and shim set into the rough frame—making sure the top frame corners stay in square and the edges are flush with the wall surfaces on each side. Then shim and nail the same way about 4 inches above the floor, so that the distance between the side jamb faces at this point is exactly the same as at the top. Drop a plumb line down the edge of each jamb and check the squareness at the top corners again; adjust as necessary. Then space three more sets of shim shingles at intermediate points along each side jamb, securing them with a pair of partly driven 8d finish nails. Use a plumb line or long straightedge and adjust the shims until both jambs are straight up and down, and at no point bow either in or out (Fig. 4-75).

Drive all the nails almost all the way in and check again for plumbness and straightness. Adjust if necessary, then drive the nails home and set the heads. Place another set of shims at the center of the head jamb for a narrow door; use two spaced sets for a wide door. Wedge the shims together until the head jamb is straight and level, bowing neither up nor down, and drive a pair of 8d nails at each point. Then install the casing on each side. The casing edges may be flush with the jamb faces, but are usually set back about $1/4$ inch or so (Fig. 4-76). If a threshold is needed, trim this to fit between the side jambs before installing the stops, and fasten it to the floor with finish nails or screws covered with wood plugs.

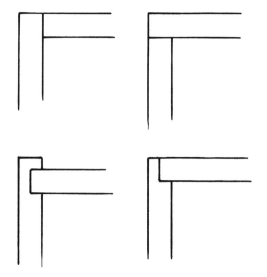

Fig. 4-74. *All of these joints can be used to join a door head jamb to the side jambs. The butt joints (top) are not recommended, the dado (bottom left) is the strongest, and the rabbet (bottom right) is perhaps the most common.*

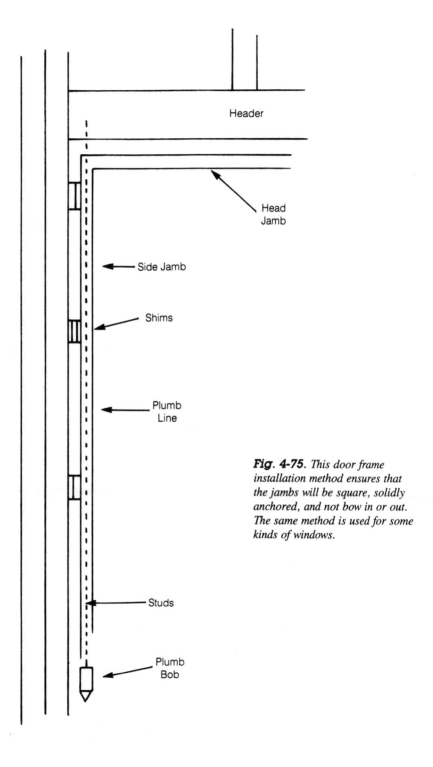

Fig. 4-75. *This door frame installation method ensures that the jambs will be square, solidly anchored, and not bow in or out. The same method is used for some kinds of windows.*

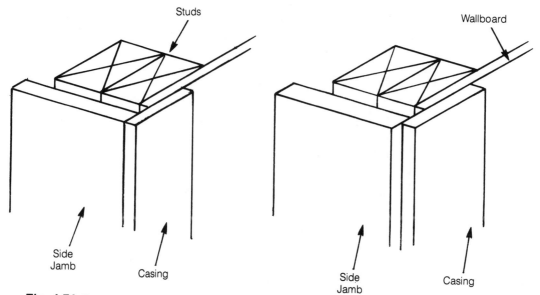

Fig. 4-76. *Door casing edges can be set flush with the jamb faces (left), but fitting and finishing is usually easier if they are set back slightly (right).*

Installing Exterior Door Frame

The procedure for installing an exterior door frame is much the same. The major difference is that a sill must be attached to the bottom ends of the side jambs with nails only (or you can add a thin coat of construction adhesive, silicone caulk, or waterproof glue). Some exterior sills are thin and lie flat on the subflooring. They butt up against the finish flooring or overlap it slightly with a rabbet joint (Fig. 4-77). In this case, the bottom ends of the side jambs are cut square and the height of the frame is calculated from the subfloor level. The sill is attached between the jamb faces, butt jointed. The outside edge of the sill extends beyond the face of the exterior siding. To install the frame, stand the unit on the subfloor and center it within the rough opening. Fasten the sill through the subflooring and into the joist below. Then square the frame up by shimming at five or six points along the side jambs, working upward from the bottom, and fasten it in place. Install a weatherstripped threshold (sometimes called a sill cap) and the casing.

An exterior door frame with a thick, wide, slanted sill is more difficult to install. The flooring and subflooring must be cut back and reinforced at the cut edges if the sill lies parallel to the joists, to accomodate the inner end of the threshold (Fig. 4-78). The header joist must be notched to bear along the lengthwise centerline of the sill. The side jambs are typically trimmed to butt against the sill surface, but can be butted against the sill ends. The sill itself is usually trimmed with "ears," the inner edges of which lie flat against the exterior sheathing. The ears extend to each side so that their ends align with the outer edges of the exterior casing (Fig. 4-79). The sill may be installed separately, followed by the jambs, or the jambs and sill may be put together and then installed.

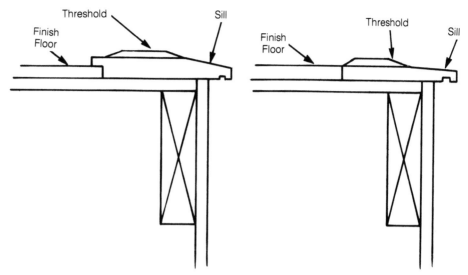

Fig. 4-77. *Some exterior door sills are rabbeted over the finish floor (left), while others butt against it (right).*

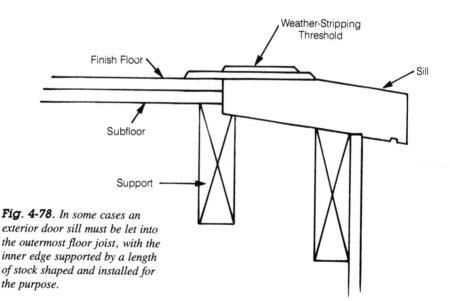

Fig. 4-78. *In some cases an exterior door sill must be let into the outermost floor joist, with the inner edge supported by a length of stock shaped and installed for the purpose.*

SPECIALTY DOOR INSTALLATION

Double-swinging (either full size or batwing), bifolding, multifolding, and accordian doors are best installed in doorways fitted with cased finish wood frames that do not have stops. Surface-sliding doors can be mounted over doorways with cased finish wood frames or those finished with plaster or gypsum wallboard surrounds that form a continuation of

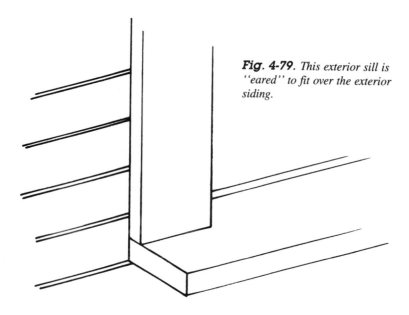

Fig. 4-79. This exterior sill is "eared" to fit over the exterior siding.

the walls. Bypass sliding doors are usually installed in a cased finish wood frame, but not always. Pocket sliding doors require their own special construction. Patio doors are prehung within a frame and are installed directly into a rough opening in the structural framing of the house.

Double-Swinging Doors

Full-sized double-swinging doors are hung on special heavy-duty pivots (Fig. 4-80) that usually come complete with installation instructions. Typically, the pivot pins are mortised into the upper and lower edges of the hinge stile of the door, close to the corners. The upper pin socket is mortised into a corresponding location centered fore and aft in the head jamb corner. The lower pin socket may be surface-mounted on or mortised into the floor.

The door frame must be square for best results. If it is out of square, the door must be carefully trimmed to the opening. The stile edges are usually rounded to a much greater degree than on single-swinging doors, to allow ample swinging clearance at the side jambs. Top clearance should be at least $1/8$ inch. Clearance at the sides is usually at least $1/4$ inch, but can be more at the pivot side, depending upon the nature of the pivots and the way they are installed. Bottom clearance is usually a minimum $1/2$ inch, but can also be more if desired.

Installation is only a matter of sizing the door to the opening, installing the pivots according to instructions, mating them, and making whatever adjustments are provided for. After finishing, push plates are usually installed on the free side stile at about chest height. Kick plates may also be installed on the bottom rail.

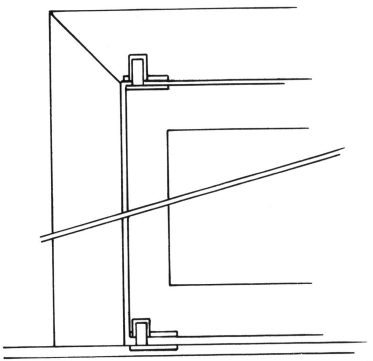

Fig. 4-80. *A double-swinging door is mounted like this, with pivot hinges at top and bottom.*

Batwing or cafe doors are smaller and lighter, so are mounted in pairs on lightweight bracket pivots that attach directly to the side jambs (Fig. 4-81). Trim the doors to fit the opening, allowing a $1/16$-inch minimum clearance gap between them at the center, plus the hardware manufacturer's specified clearance at each side jamb. Trim for an oversize fit at first, so the doors will overlap slightly at the center. After hanging the doors, retrim the meeting stiles for a final fit. If the door jambs are slightly out of plumb, you will be able to compensate during the trimming so the center gap is uniform. (There is no provision for adjustments on the pivots.)

Mount the pivots on the top and bottom edges, close to the outboard corners of each door, according to the hardware installation instructions. Determine the height of the door tops and mark one door jamb accordingly. Mount the upper and lower pivot brackets, carefully spaced to match up with the pivots, centered on the vertical centerline of the door jamb. To mount the pivot brackets for the second door, do not measure up from the floor or down from the head jamb to locate their positions. Instead, set a long spirit level or a level taped to a straightedge across the opening. Align it with the first bracket set and transfer their locations, on a level line, to the opposite jamb. Mount the second set of brackets at these locations. Make sure that the brackets are centered plumb, one above another, and are not tilted.

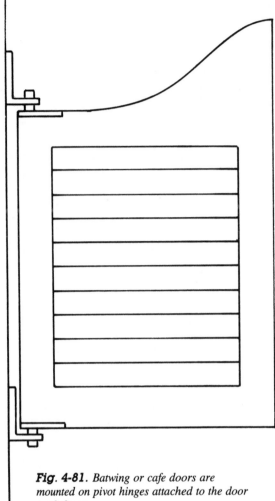

Fig. 4-81. *Batwing or cafe doors are mounted on pivot hinges attached to the door jamb faces.*

To hang the doors, tilt the top pins upward into the top bracket holes, lift the doors up to the end of the pin travel, insert the lower pivots into the lower bracket holes, and drop the doors down. Finish trimming the meeting stiles next. Then, if the doors hang a bit crooked and the meeting corners do not align, plane off the top/bottom edges until they match.

Bifold Doors

Bifold doors and door hardware kits can be bought separately or as a package. In the latter case, some of the hardware might be preinstalled. In either case, installation instructions are included. Although there are some variations, depending upon the exact type of doors and hardware, the basics are as follows.

First, the doors must be trimmed as necessary to fit the opening. If the door opening is square, all that is necessary is to trim the doors square. Leave the specified clearances at the side jambs, and between the two center meeting panels if two door sets are being installed. The bottoms should be trimmed for ample clearance over the flooring, and the specified top clearance must be allowed. Also, the top/bottom clearance is adjustable within a certain range—usually about $1/2$ inch or so—after installation. If the door opening is out of square, trimming must be done carefully, a bit at a time to compensate. This involves rough trimming first, installing the door, taking it down again and trimming some more, and so on.

If not installed, the hinges must be mounted. They may be either the surface-mounted interleaved type or the fixed-pin mortise type. Pairs are typically used, centered 10 to 12 inches down from the top edge and up from the bottom. Doors wider than 2 feet, and/or heavy ones, often are fitted with a third hinge centered between the other two. The hinge barrels align with the edges of the door and are set so there is $1/16$ inch of clearance between the doors when closed.

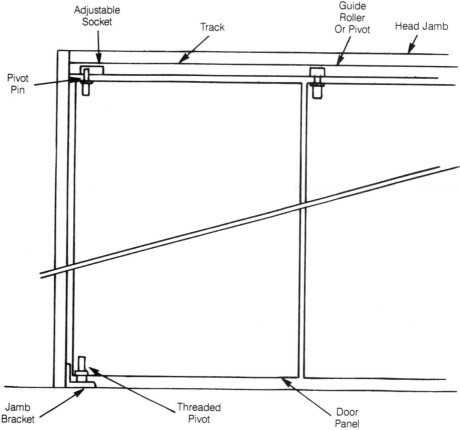

Fig. 4-82. *This is the basic arrangement for hanging bifold doors.*

The upper and lower pivots are usually mounted in holes drilled into the edges of the panel next to the door jamb, centered about $1^1/_4$ inches from the side of the door. The plain spring-loaded pivot mounts at the top, and the threaded one at the bottom. A spring-loaded guide or roller or slide is similarly mounted in the far top corner of the free panel (Fig. 4-82). The track includes an adjustable socket near one end (one at each end for two door sets) for the plain pivot and a guide pin slide and/or a bumper spring. It is trimmed to length and mounted to the head jamb with screws. The track should be set so that the faces of the doors lie flush with or are set back slightly from the outer edges of the door jambs if there is no concealment trim. If trim moldings are applied to the outer jamb edges to hide the closure cracks at the sides and the track at the top, make the setback about $^1/_{16}$ to $^3/_{32}$ inch (Fig 4-83). The socket for the lower, threaded pivot (often called a jamb bracket) is secured to the floor, against the face of the door jamb (Fig. 4-84), so that its centerline is plumb with the centerline of the track.

To hang the door, open the two panels fully and insert the top pivot into its socket. Lift the doors upward and slip the bottom pivot into its socket. Then fold the door set,

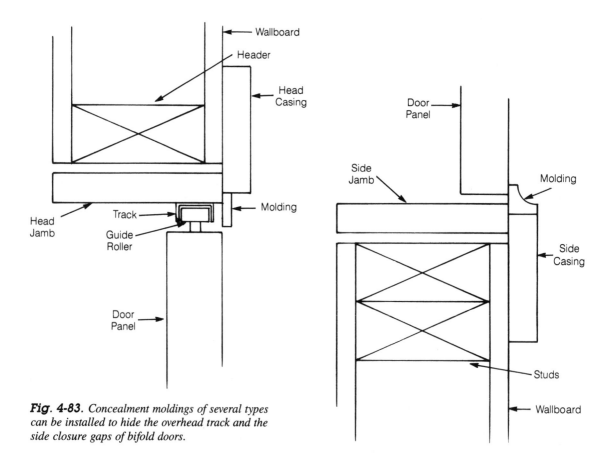

Fig. 4-83. *Concealment moldings of several types can be installed to hide the overhead track and the side closure gaps of bifold doors.*

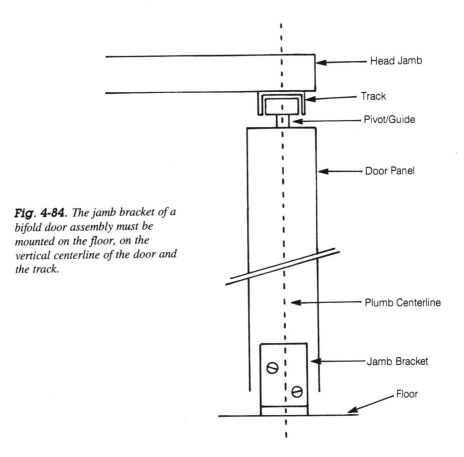

Head Jamb

Track

Pivot/Guide

Door Panel

Plumb Centerline

Jamb Bracket

Floor

Fig. 4-84. *The jamb bracket of a bifold door assembly must be mounted on the floor, on the vertical centerline of the door and the track.*

press the guide roller or pivot down, swing the free panel under the track, and pop the roller or pin into place in the track. Make the side and top clearance and door/jamb alignment adjustments by moving the threaded bottom pivot to the most suitable slot or hole in its socket. Move the top pivot socket in the track, and screw out the bottom pivot to raise the door until the clearance between the door top and the track is minimal. If you cannot make a satisfactory alignment, the doors will need further trimming in compensation. For four-panel sets, the last step is to install door aligners on the inside face of the two center meeting panels (Fig. 4-85), which keep the door edges aligned with one another when in the closed position.

Note: The hardware for bifolding doors is sized according to the size, and particularly the weight of the door panels. The heavier the doors, the stronger the hardware. When purchasing hardware separately, be sure to take this into consideration.

Multifolding Doors

To install multifolding doors, follow the same procedure as for bifolding doors. The principal difference is that there are more hinge sets to install, and the hinge barrels are positioned alternately to the inside, then the outside, from panel to panel.

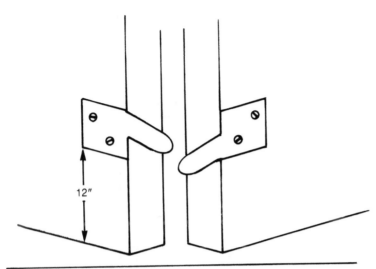

Fig. 4-85. *Aligners mounted at the meeting edges of a center pair of bifold doors slip together when the doors are closed, bracing and lining them up with one another.*

Accordian Doors

Accordian doors are usually sold complete with all necessary hardware and installation instructions. Trimming the door to closely fit an opening is not required; select a stock size that will most closely fit the existing opening. With standard doorway size, lightweight versions, the usual procedure is to trim the overhead track to fit across the opening. This is then mounted to the face of the head jamb, which is centered along the lengthwise centerline. Hang the door on the track by inserting the slides or rollers. From the inside, attach the side opposite the latch assembly to the side jamb with the hardware provided, along a plumb line. Then install the door latch assembly. There are numerous detail variations in mounting arrangements. For large, heavy, divider screen types of accordian doors, the installation is more complex; follow the manufacturer's instructions.

Surface Sliding Doors

A surface sliding door hangs over an opening rather than fitting within it (Fig. 4-86). Select a stock size that overlaps the existing opening by at least $1/2$ inch on each side, and is tall enough to extend above the top of the opening when the door bottom is at its required floor clearance level. In many cases no door trimming is needed at all, because the door track can be mounted at whatever height is necessary to provide proper bottom clearance. Select a track assembly and associated hardware to match the width and weight of the door.

Installation consists of first mounting the slides or rollers to the door top, following the manufacturer's instructions. Calculate the proper height and placement of the track so that when closed, the door fully covers the opening with no side gaps, and when open,

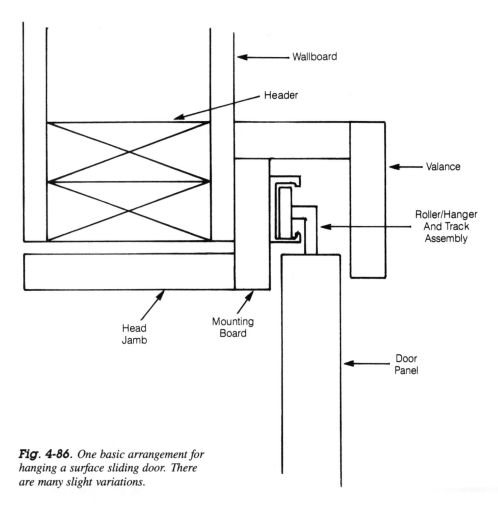

Fig. 4-86. *One basic arrangement for hanging a surface sliding door. There are many slight variations.*

provides full access through the opening. Strike a level line at the appropriate point on the wall and find the wall studs with a stud locater. Mount the track assembly; hollow-wall fasteners might be needed in some cases. Hang the door by inserting the slides or rollers in the track. Make any adjustments provided for, and install bottom guides if they are included. Cover the track assembly with a close-fitting valance or cap, and install a door pull or knob.

Bypass Sliding Doors

Bypass sliding doors are installed in pairs, one sliding behind the other, and the two overlapping slightly at the center. Although the door positions can be adjusted to some extent using the hardware, the doors must be trimmed and fitted to the opening within a small dimensional range. As with bifold and multifold doors, a concealment molding can be applied to the face of the finish frame to hide the closure gaps at the side jambs and the track at the head jamb. There are two hanging systems: The doors may suspend from an overhead double track and be kept in position by guides at floor level. Alternatively, they may travel on rollers in a floor track, with the tops guided in and restrained by an upper track, much like a patio door (Fig. 4-87).

The first step of the installation is to trim and fit the doors, incorporating the clearance allowances specified by the hardware manufacturer. Trim and fit the upper track, and mount it on the face of the head jamb so that the outermost door face is flush with or slightly set back from the outer edges of the door jambs. If concealment molding will be applied, the setback should be on the order of $1/16$ to $3/32$ inch. If a lower track is used, secure it to the floor with its centerline aligned with the centerline of the upper track. This track might be either surface-mounted on the floor or mortised in. If floor guides are used, mount them according to the instructions included. Install the guides, hangers, and/or rollers on the doors as directed, along with your choice of recessed door pulls. The pulls should be set at a height that approximately matches that of other door pulls or knobs in the room.

To install the doors, tilt the inner one up into the back track first. If it is the hanging type, move the bottom of the door over the back bottom track or against the guides. Then lower the door, and at the same time, catch the rollers in the upper track. If the door rides on bottom rollers, just set it into the back track. Make the necessary adjustments, using the hardware, to align the door with the jambs and lock it into the track. Then repeat the process with the outer door.

Pocket Sliding Doors

Retrofitting a sliding pocket door is a major construction project and installation of the associated hardware, while not especially difficult, is complex. Much depends upon the type, size, and weight of the doors selected. The size of the opening, whether one or two doors will be installed, and the nature of the existing structural framework of the house are also factors. Installation procedures vary with the specific conditions encountered and materials used. The recommended course of action here is to obtain manufacturer's literature and specifications first, then decide if the project is feasible and what will be entailed. You might find it helpful, perhaps necessary, to consult with an architect or builder before proceeding.

Patio Doors

A patio door is installed in a rough opening as a complete unit. Some doors are delivered completely assembled and ready to insert into the opening, while others are partly knocked down for shipping and must be reassembled at the job site. Either way, they are essentially prehung doors. Common practice is to install the frame alone. First, assemble it, if necessary, or remove the door and lights if they are shipped already in the frame. This makes the job easier and minimizes the possibility of glass breakage. Stand the flat-silled frame on the subfloor in an opening framed to the manufacturer's rough-opening size, then shim and fasten it to the framing as you would any door frame. Nails or screws

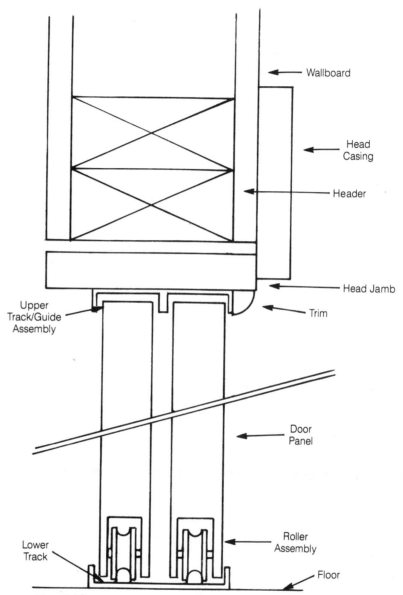

Wallboard

Head
Casing

Header

Head Jamb

Upper
Track/Guide
Assembly

Trim

Door
Panel

Roller
Assembly

Lower
Track

Floor

Fig. 4-87. *Cross section of a typical bypass sliding door installation.*

can be used to secure all-wood frames. Use nails for vinyl-clad wood frames, and screws for aluminum frames.

Once the frame is squared up and fully fastened, the fixed light(s) can be replaced and locked in, and the door reset and adjusted as necessary for easy operation. There are some procedural differences involved—depending upon the door make, the kind of frame, and whether the door is the swinging or sliding type—but none of the variations pose any difficulties. Put on the appropriate casing and finish trim, as required. Complete instructions are usually included with each unit; some are specifically designed for do-it-yourself installation.

DOORWAY BLOCK-OFF

Sometimes it is desirable, or even necessary, to block off a doorway in favor of another location, or even eliminate the doorway completely. This generally poses no problems, but be aware that the doorway wall must be refinished on both sides if it is an interior doorway. With an exterior doorway, a certain amount of exterior siding must be removed and then replaced, which can lead to matching problems.

If the doorway is fitted with a door and contains a finish frame and trimwork, remove the casing and the door and frame. (These procedures were described earlier in the chapter). If the doorway is just wraparound plaster or plasterboard, remove it from the faces of the opening. Break it away with an old screwdriver or chisel and a hammer, plus a utility knife, if necessary, to slit wallpaper or wallboard cover. Trim the edges back roughly flush with the edges and clean the faces of the rough framing. If the finish flooring is carpeting that continues through the doorway, slice that away across the opening; other continuous floorings can be left. In the case of an exterior sill that is notched into the header joist, recut the notch so that it is flat-bottomed. Then fit a piece of matching dimension stock into the slot and nail it securely. Cut a piece of matching subflooring to fit the hole in the floor and install it, along with whatever supporting and nailing blocks are needed to provide a level, tight surface. The finish flooring will also have to be matched in or replaced later.

Next, cut two lengths of dimension stock, which match the original studding thickness, to fit from side to side in the opening. Nail one flat against the flooring—secure it with anchors if a concrete floor—the other to the bottom face of the header. Use 12d or 16d nails spaced about 8 inches apart in a loose W pattern. Then, fit a full-length stud to each side of the opening. You can secure these to the existing studs with 12d or 16d nails, but long screws set with a power driver (drill pilot holes first) or $1/4$-inch lag screws will provide greater holding power. They will also virtually eliminate the chance that the studs might shift or separate slightly and open a crack in the plasterboard patch that will be installed later. Finally, fit one or two full-length intermediate studs, depending upon the width of the opening, and toenail them top and bottom with 10d nails. Space these studs for the greatest number of 16-inch centerings, to allow the easiest insulation installation with the least amount of seams, and for good sheathing support. For further stiffening, you can install solid blocking between the studs at about midheight (Fig. 4-88).

Next, cut a sheet of gypsum wallboard to fit the opening. If possible, the material should exactly match the thickness of the existing wall sheathing. If it doesn't, you can try to match in a material that is up to $1/32$ inch over or under the sheathing thickness. Otherwise, fasten shim stock to the faces of all the studs and the sole and header plates to bring an under-thickness material out flush, or nearly so, with the original surface. The shim stock might be solid (not corrugated) cardboard, 5-millimeter plywood, $1/8$-inch hard-

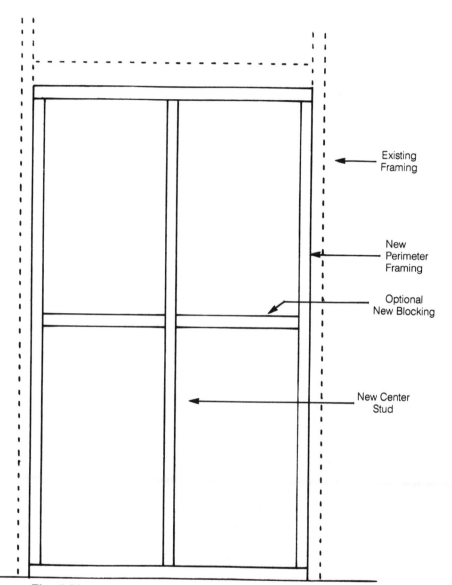

Existing
Framing

New
Perimeter
Framing

Optional
New Blocking

New Center
Stud

Fig. 4-88. A simple method of blocking off an old door rough opening.

board, or any other solid material. Use drywall screws to anchor the patch panel, and place them about 12 inches apart at all bearing surfaces. For extra strength you can run a bead of panel adhesive along the framing faces beforehand.

If the existing wall surface is plasterboard, cover the joints with mesh jointing tape and apply three coats of joint compound in the usual fashion (see "Covering Old Wall Sheathing" in Chapter 3). Then sand smooth and feather out onto both the new and the old surface. If the two surfaces are a slight mismatch, carry the feathered edges well out to each side of the joints—as much as 18 inches each way might be needed—to completely hide them. If the original sheathing is plaster, use a patch panel that is thinner by perhaps $1/32$ inch. Bevel outward and roughen up the edges of the plaster to give it plenty of "tooth," and bevel inward the edges of the patch panel. Allow a crack between the two of about $1/16$ inch. Then fill with a hard patching plaster or a plain plaster mix. Use two or three applications, feathering the edges onto the old wall 10 to 12 inches and onto the patch panel 12 to 18 inches to hide the joints and the slight mismatch. The countersunk screw heads at all intermediate attachment points must be covered as well, and this can be done with either joint compound or patching plaster.

After the final sanding, coat the repaired area with a primer or sealer that is compatible with whatever finish will be applied. Match in a new piece of base shoe and/or base molding, or replace all the original moldings. Finally, apply a finish to the wall.

All of these instructions apply to both sides of an interior wall. In the case of an exterior opening, an exterior sheathing patch panel must be put up. Use a material of matching thickness (it need not be the same kind), and secure it with appropriate fasteners. When matching in the exterior siding, it is almost always necessary to remove some of the original siding back to the joints. Then cut and fit new pieces to blend in on a random basis, or on a modular basis for plywood sheets. If you simply cut and fit pieces butt-jointed across the old opening, it will look as though you just took a door and covered the hole. The object, however, is to make the wall look as though the doorway never existed. Don't forget to take down the outside light fixture!

NEW DOOR OPENINGS

Cutting a new doorway of any width into a nonload-bearing wood-framed wall, whether interior or exterior, presents no problems. If the wall is load-bearing—either interior or exterior, but particularly the latter—and the new opening is wider than about 5 feet, it is a good idea to get professional help with the project.

To cut in a new interior passage doorway, select the approximate location and find all the wall studs in the vicinity. To do this, use a stud locater, tap the wall with a hammer, and/or drive a finish nail through the wall sheathing here and there until you strike studs. Then, if possible, adjust the location of the new opening so that one side or the other is next to a stud. Check also for electrical wires or pipes; either will have to be removed and rerouted after the wall has been opened up. Remove the base shoe and/or base molding on both sides of the wall, as well as the ceiling molding. Take down the wall sheathing on one side. Cut the sheathing flush with the face of the starting stud on one side, and flush with the face of the first stud beyond the door on the other side (Fig. 4-89).

The size of the rough opening typically equals the door width plus $2^{1}/_2$ inches and the door height plus 3 inches. However, always check this against the door manufacturer's rough opening specifications. If no door will actually be installed but a finish wood frame will, add twice the thickness of the finish frame stock to the desired finished opening width, plus 1 inch, to get the rough opening width. Calculate the height of the rough opening so that when finished and trimmed out it will match others in the room. Allow for the

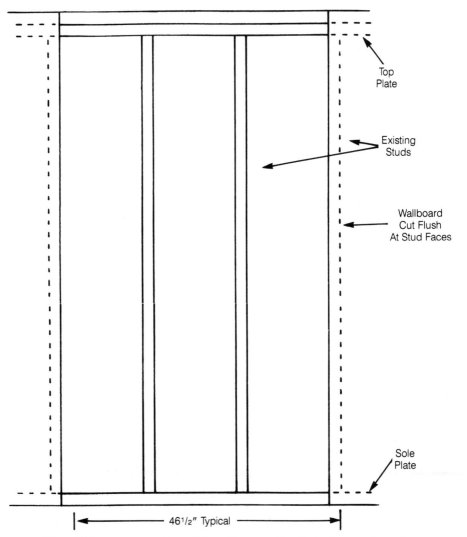

Fig. 4-89. *Opening up the wall from floor to ceiling is the first step in cutting in a new door rough opening.*

thickness of the finish frame plus $1/2$ inch of clearance between the finish head jamb and the header of the rough opening. If the finished opening will not have a finish wood frame, but will instead be covered with plaster or gypsum wallboard, add just the thickness of the finish materials to the desired width and height of the finished opening to get the rough opening dimensions.

Fasten a full-length stud to the face of the existing wall stud opposite the starting stud with 10d nails staggered about 12 inches apart to provide a nailing surface for the new wall sheathing. From the starting stud, measure out along the sole plate a distance equal to the rough opening width plus 3 inches and mark the plate. Do the same at the top plate. Cut

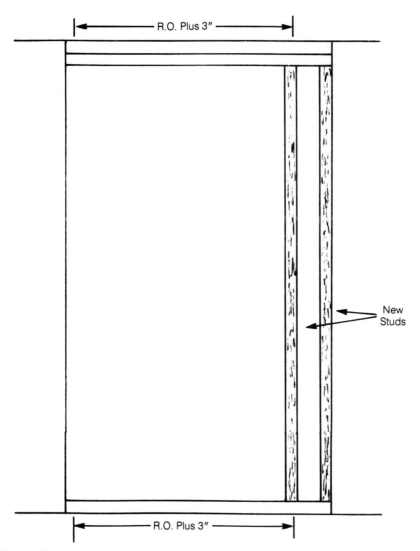

R.O. Plus 3″

R.O. Plus 3″

New
Studs

Fig. 4-90. *A new nailing stud and a side stud for the new opening have been installed.*

and fit a new full-length stud and use 10d nails to toenail it in place top and bottom with its inner face (toward the opening) on the marks (Fig. 4-90). Drive a nail through the wall sheathing from the backside at the top and bottom inside corners of this stud and the start-ing stud. Go to the other side of the wall and draw two vertical lines between the pairs of nail holes. Nail or screw the existing wall sheathing to the studs about every 12 inches, about ³/₄ inch to the outside of the lines. Draw two or more parallel lines 1.¹/₂ inches to the inside of the first ones. Cut the wall sheathing away along these lines, floor to ceiling, and remove the scrap piece. Then remove all the existing studs (and blocking, if any) that lie within the area of the new opening.

Next, build up a header to fit between the two studs. (Refer to Table 4-1 for the recommended stock size, and the section "Framing and Reframing Window Openings" for construction details.) Measure up from the floor on each side stud and mark the location of the rough opening height. Install the header between the two studs with the bottom edge at the marks (Fig. 4-91). Attach it with 16d nails driven through the stud faces and into the header ends where possible. Elsewhere, toenail the header to the studs with 10d nails, or reverse toenail with 16d nails through the side stud outer corners and into the header ends.

Cut out the sole plate section between the two side studs, flush with their inner faces. If finish flooring interferes, cut as much as possible with a saw and finish up with a wood chisel. During this procedure, protect the flooring with sheets of newspaper or cardboard. Cut a pair of trimmer studs to fit between the floor and the bottom of the header, and fasten one to each side stud with 10d nails spaced about 1 foot apart in a loose W pattern. Place a pair near the top and a pair of 16d nails at the bottom, into the sole plate ends.

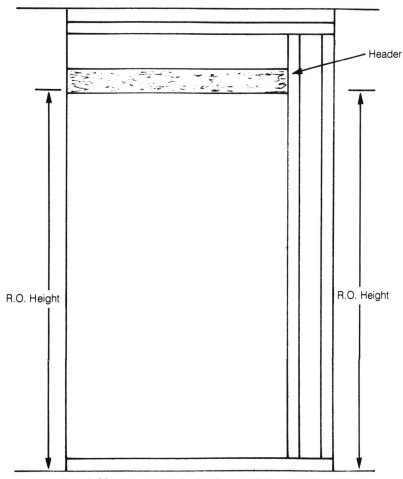

Fig. 4-91. A simple header has been nailed to the side studs.

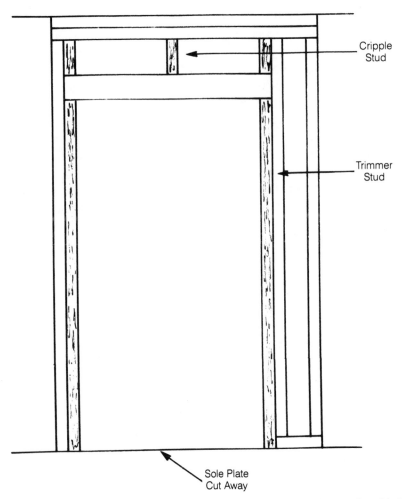

*Fig. 4-92. With the sole plate cut away, the trimmers and cripples can be added
to complete the rough framing.*

Then install cripple studs between the top of the header and the underside of the top plate,
spaced no more than 16 inches on centers. Toenail them with 10d nails (Fig. 4-92).

If necessary, fill the gap created by the absence of finish flooring by nailing in a fitted
piece of plywood or other material, using whatever thickness is required to achieve a level
surface. Or, match in a section of finish flooring to accomplish the same purpose. In many
cases the joints and/or the patch can be covered by a finish threshold installed during the
trimming-out process. Finally, install new gypsum wallboard as necessary to cover up the
framing.

If the new opening must be positioned so that an existing stud cannot be used at one
side or the other as a starting point, the framing procedure follows the pattern shown in
Fig. 4-93. Install two full-length side studs, and remove old studs and blocking from the

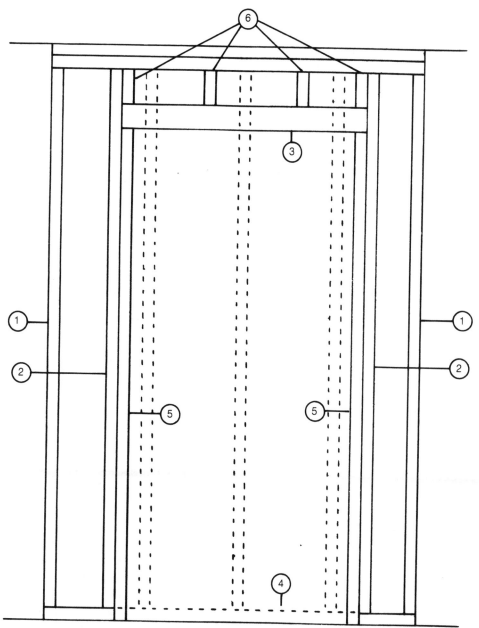

Fig. 4-93. *When no existing studs can be used in the new framing: (1) install nailing studs, (2) put in side support studs, (3) nail a header to the side support studs, (4) cut out the sole plate, (5) install trimmer studs, and (6) install cripples as necessary.*

opening area. Fit a header between the side studs, then cut out the wall sheathing on the opposite side of the wall and the sole plate section across the new opening. Install the trimmer and cripple studs, and a pair of full-length studs against the two existing outside wall studs to serve as a wall sheathing nailing surface.

To cut a new doorway into an exterior wall, follow the same steps. Work from the inside first and, if possible, install all the framing members before removing the exterior siding and sheathing. Although the framing pattern of a door rough opening is somewhat different, the procedures are the same as for cutting in a new window opening (described earlier in the chapter under "Framing and Reframing Window Openings"). Mark the outline of the opening on the exterior siding. Now, using the new framing as a guide, the siding can be removed and the sheathing cut away flush with the framing faces and renailed to the framing. The exterior siding is refitted to butt against the exterior door casing after the door frame has been installed and trimmed out.

The procedures for cutting in and framing a new rough opening wider than 5 feet in a load-bearing wall—interior or exterior—are also the same as previously described, but with an added feature. Temporary support must be installed to bear the structural load as the existing wall is modified and a new header put in. The rough opening size must conform to certain specifications if a door set such as a patio door unit or a bypass slider is to be installed. If an open doorway is desired, the exact dimensions do not matter. In either case, the result is the same as removing part or all of an existing load-bearing wall. The reframing required in that instance is also the same as for framing a door rough opening. Refer to the section "Removing Walls" in Chapter 3.

RESIZING DOORWAYS

Reducing the width of standard passage or entry doorway is seldom a good idea— most need to be wider, not narrower. However, there are sometimes good reasons to make a wide doorway narrower during a remodeling project. Blocking down a 6-foot opening to a 3-foot passageway, for example, increases usable wall and floor space in two rooms.

The first step is to remove the casing and finish door frame. Or, strip the plaster or gypsum wallboard away from the rough framing faces if the doorway does not have a finish wood frame, as explained earlier in the chapter. To reframe a smaller doorway using one side of the original as a starting point, determine the required width of the new rough opening. If a new door will not be installed, any workable measurement can be chosen. Cut two adjacent pieces of dimension stock that match the wall studding to a length equal to the width of the existing rough opening, minus the width of the new rough opening, minus 3 inches. Fasten one piece to the face of the existing header and the other to the floor, with both pieces butted to the original trimmer stud opposite the starting side. Use 10d nails in a staggered pattern.

Now, fasten a full-height stud to the ends of the pieces with 16d nails. Attach a second full-height stud to the face of the first with 10d nails staggered about 1 foot apart. Fasten a third full-length wall stud in the same way to the face of the existing wall stud opposite the starting side. This will provide a nailing surface for the new sheathing. If necessary, install

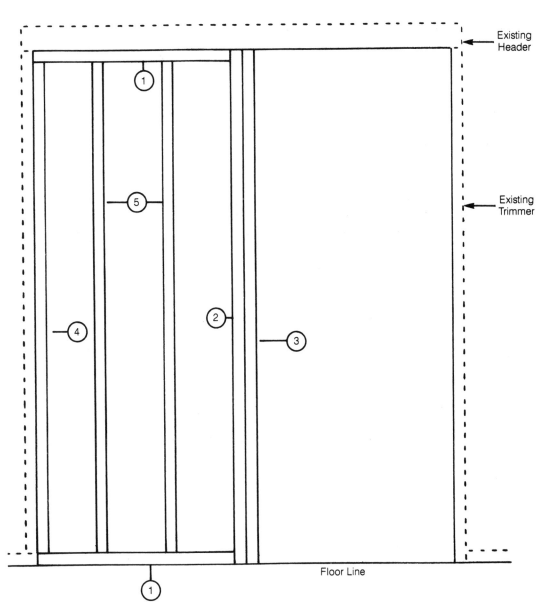

Fig. 4-94. *The framing sequence for making a large door opening smaller. In some cases it will not be necessary to install trimmer studs.*

new full-height intermediate studs on centers no more than 16 inches apart by toenailing them to the upper and lower plates with 10d nails (Fig. 4-94). Cover the new framing on both sides with a wall sheathing material that matches the old (see "Installing New Wall Sheathing" in Chapter 3) and then build up a new finish door frame (covered earlier in this chapter).

If the new doorway is to be positioned away from both sides of the original rough opening, follow the same procedures. However, install two upper and lower plates and

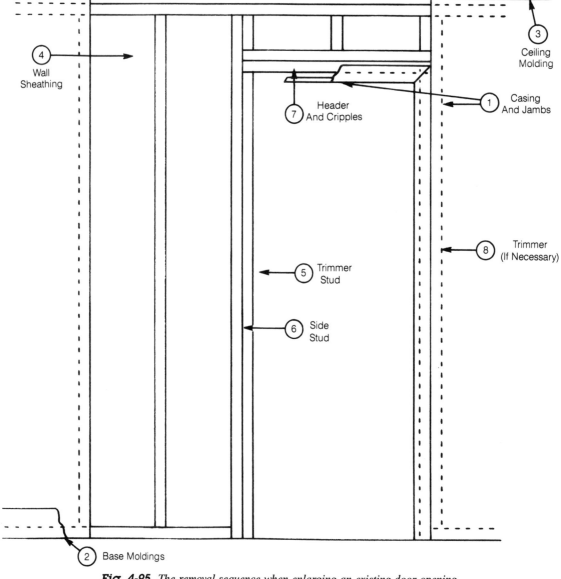

Fig. 4-95. *The removal sequence when enlarging an existing door opening.*

two sets of double studs at the sides to place the new rough opening at the desired spot within the old opening.

Enlarging an existing opening is a bit more difficult. Start by removing the old casing and finish frame. Remove base moldings on the side that will be extended, and the ceiling molding above the doorway. Take the wall sheathing off on both sides of the wall above the doorway, and along the side that will be opened up to expand the doorway, back to the first wall stud beyond the new rough opening location. Pry out the trimmer stud on that side, and knock out the full-length stud as well. Then remove the original header, which might require removing the opposite trimmer stud as well. If there are other studs in the way of the expansion, remove them as well (Fig. 4-95).

From the side of the original rough opening framing that will remain, measure along the floor toward the opposite side a distance equal to the width of the rough opening, plus

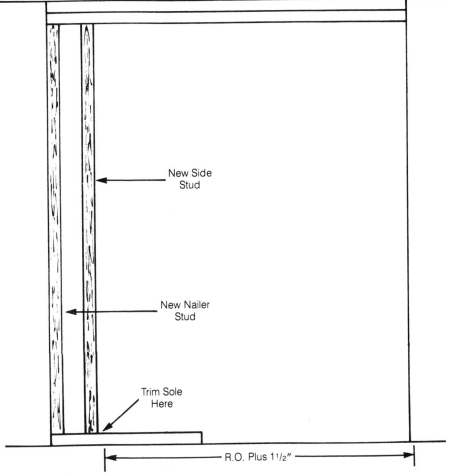

Fig. 4-96. *With the old material out of the way, a new nailing stud and side stud can be installed.*

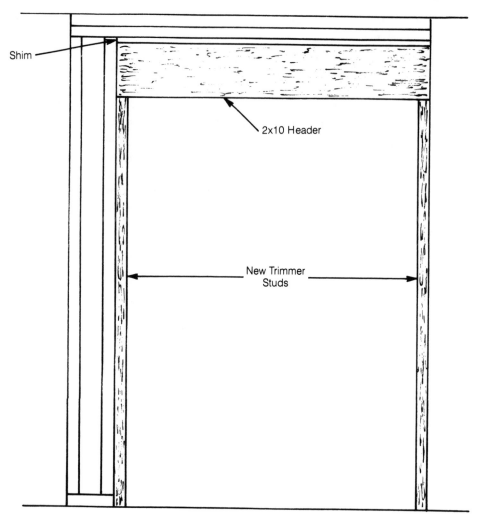

Shim

2x10 Header

New Trimmer
Studs

*Fig. 4-97. Installation of a full header (or a simple header and cripples) and
trimmers completes the job.*

$1\frac{1}{2}$ inches, and mark a square across the top of the sole plate. If you removed the trimmer
stud on the side that will remain, add $1\frac{1}{2}$ inches to that distance (3 inches total). Install a
full-height stud here with its inner (toward the opening) face on the mark. Toenail it to the
sole and top plates with 10d nails. Trim the sole plate off flush with the inner face of the
stud (Fig. 4-96).

   Make up a new header (see "Framing and Reframing Window Openings" earlier in
this chapter) and install it between the side studs in place of the old one. Secure it with 16d
nails driven through the face of the accessible side stud into the header end. At the other
end, secure by toenailing into the side stud with 10d nails or reverse toenailing through the
corner of the side stud into the header end. Then cut a trimmer stud for each side to fit

between the bottom of the header and the floor. Fasten them to the faces of side studs with 10d nails staggered about 1 foot apart. In the same way, add another full-height stud to the furthest wall stud to act as a nailing strip for the new wall sheathing (Fig. 4-97). Match in new wall sheathing over the open framing, and finish out the rough opening either with wallboard or by installing a finish door frame and trimwork.

Expanding an existing exterior doorway in order to install sidelights, a double entry door, or a patio door assembly is done in much the same way. If the new opening is very much larger than the old and will be installed in a load-bearing wall, install temporary support to bear the structural load until the new rough framing is completed. Get professional advice if you are not sure how to proceed. The major difference between interior and exterior doorway resizing is that the exterior wall sheathing and siding must be removed and later refitted to the outside door trim. It is also a good idea to provide air infiltration and vapor barriers between the rough and finish framing, as is done with windows.

5
CHAPTER

Ceilings

YOUR CEILING MIGHT BE SLANTED, ATTACHED TO THE UNDERSIDE OF A PITCHED ROOF frame, but more likely it is a broad, flat expanse that lies across the top of the room and separates you from the attic or the second floor. You probably don't pay much attention to the ceiling—unless, of course, it gets damaged in some way, or begins to sag or fall apart from age, or becomes grimy and fly-specked. Then the ceiling has to be fixed up somehow, or given a quick coat of paint.

In many homes, the walls and woodwork are refurbished three or four times and the floors twice, before the ceilings are redone. Most ceilings do not, in fact, add much to the overall impact of the decorative scheme, and most homeowners do nothing to change that. There is nothing wrong with plain, blank ceilings if this is what you like—but they don't have to be that way. Consider Michelangelo's timeless frescoes on the ceiling of the Sistine Chapel.

This chapter will cover repairing, re-covering, leveling, painting, and generally fixing up ceilings. In addition, a number of ceiling treatments both decorative and functional will be presented, some of them out of the ordinary. As you will see, the ceilings of your house can become an integral part of your interior decorating scheme, or augment it substantially, or even become a focal point.

REPAIRING CEILING SURFACES

Damage to ceiling surfaces—if it is localized, not extensive, and does not involve hidden structural damage or faults—can usually be repaired without much trouble. Hopefully the repair work will also be permanent. The most common problems are: stains from water leakage, water damage, cracks from aging and building shifting or settling, nail heads popping through from framework expansion and contraction, and mechanical damage from being struck.

Plaster

Stains in plaster, especially old hair plaster, can be troublesome. If the surface is to be covered with some solid material that is mechanically fastened, the stains can be ignored. However, if the surface is to be painted or covered with a glued-on material, the stains should be treated. Scrape the surface of the stained area gently to remove any loose paint or crumbled material. Then apply at least two coats of shellac—or better, an oil-based sealer formulated for the purpose.

If the plaster has been softened by water and is crumbly, or if it is loose and coming away from the lath, dig all the plaster away back to solid material. Then fill the area with patching plaster or spackle, depending upon the area size. These repairs are made in the same way as wall repairs (see Chapter 3).

Mechanical damage such as nail or screw holes, gouges, and chips are easy to fix. Scrape out any loose plaster, then vacuum the dust out of the damaged area. Fill with one or two applications of vinyl paste spackle for small defects, patching plaster for larger ones. Finally, sand the repairs smooth and feather the edges out onto the existing plaster. To fill cracks, use the same procedure. Check all along the crack edges first to make sure the plaster is not coming loose from the lath. If it is, dig the plaster away back to sound materials. If not, fill larger cracks directly, after cleaning the dust out of them. If the cracks are fine or hairline, draw the tip of a knife blade along them and gouge out a small groove to give the spackle a good footing to which it can adhere. Allow plenty of time for the spackle to cure, then sand the patches smooth.

Gypsum Wallboard

Stains on gypsum wallboard that has not been primed or painted cannot be removed; the stains become embedded in the paper facing. Some kinds of stains on painted wallboard can be removed, but covering them is usually easier and more effective. If the stains have been caused by moisture or free water, check first to see if the gypsum core is still solid. If the material feels soft or punky when you press your thumb on it, the damaged piece of board should be cut away and replaced with a patch. This is done in the same way as wall repairs. To cover the stain, whether on uncoated or painted wallboard, apply at least two coats of an oil-based sealer. If the stains were caused by moisture, you might find that the paper wallboard facing has swollen and then dried, leaving an uneven surface or perhaps a ridge around the spot. This can usually be removed by sanding gently with fine sandpaper. Apply a coat of sealer, then lightly sand again, and apply another coat of sealer.

Cracks in gypsum wallboard will almost always occur at joints between pieces. If damage is severe enough to open cracks in the field of a sheet itself, it is likely that much more than surface repair will be needed. The cause of most joint cracks is that joint compound was applied incorrectly, or not enough was applied in the first place. Slight shifting or settling of the structure is also a factor. Sometimes a joint will open slightly because the nails loosen up from expansion and contraction or vibration. Always check to make sure the nails are fully set and the piece is snug against the joist or nailing strip before making repairs.

If the crack runs along the edge of the joint tape, the tape will be loose. Pry it up and tear it off back to solid material. If the bad spot is only an inch or so long and the joint itself between the two pieces seems solid, patch with spackling compound and sand the patch smooth after it cures. If the bad spot is fairly large, or extends across the joint, cut and scrape away enough compound and old tape so that you can fit a short new piece of tape in place without having it bulge out from the surrounding surface. Apply a length of self-stick mesh joint tape in line with the original tape, then coat it with at least two applications of joint compound, feathered out onto the surrounding surfaces. A third, touch-up coat might be needed; vinyl paste spackle can be used for this. Sand the patch smooth after the compound has cured.

If the crack is directly over a joint between two pieces, the tape itself probably will have split. Simply filling the crack with spackle will not work; it will crack again. With a shave hook or a heavy-duty paint scraper, scrape away the compound and tape, down to the surface of the wallboard. Cover the joint with self-stick mesh joint tape, and coat it with two or three applications of joint compound or patching plaster. Feather the compound well out to the sides of the joint, and sand the patch smooth after the compound cures. If the crack recurs after a short time, the problem is structural and further surface repairs are pointless.

Nails that work back out of the joists, pushing the covering joint compound and finish outward, are a common problem. Driving the nails back in and re-covering them might or might not work. There are two ways to make repairs: First, drive the nail head about halfway into the gypsum using a punch, then set two more nails about 1 inch above and below the first. Drive the nail heads just below the wallboard surface, creating shallow dimples without breaking the paper facing (Fig. 5-1). Fill the dimples and nail holes with vinyl paste spackle. The second method is to carefully remove the popped nail with a pair of electrician's slim-nosed diagonal cutting pliers, then drive a bugle-head wallboard screw into the hole, countersinking it slightly. Cover the indentation with spackle. In either case, at least two applications of spackle will be needed. Sand these smooth after the compound cures. Popped nails usually occur at random; if you find several in one area, or a whole row of protruding nail heads, look for structural damage or internal problems.

Minor mechanical damage such as scrapes or gouges are easy to repair. Cut away any loose crumbs and bits of gypsum and peel off fringes of paper facing. Smooth on a layer of vinyl paste spackle, let it cure, then sand gently around the edges to smooth off scuffed paper fibers. Recoat at least once with spackle and sand the patches smooth. Nail and tack holes often are surrounded by a ridge of paper facing, and small holes sometimes cannot be adequately filled because of the paper fuzz partly covering the top of the opening. In that case, center a broad-faced prick punch or a $4/32$ nail set over the hole and tap it gently with a hammer, leaving a larger indentation about $1/16$ to $1/8$ inch deep. This will leave a clean surface edge and drive the paper fuzz out of the way, and you can fill the holes easily with spackle.

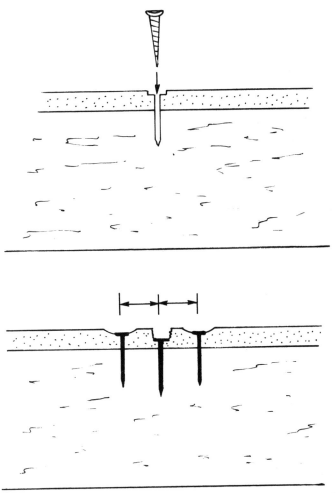

Fig. 5-1. *Ceiling nails that have popped can be repaired by driving adjacent nails (left) or removing the nail and substituting a wallboard screw (right).*

Ceiling Tile

Stained ceiling tile usually cannot be satisfactorily cleaned. If the ceiling is to be painted, seal the stained areas with one or two coats of oil-based sealer. Alternatively, replace the stained tile with a new one of the same pattern. The replacement might be obvious, though, if the ceiling has been in place for a while, because the original tile coating will have changed color slightly from aging and an accumulation of grime.

To replace a tile, use a fresh single-edged razor blade to cut through the mounting flanges of the tile all around its perimeter (Fig. 5-2), until the tile falls away of its own weight. Take care; the tiles break easily. Fish the remaining loose bits of tile tongue out of the grooves of the surrounding tiles. Leave one mating tongue or flange on the new tile in

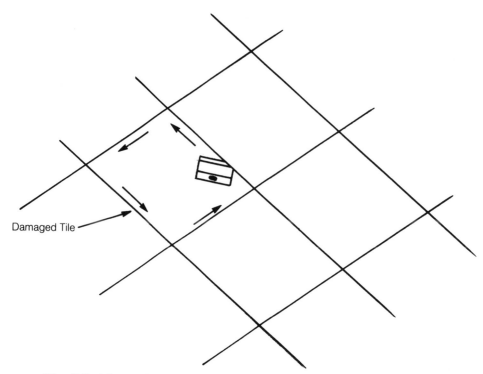

Fig. 5-2. *A damaged ceiling tile can be removed by cutting through its flanges with a razor blade along the seams between the tiles.*

place, and use a razor blade to cut away the remaining ones. Lock the flange or tongue of the new tile into one of the existing tiles and set the tile up into the opening. You can secure it to the furring strips or a metal runner with a bead of construction or tile adhesive; hold the tile firmly in place until the adhesive sets up. You can also drive two or three small brads or wire nails through the tile into a furring strip. The nails can often be set into the small surface holes present in most acoustic tile, and they will not show if you work carefully.

Tiles that have been damaged by water may no longer be solid enough to stay properly in place, or may have swollen or buckled into an unsightly mess. In that case, remove the old tiles and replace them with new ones, using the same procedure as just described. Gouged or dented tiles usually cannot be successfully patched because the patching compound does not hold well and sanding smooth is a problem. If the defect is very small, such as a nail hole, fill with vinyl paste spackle and immediately scrape the surface smooth. Use a small flexible blade such as an artist's palette knife. Then wipe the surface clean and smooth with a damp cloth; do not let the patch dry and then attempt to sand it smooth. For larger defects, replace the tile. In cases where the new tile color does not match well, the entire ceiling might have to be washed and painted.

Ceiling tiles will sometimes sag or droop slightly if the staples loosen or a few adhesive dots fail. This creates an uneven surface. If you can easily push the offending tile back up into the correct plane, the problem is just looseness and not a failure of the support system. There will be a mounting strip along two opposite sides of the tile, and none on the other two sides. To find out where the strips are, stick a fine needle up through the tile close to one edge. At the two edges where you encounter resistance, drive two or three 1-inch fine brads up into the supports to secure the tile in its proper position.

Wood

Extensive exposure to water can damage plywood ceiling material by separating and buckling the plies. This involves a major replacement or cover-up project. Otherwise, moisture or even free water is unlikely to cause permanent damage to other kinds of wood ceilings, unless it is present continually and causes rot. The resulting stains, however, can cause problems. On unfinished or natural finished surfaces, stains are difficult and sometimes impossible to effectively remove, depending upon the nature of the staining agent. The first step is to try sanding with a medium-grit paper, followed up with a fine grit. If there is a finish coating, that will have to be cut through into the wood. If sanding does not do the job, the only recourse is to try bleaching. First try several applications of a plain laundry bleach, keeping the surface well dampened but not dripping wet. If this does not work, try a stronger bleach. There are several commercial preparations made for the purpose; consult your paint supply dealer.

If the stain appears on a painted surface, first scrape or sand away any loose or flaking paint, as well as the surface portion of the stain. Then coat the area with two or more applications of an oil-based sealer, as necessary.

To repair mechanical defects in wood ceilings—such as nail holes, scratches, or gouges—fill them with an appropriate material. Then make repairs as you would to wood flooring or wall coverings.

FIXING CEILING SAGS

Sags in open construction floor/ceiling or roof/ceiling systems are rare. Whan they do occur, the problem is deflection of the basic structural members of the building and is not readily correctable. Sags that occur in a grid-type suspended ceiling are usually caused by incorrect adjustment of the hanging wires, or because one or more of the hangers has loosened up or was not originally made secure. The solution is simply to adjust the hangers until the ceiling grid is level.

Sags in conventional lath-and-plaster or gypsum wallboard ceilings attached to joists or subframes are common, and there are a number of possible causes or combinations of causes. The sag might simply be a gradual bellying-down toward the center of the room from wall to opposite wall, it might alternate and create a ripple effect, or there might be one or more isolated areas in the ceiling field that droop. Sometimes these problems can be remedied without much trouble, and sometimes major renovation might be required.

Occasionally the best course is simply to build a new ceiling below the old one. In all instances, however, it is wise to investigate thoroughly to find the root of the problem and either cure or alleviate it, not just cover it up and hope for the best.

Isolated sagging areas in a ceiling that is otherwise relatively flat and intact can usually be traced to one of two causes. The first is moisture or free water that has softened and degraded the ceiling material and caused it to fall away from its fasteners. The second is a loosening of either the ceiling or the ceiling support fasteners because of inadequate fastening for the ceiling weight involved, expansion and contraction of the various materials, or perhaps excessive vibration from the floor above (appliances or equipment) or throughout the structure (heavy street traffic nearby).

If the topside of the ceiling is accessible from above, you can easily see if the ceiling or the ceiling supports (furring strips, for instance) have pulled away from the joists or if old plaster has broken away from the laths. If you can readily push the sagging material back up level or it flexes fairly easily, you can also assume this is the case. If the ceiling material is solid and intact, you can force it back into place. If it is breaking up or crumbling, you will have to remove that section and replace it.

The first step in fixing this problem is to locate the joists or furring strips above the ceiling. If you cannot see the nails, use a stud locator to find the nailing points. Draw some light guidelines on the ceiling with a pencil to indicate where the fasteners are. Cut a prop of 1 × 4 or other handy stock about 3 inches longer than the distance from the bottom of the ceiling sag to the floor. Then nail a crosspiece to one end of the prop, leaving it a bit loose, to form a T. Place another scrap of wood on the floor, and wedge the prop in place on a slant under the sag and also beneath a joist. Then tap the bottom of the prop towards the vertical position, forcing the ceiling upward (Fig. 5-3). Do this carefully to avoid damage. Then drive a series of bugle-head wallboard screws through the ceiling and into the furring or joists. Move the prop as necessary in order to drive the screws into all areas of the sag, until the whole ceiling section is flat again. Then fill over the screw heads with vinyl paste spackle or patching compound.

If the sag was caused by free moisture and the ceiling is made up of sheet material like gypsum wallboard or plywood, it might have warped and buckled into a firm set. If so, there usually is no way to force it back into place; replacement or re-covering is necessary.

A series of sags across a ceiling that results in an undulating or ripple effect is usually caused by bows in the joists. This condition is most evident at the center of the ceiling, or the center of the joist span. It can occur if some joists were originally installed crown (bow) upward and some downward. Joist bottoms can be out of line with one another as much as 2 inches or more from the highest point to the lowest. There is no practical way to force the joists into proper alignment, even if the entire ceiling is removed. The remedy consists of installing a new ceiling and ceiling support system, with the joists left in place.

If all or most of a ceiling sags, with the low point near the center of a room or area, a structural problem is likely. All the joists, ceiling subsupports, and ceiling covering are sagging together. A possible cause is that the weight of the ceiling, or the combined weight of the ceiling, flooring above, and objects placed upon the floor, is too much for the size

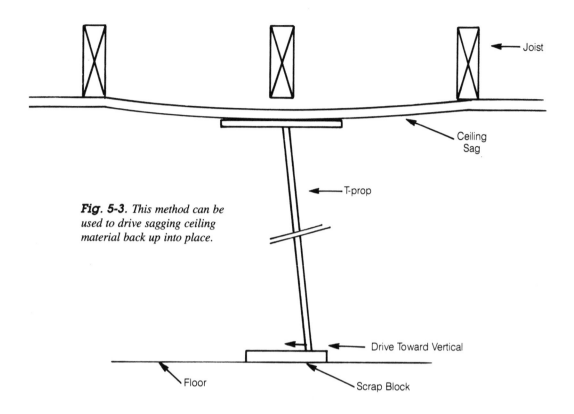

Fig. 5-3. This method can be used to drive sagging ceiling material back up into place.

and/or spacing of the joists—the safe load capacity has been exceeded. The sag can result over a long period of time, slowly forcing the unsupported midspans of the joists downward. If the condition is not corrected, the floor frame might eventually fail.

There are three remedies, all requiring removal of the existing ceiling (covered later in the chapter). One is to remove the entire ceiling/floor system and build in a new one, using heavier joist stock, closer joist spacing, or both. The second is to add a new joist to each of the old ones, while at the same time leveling all the new lower joist edges to eliminate the sag. If there is flooring above, that must be removed too, and both upper and lower joist edges leveled. The third remedy is to install a girder directly beneath the lowest point of the sag at right angles to the joists to support them. The girder is tied into and supported at each end within the walls of the room, and is boxed in and/or decorated to suit the room decor. Then the ceiling framework must be leveled. All of these solutions are major renovation projects and are best undertaken with the help of a professional. A fourth fairly simple method can be employed if there is no floor above but there is sufficient working space. Build in a strongback of dimension stock as shown in Fig. 5-4. This will alleviate the problem but will not cure the sag; to do that, the ceiling must be leveled with a subframing.

Sometimes a ceiling sags off toward a wall. If it is toward an outside wall, the problem might be foundation settling or shifting. If it is toward an inside wall, the support

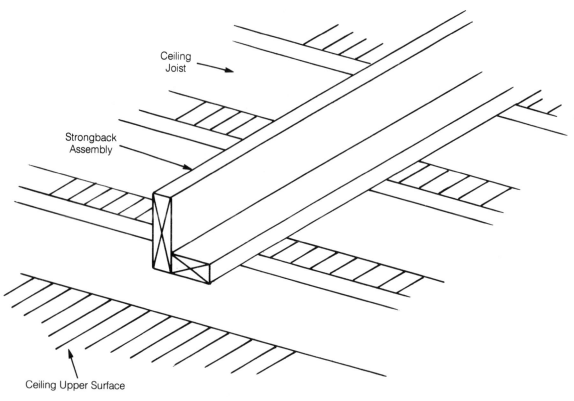

Fig. 5-4. *Installing a strongback is one way of strengthening a weak ceiling structure.*

beneath that wall might be inadequate or settling. These are serious structural problems, and should be assessed by a competent professional builder or engineer before any correction attempts are made.

RE-COVERING CEILINGS

In some cases it is easier to cover over an existing ceiling surface and do your redecorating on a smooth, clean, fresh surface than it is to struggle to prepare the old surface for a new finish. Some examples are: a plaster ceiling that is beginning to crack and crumble, a ceiling that has had multiple coats of paint and is now peeling and flaking, or a mildly textured surface that is grimy and unattractive.

If the original surface is flat, reasonably smooth, and mostly intact and solid, it can be re-covered to provide a clean surface. Defects, cracks, popped nails, and even crumbling or loosened plaster can be covered right over. There is one added consideration, however. The ceiling subframe and/or joists must be sufficiently stiff, strong, and well secured to carry the weight of the new ceiling material. In houses with 2- × -8 or heavier joists on 16-inch centers, this is seldom a problem. If modest strengthening seems necessary, a strongback installation (see Fig. 5-4) might serve the purpose if there is no flooring above.

The usual choice for re-covering a ceiling that is to be painted, textured, or papered is gypsum wallboard. If the existing ceiling is flat and smooth, $1/4$-inch-thick wallboard can be used, but $3/8$-inch is less flimsy and will do a better job. If the ceiling surface is rough and/or slightly uneven or loose in a few spots, $1/2$-inch wallboard is a better bet. Unless the ceiling is wavy or ripply, the wallboard can be installed directly against the existing surface.

Preliminaries

The first step is to turn off the electricity feeding ceiling lighting fixtures, remove the fixtures, and cap off the wires with wirenuts. If there is a molding at the wall-ceiling joint, remove it. If the pieces are salvageable, number and key each one for easier replacement later. Then find all the joists or furring strips with a stud locater. Snap chalklines on the ceiling along the centerlines of each.

Procedure

The process of cutting, fitting, and finishing the wallboard is basically the same as for re-covering walls (refer to the section "Covering Old Wall Sheathing" in chapter 3). Start in one corner of the room and cut the first piece of wallboard to fit snugly into the corner, with the outer edge of the piece set against an appropriate chalkline. Nail a temporary cleat of 1 × 2 to the wall to hold one end of the sheet. Boost one end of the sheet up and set it on the wall support. Then push the other end up and hold the sheet in place with one or more T braces made up of 1-×-4 stock (Figs. 5-5 and 5-6). Secure it with bugle-head wallboard screws of sufficient length to extend into the joists or furring at least $3/4$ inch. Drive the screws with a power driver (which you can either buy or rent) until the heads are just below the paper surface of the wallboard. Do not tear or run through the paper.

Repeat the procedure with successive sheets until the ceiling is covered, making the necessary cutouts for electrical boxes as you go along. If possible, use sheets long enough to span the entire ceiling, in order to avoid end joints. These are squared off and harder to tape and cover than the tapered-edge side joints. If end joints are necessary, stagger them so that no two are adjacent; each must fall on a joist or furring strip centerline. Leave a gap between end joints of about $1/16$ inch. Finally, tape and fill the joints and cover over the screw heads, then sand the compound smooth.

If the existing surface is somewhat ripply, to the extent that securing the new covering directly against it is likely to result in ripples in the new covering as well, you might wish to use a different method. Locate the joists or furring strips as previously explained and set the chalklines. Then fasten new 1-×-4 furring strips to the existing joists or strips at right angles (Fig. 5-7). You can nail them up, but fastening them with long bugle-head wallboard screws and a power driver is easier, faster, and will hold the strips more securely. Use two screws at each mounting point.

Wherever gaps occur between the strips and the ceiling at the mounting points, shim with pieces of thin wood or bits of cardboard to achieve a plane surface. (Do not use corrugated cardboard—it compresses too much). If the covering will be gypsum wallboard,

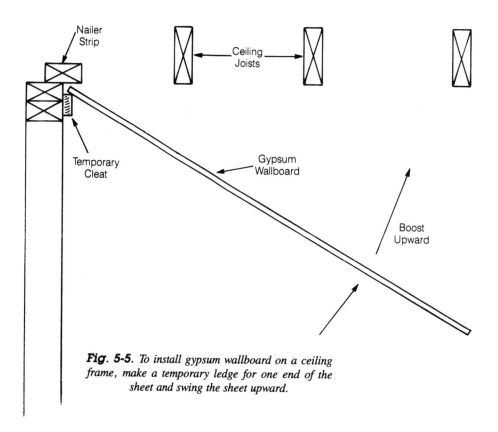

Fig. 5-5. *To install gypsum wallboard on a ceiling frame, make a temporary ledge for one end of the sheet and swing the sheet upward.*

use material at least ³/8 inch thick and space the furring strips on 16-inch centers. If you plan to put up ceiling tile, space the strips on 12-inch centers. When all the strips are in place, install the new covering in the usual fashion. Electrical boxes can either be fitted with box extensions or removed from the existing ceiling and reset in the new covering.

An existing shabby ceiling can also be directly covered with some kinds of finish wood planking, which is described later in the chapter.

CEILING REMOVAL

Circumstances occasionally dictate that an existing ceiling be removed completely. If, for example, a plaster or wallboard ceiling is in such bad condition that repairing or redecorating would be fruitless, it must be torn down and a new ceiling put up. Or, there might be reason to change the level of a suspended ceiling, or remove it altogether in favor of a framed-up ceiling. When restoring an old house, it might be necessary to remove a false or dropped ceiling in order to get at the original one. The original ceiling itself might then have to be replaced.

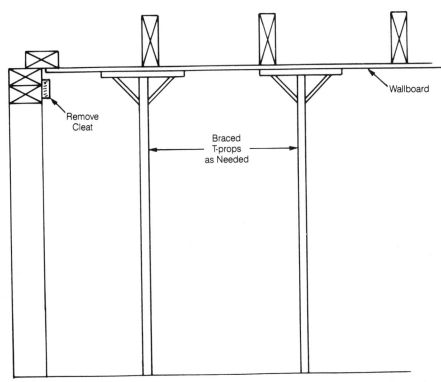

Fig. 5-6. *Hold the sheet in place with T-props while it is being secured.*

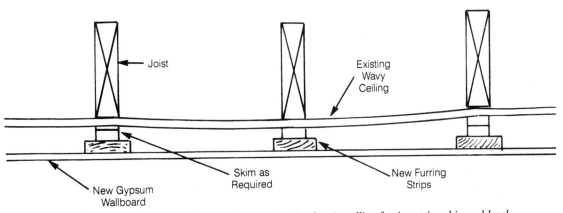

Fig. 5-7. *An existing wavy ceiling can be smoothed by first installing furring strips shimmed level, then applying new sheathing.*

Suspended Ceilings

To more or remove a suspended ceiling, first take out all of the ceiling panels that do not have lighting fixtures mounted within them. This will allow free access to the above-ceiling area. The panels are a bit fragile, so if they are to be reused, handle them carefully and stack them on edge in a safe place. To remove lighting fixtures, first turn off the electrical circuits that feed them, and tape off the circuit breaker handles or remove the fuses. Disconnect the wires at the junction boxes or in the fixtures and secure the cables up out of the way. Fixtures mounted to ceiling panels—such as cans, highhats, or dropped lens types—might be bar-mounted, or they might be set into locking frames. Undo the mounting hardware and remove the fixtures from the panels, then take the panels out of the gridwork. Troffer lighting fixtures usually just rest upon the gridwork. Take the fluorescent tubes out, then push the fixtures straight upward, tilt them to clear the gridwork, and lower them endwise through the opening.

Take out the border cross tees's first, then unlock and remove the full-length cross tees. Then remove the main runners one by one, clipping off or untwisting the suspension wires as you go. The wall molding will be nailed in place. To avoid damage to the walls, pry the nails away from the molding with an old screwdriver or a very small pinch bar, grip them with electrician's diagonal pliers, and pull them free. If minor wall and/or molding damage is of no consequence, just drive a pinch bar up under the molding and pry outward—the strips will come away easily. Remove any remaining suspension wire and fasteners, and the job is done. (If the same ceiling is to be set to a different level, leave both in place.)

Preparations

The first step in removing other types of ceilings is to turn off the electricity feed to ceiling lighting fixtures, take down the fixtures, and cap off the wires in the outlet boxes. Also remove wood moldings at the wall-ceiling joint, and number and key the pieces for replacement later, if that is part of the redecorating plan. Clear all the furnishings out of the area, remove paintings or other hangings from the walls, and cover the floors with drop cloths or plastic sheeting. It is also a good idea to seal off doors into other rooms as much as possible to prevent dust from spreading. This is imperative if the ceiling is plaster, and/or if the house is old. Wear old clothes, a hat, and safety goggles. Depending upon the ceiling material, you might also want to wear gloves and a dust mask. The mask is essential if plaster is involved.

Gypsum Wallboard Ceilings

To remove a gypsum wallboard ceiling, punch a starter hole through the wallboard between two joists at any convenient spot. With the crook of a wrecking bar, pull chunks of the material down, working from bay to bay between the joists or furring strips. Be careful not to hook onto pipes or wires and damage them. Then clean the remaining scraps and nails off the joist or furring strip faces with a pinch bar or claw hammer. If furring is in good condition, it can be reused for the new ceiling. Check the nailing and renail wherever necessary. If the furring is in poor shape or will not be reused, just pry the pieces off the joists with a pry bar or claw hammer, bend the nails over, and discard the pieces. It's a good idea to pick up the worst of the debris as you go along, so you don't trip over it or get a nail in your foot.

If the wallboard is attached to a dropped ceiling frame that is to be left intact to hold a new ceiling, proceed with some caution when you pull down the wallboard to prevent damage to the frame. Check the entire framework for loose fasteners or joints, and add bracing if it seems necessary. If the whole system is to be removed, peel the bulk of the wallboard off first. Detach or remove any wires or pipes that might be secured to the frame. Then attack the frame section by section, taking apart the smaller and lighter members and bracing to free up the more major components. Work carefully as you pull away members attached to the walls, to avoid damage there. A logical sequence of disassembly will usually be apparent, and in some cases it might be easiest to cut some of the longer members in pieces during the process, or remove the framing in chunks.

Plaster Ceilings

Removing a plaster ceiling is a more difficult and dirtier process than others. If the plaster is on wood lath, it will come away in great showers of dust and chunks. In an old house, the cut nails in the laths will probably be rusted tightly in place, and the lath is likely to very tough and springy. In addition, you can count on quantities of dirt, dust, mouse droppings, and all manner of indeterminate detritus falling down around you.

A good place to start is at a loose or broken spot, where an electrical outlet box has been cut into the ceiling, or any place you can get the point of a wrecking bar beneath a lath or two. Failing that, you will have to drive a hole into the ceiling between a pair of joists, which is easier said than done. You can break out a small section with a heavy hand sledge, or find a crack between laths and force a pinch bar through. Or, you can cut a hole with a reciprocating saw. However, use a power saw as little as possible, because the plaster dust is extremely hard on saw workings and blades. Once you have a starter hole, you can yank the lath and plaster down with the crook of a long wrecking bar, working from bay to bay between the joists and following the path of the laths.

If the plaster is on metal lath, you can either start at an open edge, or poke a hole through at some point with the tip of a wrecking bar. Then curl the crook of the bar into the lath and peel it away from the joists. This material is tough, springy, and full of sharp edges; wear heavy gloves. At many points the lath will have to be pried and worried loose from its fasteners, and you might have to cut chunks apart with tin snips or wire cutters if the pieces become too cumbersome or tangled to work with easily.

The plaster might have been applied over gypsum lath (also called rock lath), which is much like gypsum wallboard but in smaller sheets. This material is usually 1/2 inch thick and attached with wallboard nails. The approach to removal is the same as for gypsum wallboard. The total ceiling thickness, though, is likely to be between 3/4 and 1 inch (perhaps a bit more), so it will break up and come free with more difficulty.

Acoustic Tile Ceiling

Removing an acoustic tile ceiling is simple if the tiles have been stapled to joists or, more likely, to furring strips. Drive a screwdriver into the midpoint of a joint between two tiles. If the blade bottoms after about $^1/_2$ inch of travel, you have hit the furring—move to an adjacent joint. Shove the screwdriver up and in, and pry down. The tile should easily break apart. Then all you have to do is grab each successive exposed tile edge with your fingers and pull the tiles down. The bits and pieces left on the furring can be stripped off with the claws of a hammer.

If you wish to salvage the tiles, the approach is a little different. Check the room corners and find a corner tile with visible staples or small nails along its wall edge that were once hidden by the ceiling molding. Pull the tile down carefully; this first one might break up. The staples will now be exposed in the next tile back, which you can pry out with a small screwdriver and pull away with pliers. Ease the tile tongues out and away from the neighboring tile grooves. By following this procedure back along the tile lines, you should be able to take down nearly all of the tiles undamaged and reusable.

If the tile was secured to the furring with glue dots, the tiles will come away just as easily, but the glue and chunks of tile will remain on the furring. This residue can usually be scraped off with a heavy-duty paint scraper or knocked off with a hammer and a broad-bladed wood chisel. A faster method is to loosen the glue with a heat gun. Work carefully, and as soon as the glue softens, scrape it away with a putty knife.

If the tile was glued directly to a gypsum wallboard substrate, work a stiff, fairly broad-bladed putty knife between the tile and the glue dots and pry the tiles downward. The tiles will break up, and some of the tile material and glue will stay on the wallboard in places; in others, the paper facing of the wallboard will tear out with the tile pieces. The wallboard surface probably will not be reclaimable for painting, but it might be texturable. Often as not, it will have to be re-covered or removed. Also, glued-up tiles are not usually salvageable.

Ceiling tiles that were installed on ceiling-mounted runners or tracks are easy to remove, and nearly all of them should be reusable if you wish. Pry out one or two tiles next to a wall at row ends to see how they are secured and fitted to the support runners. Determine which tiles were installed last—usually at one of the room corners—and start there. Unlock or unclip each individual tile and remove it, working back to the original starting point. Then pry the runners or tracks loose from the ceiling. If this is carefully done, the tracks will also be reusable. Or, new ceiling tile can be installed in the tracks, provided the two are compatible.

Wood Ceilings

Removing a wood ceiling is usually not a difficult job. The starting point is almost always the first strip or plank alongside a wall and it runs parallel with it. The key is to get the first piece out without causing damage to the wall. Rather than attacking it with a wrecking bar, locate the nails that secure the piece. Using a cat's claw (Fig. 5-8), tap the prongs under the nail heads, then pry the nail heads up until you can get the claws of a hammer under them. Remove all the nails and pull the piece down intact. Alternatively, make a plunge cut with a reciprocating or saber saw, and cut a section out of the piece to give you a starting point for prying. Once you have your starting opening, just pry succes-

Fig. 5-8. *This demonstration shows how a cat's claw is used to easily pull headed nails.*

sive planks loose with a pinch bar. Separate or break off the tongue-and-groove flanges, if necessary.

LEVELING CEILINGS

Two considerations are involved in leveling a ceiling. The first is to achieve a reasonably plane surface—that is, one that is approximately flat in all directions. Approximately, because the tolerances for ceiling flatness are loose, and a perfectly flat surface would be impossible to achieve. Depending upon the nature of the finish ceiling surface and the kind of lighting in the room or area, a degree of flatness of about 1/4 inch in 4 feet in any direction is sufficient.

As to the second consideration, a satisfactorily flat ceiling might not necessarily be level. The plane of the ceiling might be tilted slightly from one wall to another, run off toward a corner, or be off-level in a compound fashion. So, leveling a ceiling might consist of removing sag or ripples and achieving a plane surface, or removing any tilt or run-off from the ceiling and making the whole affair perfectly horizontal, or both. You have a choice of two approaches, depending upon the kind of finish ceiling you intend to install. If actual levelness is not a consideration and you are only after a smoother, flatter surface, you can leave the existing ceiling in place and work from it. If the ceiling must be made horizontal, and/or sags and dips must be eliminated, the existing ceiling must first be removed (see the previous section), whether it is secured to ceiling joists or is the dropped variety.

There is a third approach, too, useful if the existing ceiling is at least 8 feet high. Ignore the existing ceiling altogether and build a dropped ceiling or install a suspended ceiling. Both should be made horizontal and plane, regardless of the conformation of the structure to which it is attached.

Procedure

To install a new, flat ceiling over an old one without regard to the actual levelness of the entire system, you must first find the lowest points of the old ceiling. If there is a molding along the wall/ceiling joint, remove it. With a stud locater, find the ceiling joists or furring strips. Find the centerline of each joist or strip at each end, next to the walls. You can do this by tapping a finish nail or an old screwdriver up through the ceiling and finding the edges of them. Then snap centerline chalklines across the ceiling to act as nailing guides. Next, snap a series of chalklines at right angles to the first series, from wall to wall and on 16-inch centers (12-inch centers for ceiling tile).

The next steps are tedious but accurate. Nail 1-×-2 furring strips to the walls, snug to the ceiling. Don't drive the nails all the way in, to allow for easier removal later. Drive 6d nails partway up into the bottom edge of these strips on the walls that are at right angles to the joists or ceiling furring strips, and in line with their centerlines. Then stretch a series of taut lines from nail to opposite nail, directly under the joist or furring strip centerlines. Push the string loops up snug to the underside of the wall furring strips, and stretch the lines tightly so there is no sag (Fig. 5-9). Then drive a similar series of nails into the bot-

Joist

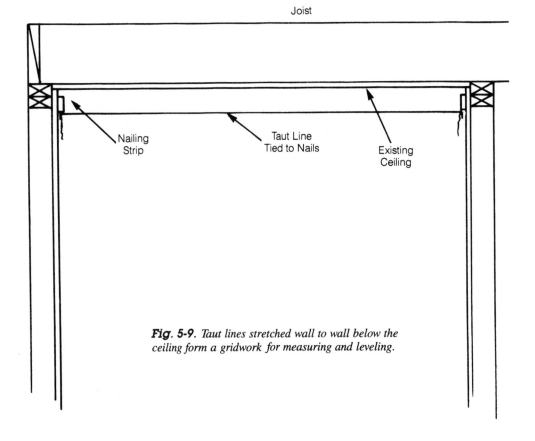

Nailing
Strip

Taut Line
Tied to Nails

Existing
Ceiling

Fig. 5-9. *Taut lines stretched wall to wall below the ceiling form a gridwork for measuring and leveling.*

tom edges of the other wall furring strips, beneath and in line with your 16-inch (or 12-inch) centerlines. Stretch another taut line between the first pair of nails. Adjust the height of this line on the nails so that it does not push any of the crossing lines upward.

Now, at each intersection of lines, measure the distance from the crossing line to the ceiling, straight up. Note the measurement to the nearest $1/16$ inch, and write it on the ceiling. Take the line down, move it to the next pair of nails, and repeat the process. Continue until you have gone across the room. When completed, you should have a gridwork pattern on the ceiling with a measurement figure at each chalkline intersection. Take down the taut lines and remove the wall furring strips.

The smallest measurement you recorded is the lowest point of the gridwork. It is possible that some point between the grid lines is lower. If so, it will show up as you set the new furring strips. The easiest remedy is to furrow out the small section of ceiling that interferes with the proper lie of the furring strip, rather than to adjust the entire grid.

At the lowest point of the grid, the new furring will lie tight to the old ceiling. At all other points, shims must be slipped between the furring and the ceiling so that all the strips remain in the same plane, or nearly so. The difference between the lowest-point measurement and the measurement noted at any given grid intersection equals the thickness of the shim needed at that point.

For example, assume the lowest-point measurement to be 1 inch. At the first intersection along the centerline of a new furring strip, which is the first nailing point out from the wall, the measurement is $1^3/4$ inches. The difference is $3/4$ inch, so a shim $3/4$ inch thick will be needed here between the furring strip and the ceiling. The next measurement is $1^1/2$ inches, so a $1/2$-inch shim is required. The third measurement is 2 inches, so a 1-inch shim is needed, and so on.

The easiest place to start putting up the new furring strips is usually along the centerline in the grid where the lowest point lies. Cut a length of 1-$\times$-4 furring to reach from the wall to the centerline of a joist (or to the opposite wall). Align it at right angles to the joists along its centerline marked on the ceiling, and nail it up at the lowest point. Then move to an adjacent grid intersection and insert a shim of the required thickness, and secure both shim and furring to the joist (Fig. 5-10). Continue until that strip is fastened in place, and go on to the next. To start a strip that you cannot attach directly to the lowest point, first nail a shim of the required thickness to the ceiling, then fasten the strip up through the shim (Fig. 5-11).

The shim stock can be any solid material that happens to be handy. Scraps of $3/4$-inch board, pieces of plywood of various thicknesses, scraps of flat wood molding, and bits of solid (not corrugated) cardboard can be stacked in combinations to gain whatever thickness you need. The nails should be long enough to go through the old ceiling and extend at least $3/4$ inch into the joists or old furring strips above. Or, for greater holding power and less vibration during installation, you can use long flat-head wood screws and drive them with a power screwdriver. For furring strips, use kiln-dried 1 $\times$ 4s. As you put the strips up, it is a good idea to check them for flatness and alignment using a 4- or 6-foot spirit

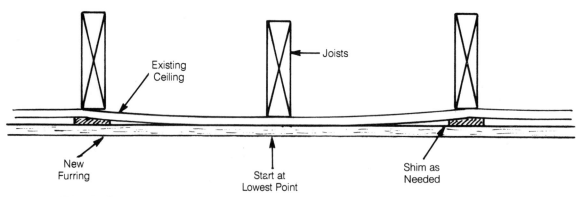

Fig. 5-10. *The first furring strip is nailed directly to the lowest point on the ceiling, with shims inserted elsewhere is required.*

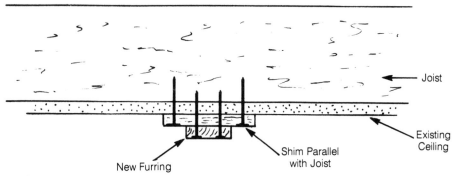

Fig. 5-11. *When thick shims are needed, nail them to the joist bottom first, then nail the furring to the shims.*

level, both along the strips and across them. If one becomes misaligned at some point, you can loosen it and add or subtract from the shimming.

Leveling a new ceiling as well as getting it into a reasonable plane when working from an existing ceiling might or might not be practical. If the ceiling level is off by only an inch or so over the length or width of the room, the discrepancy is unlikely to be noticed much, if at all. If it is off level by a considerable amount, shimming the new furring strips and anchoring them solidly might pose some problems. However, if the ceiling is removed to expose the framework, establishing plane support is a bit easier than if the ceiling were in place, and making it level involves little extra work.

If there is old furring attached to the joists, remove it. Then you must find the lowest joist point. On the bottom of the first joist parallel with a wall, mark off 16-inch centers (12-inch for a ceiling tile installation). Do the same along the joist nearest the opposite wall, and snap a chalkline across the joist bottoms from each mark to opposite mark. Tack nail temporary 1-×-4 nailing strips to the walls, running parallel with the joists. The top edges must be flush with the top of the wall plates. Drive nails partway into the bottom edges of the strips, directly in line with the chalklines. Stretch taut lines between opposite pairs of nails, as explained earlier, to set up the measuring grid. At the intersection of each taut line and joist, measure the distance from the line to the joist bottom (Fig. 5-12), straight up, and write the figures on the joist sides. The smallest measurement will be the lowest point on the grid lines. When this is done, take down the temporary nailing strips.

Next, check the room corners to see if any one of them is lower than the lowest joist point. Drive a nail into that lowest joist point, attach a taut line to it with the loop pushed up against the joist bottom, and thread a line level onto the line. Take the other end of the line to a corner and, stretching it tight, hold it flush against the upper edge of the wall plates. Have a helper run the line level to the approximate center of the line and then read it for level. Adjust the line in the corner up or down until it reads dead level. If the line ends up below the top of the wall top plate, the corner is high and needs no further attention. If the line ends up above the top of the top plate, the corner of the top plate is the lowest point at ceiling level; all the joist bottom edges are higher.

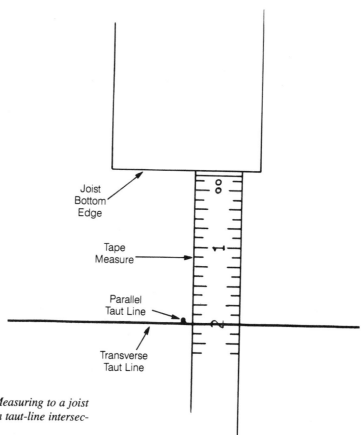

Joist
Bottom
Edge

Tape
Measure

Parallel
Taut Line

Transverse
Taut Line

Fig. 5-12. Measuring to a joist bottom from a taut-line intersection.

If all of the corners are higher than the lowest joist point, swing the taut line around and carry it to the nearest wall that is at right angles to the joists. Hook the line over the top of your thumbnail and set the line against the wall, pulled tight. Adjust up or down against the wall until the line level reads dead level, and make a mark there (Fig. 5-13). Set the top of a 4-foot or preferably 6-foot spirit level against this mark. Adjust it until it is perfectly horizontal. Draw a line on the wall, using the level as a straightedge. Continue this all the way around the room. In theory, the end of the line should exactly meet the starting point, but this seldom happens. Go back and check your level accuracy and adjust one or two lines as necessary, so that you come out even. This line marks the position of the upper face of a new plane and level ceiling assembly.

If a corner should happen to be the low point, make a mark on the wall to indicate where the taut line rests when dead level. Tuck the end of the long spirit level into the wall corner with the mark just visible above its top surface, and draw the perimeter line around the room as just discussed.

The next step is to fasten leveling boards to one side of each joist. The bottom edge of each board at each end will lie just on the line you have drawn on the wall. Use only

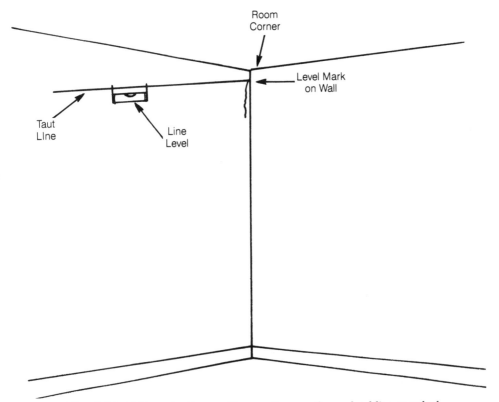

Fig. 5-13. *Making a leveling mark at a wall corner from a level line stretched from the lowest joist point.*

boards that are straight-edged, not bowed or crowned or crooked. The boards need not be completely flat, because they will be pulled into line as they are fastened. A minimum of 2 inches of the board must lie against the joist side. Usually nominal 1 × 4s will do the job, but if the level of the joist bottoms is seriously off, 1 × 6s might be needed in some spots.

If the boards are long enough to reach from wall to wall, cut each to fit, leaving about ¹/₁₆ inch of clearance at each end. Have a helper hold one end against the line on one wall while you set the other to that line and tack it in place with a 6d nail. Then tack the other end in place and check the alignment. Fasten the board to the joist with a series of 6d nails applied in a broad W pattern along its entire length. It would be even more effective to drive 1¹/₂-inch bugle-head wallboard screws into the boards with a power screwdriver. After attaching leveling boards to all the joists, finish the job by securing 1-×-4 nailing strips to the walls. They run parallel to the joists, with their bottom edges on the line. When you have finished, you should have a series of mounting surfaces for the new ceiling covering that are all level and in the same plane (Fig. 5-14).

The last step is installing the furring strips to which the new finish ceiling will be attached. You can transpose from the lines made earlier and snap new chalklines across the bottom edges of the leveling boards on 16-inch centers (12 inches for ceiling tiles). Alternatively, simply measure out from a wall and mark each board at the correct inter-

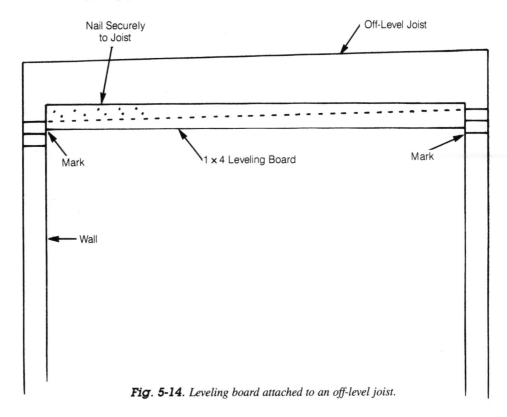

Fig. 5-14. *Leveling board attached to an off-level joist.*

vals. Using these marks as a guide, put up 1-×-4 furring strips at right angles to the leveling boards. Secure them with a pair of 8d nails or 1½-inch bugle-head wallboard screws at each mounting point. Wherever a joint must occur, solidly attach a short piece of 1 × 4 to either the leveling board or the opposite side of the joist. Fasten the end of the first furring strip to one, and the end of the second one to the other (Fig. 5-15).

CEILING INSULATION

Thermal insulation is not usually installed in a floor/ceiling system that separates two levels of mechanically heated or cooled living quarters. There are exceptions, however: if temperatures of the two levels are frequently or always kept 10° Fahrenheit or more apart, or if one or the other of the two levels is closed off and unheated or uncooled for significant periods of time. For either of these cases, refer to "Floor Insulation" in Chapter 2. A third exception: If thermal insulation is installed in a floor/ceiling system for sound deadening purposes.

A ceiling that separates any mechanically heated or cooled area from an unfinished, unheated, or uncooled area above—whether a full attic, a crawlway, or the roof structure itself—should be thermally insulated. The recommended minimum R-value for such insulation ranges from R-19 in warm climates, where mechanical cooling is the principal consideration, to R-38 in the very cold parts of the country. Recommendations or requirements might be different in your locale.

Types

There are several types of thermal insulation that can be used for the area just described. They include mineral wool blanket in roll or batt form; loose-fill insulants like

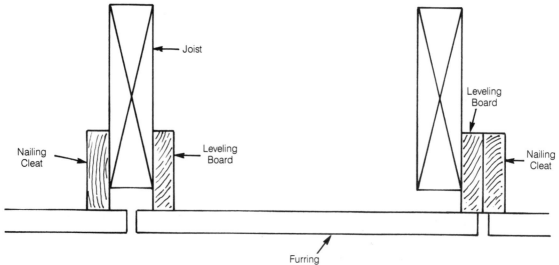

Fig. 5-15. *Two methods of providing attachment points on leveling boards for furring strip ends.*

mineral wool, cellulose, perlite, or vermiculite; and various rigid sheet products such as polyurethane, polyisocyanurate, polystyrene, and phenolic. The insulant you select depends upon local costs and availability, the total R-value you need to achieve, and the installation conditions—especially physical accessibility of the area to be insulated. To determine which insulant might work out best for your particular circumstances, consult with your building supplies dealer, or a heating/cooling engineer. A local building contractor or the municipal building department might also have some valuable suggestions to offer.

Note: When putting in loose-fill or blanket insulation, be sure to wear a dust mask; goggles and gloves also help.

When an existing ceiling is to be left in place, obviously insulating must be done from above. If there is full access to all parts of the ceiling framework, either loose-fill or blanket insulation can be installed. If no insulation exists, you can install mineral wool batts or lengths cut to fit from rolls. Fit the pieces snugly between the ceiling joists with the foil or kraft paper facing down (Fig. 5-16), and abut them tightly together by wiggling them. Make sure the material is fully fluffed up, and fill all cracks and small openings with chunks of scrap. Fill gaps around masonry chimneys or metal stacks with insulation that has all facing removed. Alternatively, fill the bays between joists with a loose-fill insulant, pouring it in and raking it smooth (Fig. 5-17). This material is easier to push into the cramped spaces under eaves, but has the disadvantage of having no integral vapor barrier. However, insulation should not rest directly upon acoustic ceiling tile.

Replacing Old Insulation

If there is old insulation present, check its condition and thickness. If it is thin, compacted, damaged (from water leaks, for example), or an old type with dubious insulating

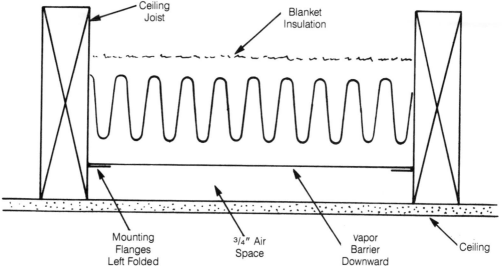

Fig. 5-16. *Faced blanket insulation installed from above a ceiling.*

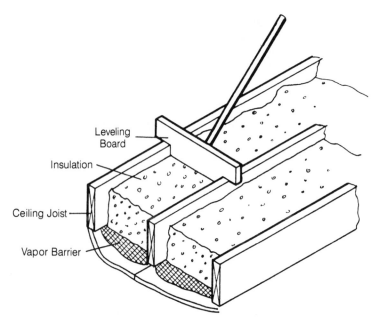

Leveling Board

Insulation

Ceiling Joist

Vapor Barrier

Fig. *5-17. Loose-fill insulation can be raked in and leveled between ceiling joists.*
(Courtesy USDA.)

value, you might want to remove it and start afresh. If the insulation appears to be still useful, simply add another layer on top, filling the bays between joists. Check first, though, to see if there is a vapor barrier present. If there is, you can add loose-fill material or unfaced mineral wool blanket. If only faced blanket material is available to you in the thickness you need, peel the facing off the insulation before installing it.

If the joists are not set on 16- or 24-inch centers, but are random or oddly spaced so that standard blanket insulation will not fit properly in the usual fashion—lengthwise between joists—you have two choices: Use loose fill instead, or cut short pieces to fit sideways between the joists, and install them side by side like paving blocks (Fig. 5-18). You could also trim standard 24-inch insulation lengthwise to fit narrower joist bays, but that can create considerable waste.

Adding to Existing Insulation

If there is decent insulation present and the joist bays are well filled, but you wish to add some more to upgrade the overall R-value and thermal efficiency, the best bet is to install unfaced 6-inch-thick mineral wool in 4-foot batts. Start at the corners of the area and lay the batts on top of the joists at right angles (Fig. 5-19). Push them into the eave areas as necessary, but don't compress the batt edges against the roof rafters by more than about half thickness—even at that, you will lose some thermal efficiency. Be sure not to block vent ports in the eave soffits. Cover the entire area with batts wiggled snugly together. Cut pieces to fit over the access hatch, and pull them into place as you leave the

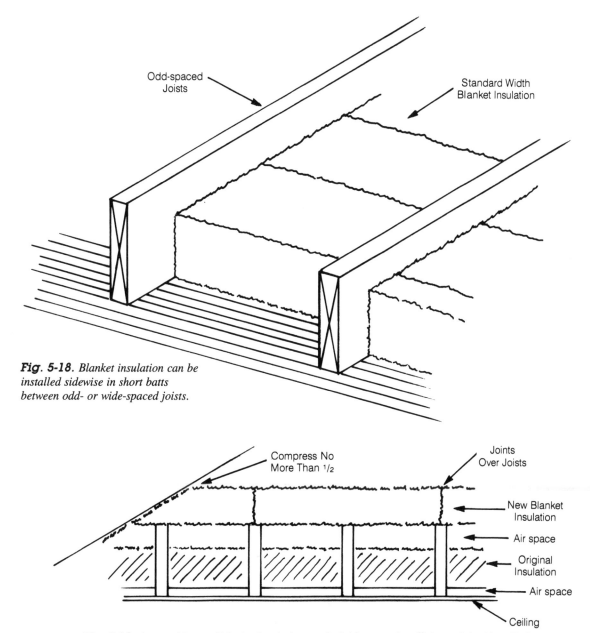

Fig. **5-18**. *Blanket insulation can be installed sidewise in short batts between odd- or wide-spaced joists.*

Fig. **5-19**. *A second layer of blanket insulation can be laid over an insufficient original installation, but should not include vapor barrier facing.*

attic. If access is by means of a doorway or a pull-down ladder, these can be insulated as well.

If the space above the ceiling is not fully accessible or there are cramped areas that cannot be reached, manually installing insulation is neither practical nor possible. The solution is to have cellulose or mineral wool loose fill blown into the ceiling spaces. Contact an insulation contractor for this.

Installing Insulation Prior to New Sheathing

If an existing ceiling is to be covered with new sheathing, there is another useful approach to retrofitting insulation. You can fasten a rigid sheet form of insulant to the old ceiling. In order to gain a worthwhile degree of added thermal efficiency and R-value without excessive thickness, use polyisocyanurate or phenolic insulants. Of the two, phenolic is the better choice because it is not flammable and it also has a somewhat higher R-value per inch of thickness. These materials are available with an integral vapor barrier.

To install this insulation, first locate the centerlines of the ceiling joists or furring strips with a stud locater, and snap chalklines on the ceiling as a nailing guide. Boost the sheets of insulation in place—they are lightweight—and nail them to the joists or furring strips. This process is similar to re-covering an old ceiling with gypsum wallboard, as described earlier in the chapter. Cover the joints with a recommended sealing tape. Then install the new ceiling sheathing, tiles, etc. right over the insulation or on new furring strips.

Installing Insulation Prior to New Ceiling

If the old ceiling has been taken down, you can either insulate from above after the new ceiling has been put up, as just explained, or you can install mineral wool blanket insulation from below. You can use either faced or unfaced batt or roll material of suitable thickness. Push the material up between the joists with the facing downward, fluff it, and run your fingers along the sides and into the corners to make sure that it lies snug and there are no gaps. When you cut pieces to fit odd spaces, make them about $1/2$ inch oversize in all directions. Unfaced material will stay in place by friction. Secure faced material by pushing it up about $3/4$ inch above the bottom edges of the joists. Then fold the facing flanges down and staple them to the sides of the joists (Fig. 5-20). Staples $1/4$ inch long are ample, spaced about every 16 inches or so.

If you wish to put in a double or triple layer of mineral wool blanket for a greater R-value, start by setting the topmost layer on top of the joists, at right angles. Shove unfaced batts up between the joists and lay them out, shifting them snugly into place from below. Then install the next layers between the joists in the usual way. Only the bottom layer should have an integral vapor barrier, and that should be below the insulation, at ceiling level. You won't need this barrier if a separate vapor barrier is to be added (see the following section).

When the ceiling is the interior face of a roof system and is attached to the upper edges of the roof rafters—with the rafters exposed as part of the decor (a purlin or log-rafter assembly in a log house, for example)—insulation is installed between the ceiling and the roof weather surface. Retrofitting or adding insulation can only be done from the outside during a roofing rebuild. However, in the more conventional arrangement, where the ceil-

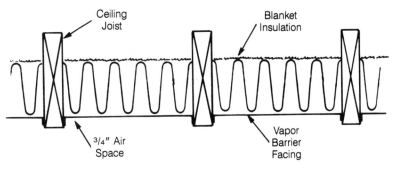

Fig. 5-20. *Install blanket insulation from below with the vapor barrier facing downward. Secure it by stapling through the mounting flanges.*

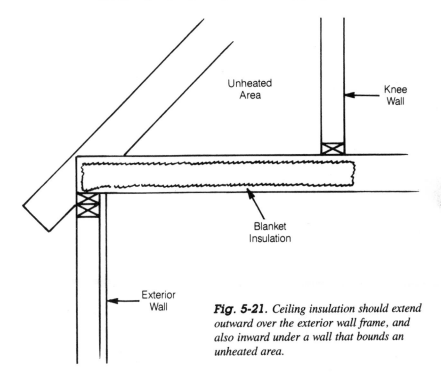

Fig. 5-21. *Ceiling insulation should extend outward over the exterior wall frame, and also inward under a wall that bounds an unheated area.*

ing material is attached to the inside or bottom edges of dimension stock roof rafters or beams, new or added insulation can be installed.

The usual choice, where an existing ceiling has been removed or one is to be installed—converting an unfinished attic to living quarters, for instance—is mineral wool blanket. The material is installed in just the same way as in any conventional ceiling (or wall) framework. If more than one layer of insulation is installed, only the innermost layer should be fitted with a vapor barrier facing, and that should lie inward, just above the ceiling. Install the material so that the lowermost end of each strip extends to the outer edge of outside walls, or just beyond the outer edge of knee walls (Fig. 5-21). If, as might be the case in an attic conversion job, there is existing insulation in the floor/ceiling system and perhaps a vapor barrier as well, these can be ignored. No harm will come from their presence, and neither need be removed.

When installing insulation between rafters, leave at least 1 inch of free airway between the upper face of the insulation and the lower face of the roof sheathing (Fig. 5-22). Air circulation should be provided by installing vent ports in the roof soffits. If the

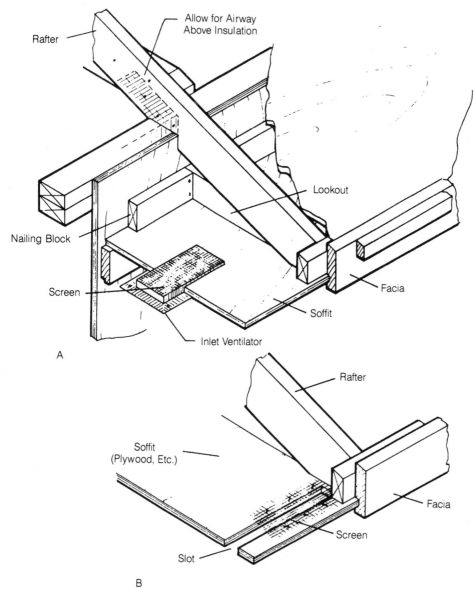

Fig. 5-22. *Make provisions for ample through ventilation above thermal insulation in a roof or attic area. (Courtesy USDA.)*

insulation extends all the way to the peak or uppermost points of the roof, ridge vents should be installed there to allow full venting (Fig. 5-23). If the insulation carries up only to collar ties, leaving an open space in the gable peak, louvers should be installed at each gable end (Fig. 5-24). For this kind of arrangement, the amount of ventilation needed is usually 1 net square foot of ventilator area for every 900 square feet of ceiling area. However, a greater amount does no harm and results in a more positive air flow.

If there is an existing ceiling and there is no particular need to remove it, insulation can be added by attaching sheets of rigid insulation through the ceiling of the rafters, as explained earlier. The insulation is then covered with new ceiling material.

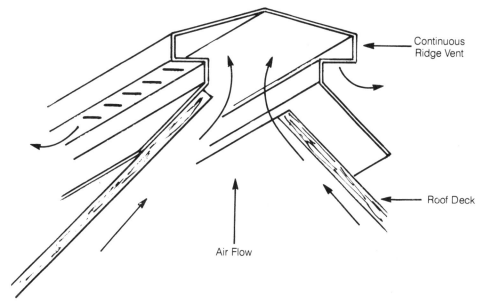

Continuous Ridge Vent

Roof Deck

Air Flow

Fig. 5-23. *In some constructions, ridge venting with a specially made metal ridge cap is necessary.*

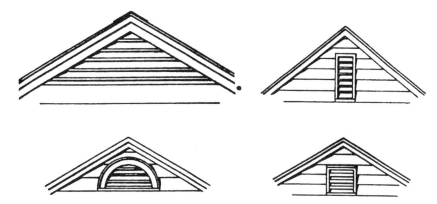

Fig. 5-24. *Several designs for gable louver vents. (Courtesy USDA.)*

Note: There is no point in installing thermal insulation above a conventionally framed dropped ceiling. It will have little if any effectiveness. There is one exception, and that is when the insulation entirely fills a shallow space between the dropped ceiling and the original one, or the joists themselves. The insulation must be positioned at or above the wall tops. Trying to install insulation directly above a suspended ceiling is likewise fruitless, for the reason just given—and also because, physically, it is next to impossible.

CEILING VAPOR BARRIER

A ceiling that separates heated living quarters from an unheated area should be fitted with a vapor barrier. This is true whether the ceiling is attached to the joists of a ceiling frame, floor/ceiling frame, or the rafters of a roof system. Its purpose is to eliminate harmful condensation within the building sections when moisture migrates outward from within the house. In addition, a vapor barrier reduces air infiltration/exfiltration, which in turn reduces heat loss.

The vapor barrier should be installed between the ceiling and the thermal insulation (Fig. 5-25). If the ceiling material is attached to furring strips, the barrier can lie either above or below the furring. If the ceiling material is mounted to resilient channels or metal runners, the barrier must first be attached to the joists or furring, above the channels. In

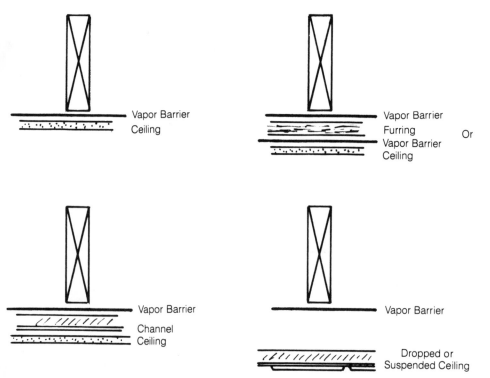

Fig. 5-25. Vapor barrier positions for various ceiling constructions.

the case of a dropped or suspended ceiling, the vapor barrier is attached to the framing directly beneath the thermal insulation, or can be spread across an existing ceiling above.

The vapor barrier that is present in faced mineral wool blanket insulation should not be depended upon; it is seldom completely effective. A separate barrier is always installed if good construction practices are being employed. The vapor barrier affixed to rigid insulation—usually aluminum foil—is effective provided the insulation joints are sealed with a recommended tape and the edges are closely fitted to the structure. No further barrier is needed. If you do not know whether a vapor barrier is present in a cold-side ceiling system, or are uncertain as to the effectiveness of an existing one, there is no harm in adding a second barrier. It must, however, be installed below the insulation.

The most commonly used material for this purpose is 4-mil or 6-mil polyethylene sheeting, available in rolls of various sizes. To install the sheeting as a ceiling vapor barrier, spread it across the old ceiling surface, the joist lower edges, or the furring—whichever applies. Pull the sheeting taut, and secure it with 1/4-inch-long staples. Overlap all edge and end seams by at least 6 inches, or overlap by one full opening between joists or furring strips. Seal the seams with duct tape. Let the sheeting lap down the walls several inches. When the ceiling material or ceiling molding is installed, sandwich the flap of material tightly against the walls and trim off the excess (Fig. 5-26). Cover over lighting fixture outlet boxes, then slit an X across the box opening and tuck the flaps around the box.

If the existing ceiling is to remain in place and only will be repainted or papered, but you wish to incorporate a vapor barrier, there is another possibility. Several vapor barrier paints that do a good job of minimizing water vapor transfer are commercially available. They come in numerous colors and can be left as a finish coating, or they can be painted over again with an ordinary ceiling paint. They can also be coated with a compatible texturing material, or papered over. Consult with your paint supplier for an appropriate product.

CEILING SOUND CONTROL

If there are no living quarters above a ceiling, or if there is no chance of converting or finishing this area, then sound control at the ceiling level is not a consideration. However, if the above-ceiling area might at some future date be made over into living quarters, it is best to incorporate sound control measures into a current ceiling remodeling job. If living quarters already exist over a ceiling and noise transmission from above is a problem, follow the steps given here during a ceiling renovation. If work is done at ceiling level, a new finish must be put up.

Noise reduction measures that concern only the ceiling will have minimal effectiveness. For significant noise reduction, the ceiling/floor system must be considered as a unit, and sound control undertaken at both levels. Refer to the section "Floor Sound Control" in Chapter 2.

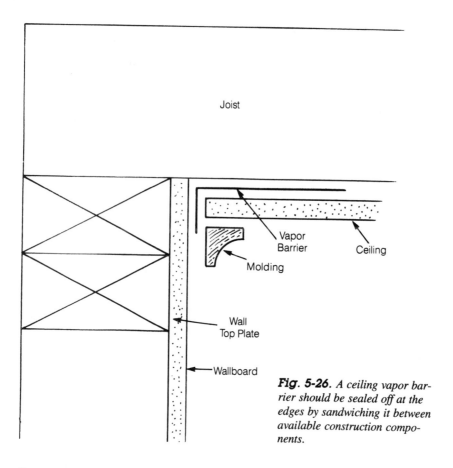

Joist

Vapor
Barrier

Ceiling

Molding

Wall
Top Plate

Wallboard

Fig. 5-26. A ceiling vapor barrier should be sealed off at the edges by sandwiching it between available construction components.

Install Acoustic Tile

The simplest, and also the least effective, ceiling sound control is acoustic ceiling tile. Some tiles have a slightly better acoustic rating than others, but in this sort of installation there is little practical difference. The ceiling tile diminishes noise from above slightly; mostly it will dampen reverberation and hush harsh noise that originates in the room itself or that travels into the room through its doorways. Carpeting, upholstered furniture, fabric drapes, wall hangings, and furnishings in general will also help to dampen local sound. Ceiling tile can be put up in any of several conventional methods; this is covered later in the chapter under "Acoustic Tile Ceilings."

Install a Dropped Ceiling

If an existing ceiling is sufficiently high, installation of a dropped ceiling on its own framework, which has no physical connection to the existing ceiling, will serve to lessen sound transmission from above. Likewise, a suspended ceiling consisting of a metal grid and acoustic panels will help, especially if the grid hangers are not connected to the existing ceiling framework.

Install New Ceiling Support System

Another sound-dampening measure is to remove the existing ceiling and install a new and separate ceiling support system. This involves setting a series of new ceiling joists halfway between the existing ones and parallel with them, as shown in Fig. 5-27. Usually 2-×-6 or 2-×-8 dimension stock is used, depending upon the span and the weight of the new ceiling that will be attached to the joists. The bottom edges of the new joists are set 1 to 2 inches below the bottom edges of the existing ones; at no point should they touch or be braced to them. In a retrofit, the joist ends can be notched slightly to sit upon the wall top plate (Fig. 5-28), then toenailed to the plate. Or, if the drop below the top of the plate is sufficient, they can be attached to nailing blocks, which in turn are nailed to the plate (Fig. 5-29). Furring or the new ceiling material is then fastened to the new joists.

Mount Resilient Channels

The most efficient sound control measure involves mounting specially made resilient channels to the bottoms of the existing ceiling joists or furring (Fig. 5-30). The channels can be mounted lengthwise along the joist bottoms—if the joist spacing is compatible with the size of the ceiling material units to be installed—or at right angles across the joist bottoms just as furring strips are mounted. The latter is the most common arrangement. The new ceiling material can be attached directly to the channels, or to furring strips attached to the channels. Specific installation details vary slightly depending upon the product.

REPLACING CEILING SHEATHING

By far the most common material employed as ceiling sheathing is gypsum wallboard. A lath and plaster ceiling is also an option, but this must be professionally installed. Gypsum wallboard can be decorated with paint or paper, or it can serve as a backer for tile of various sorts, texturing compounds, wood planking, hardwood plywood, or a false beam arrangement.

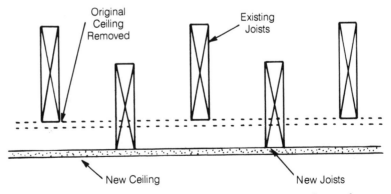

Fig. 5-27. *This ceiling construction system is an effective sound control measure, and can be retrofitted in an existing room.*

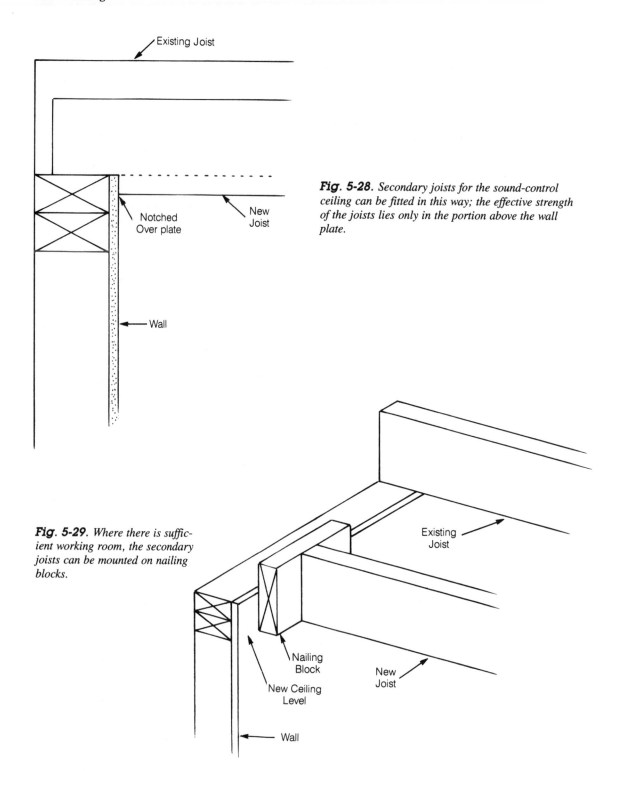

Fig. 5-28. Secondary joists for the sound-control ceiling can be fitted in this way; the effective strength of the joists lies only in the portion above the wall plate.

Fig. 5-29. Where there is sufficient working room, the secondary joists can be mounted on nailing blocks.

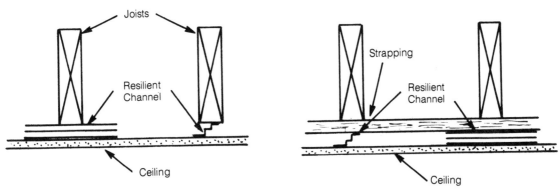

Fig. 5-30. *Resilient channels for mounting a sound-control ceiling can be attached either parallel with or at right angles to ceiling joists or furring strips—even through an existing ceiling (not shown here).*

Standard 4-foot-wide, ¹/₂-inch-thick wallboard sheets with tapered edges are the usual choice. Sheets up to 14 feet long are available, and using the longest sheets manageable limits the number of squared end joints that must be filled and disguised. Wall-to-wall installation is ideal.

When the new finish ceiling is to be a rigid material such as nominal 1-inch planking, there is no need to tape and fill tapered-edge wallboard joints, nor are end joints of any concern; everything will be covered up anyway. This introduces several other options for ceiling sheathing. One is gypsum lath, which is round-edged; comes in ¹/₂-inch-thick panels that are 16 or 24 inches wide and 4, 8, or 12 feet long; and is easier to handle than the larger sheets. Another is ¹/₂-inch-thick gypsum sheathing, which comes in widths of 2 and 4 feet, and lengths of 8 and 9 feet. It is either square-edged or tongue and groove. Gypsum backing board could also be installed.

If the finish ceiling is to be semi-rigid material such as thin wood planking or plywood sheets, or small units like ceiling tile with its multiple joints, the joints of tapered-edge wallboard should be taped, filled, and smoothed off. Otherwise, the depressions along the tapered edges can cause problems when the finish material is installed (Fig. 5-31). An alternative is to use the products that do not have tapered edges, which will result in a smooth surface.

Any of these gypsum board products can be mounted directly to the ceiling joists or to furring strips. If you have taken down an old ceiling and there are furring strips present, they might be reusable. Check them for level and planeness at numerous spots by setting an 8-foot straightedge against them, both lengthwise and crosswise. A straight board with a spirit level taped to one edge works well for this. If the strips themselves are not cupped or warped, and the ceiling plane is satisfactorily flat and level, fasten the new sheathing back on them. If there are some off-level or out-of-plane spots, you might be able to pry the strips down at high spots to closely match the low ones and insert shims to even them out. This prying will reduce the holding power of the nails, so as you reset a strip, drive one more ring-shanked nail or a drywall screw.

If there are no furring strips or the old ones are not reusable, check the joists with a straightedge and spirit level for level and planeness. If all is satisfactory, you can attach the sheathing directly to the joist bottoms. If not, you will have to level and flatten the ceiling plane by installing new furring strips or leveling boards (refer to "Leveling Ceilings" earlier in the chapter).

Once the ceiling framework is squared away, installation of the sheathing can begin. The process is essentially the same as discussed in "Re-covering Ceilings."

PAINTING CEILINGS

The most common finish for a ceiling is paint. The preparation for painting and the kind of paint to use depends upon the nature of the surface and the desired end result. Suitable finishes for ceilings include oil-based paint or enamel, or a latex, resin, or casing type of emulsion paint. The most widely used are latex and acrylic latex paints. Flat, semi-gloss, and gloss finishes are available. For most purposes, a flat finish is desirable and has the greatest hiding power. Minor surface imperfections as well as roller or brush marks are imperceptible, and the surface diffuses rather than reflects light. Semi-gloss finishes are sometimes used in kitchens, bathrooms, and laundries. The harder skin makes these finishes more spot- and stain-resistant, they do not absorb moisture, and they are more durable and cleanable. High-gloss finishes are the most durable, but also the hardest to apply. Minor imperfections in the surface show up readily, and the surface reflects light and tends to have a harsh, cold look.

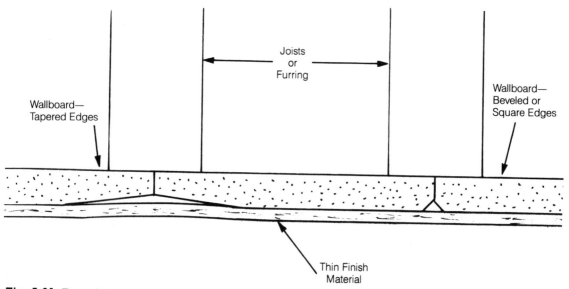

Fig. 5-31. *Tapered joints in wallboard ceiling sheathing might show through thin finish material as dips or waves when the material is nailed up. Bevel-edged or rounded-edge sheathing causes no problems.*

An enormous array of colors are available. White or off-whites are the most popular choice; pastels or light tints are also popular. The darker or the more intense and vibrant a ceiling color, the lower the ceiling appears to be, while a white or very pale tone makes the room look larger. Rather than simply assuming that a ceiling should be always white, as most are, you might wish to discuss other possibilities with your paint dealer or an interior decorator. The often neglected ceiling can become a more important and emphatic part of your overall decorating scheme with little extra effort or cost.

Room preparation is the same for any kind of ceiling painting operation. Remove as much furniture as you can, to allow a maximum of free working space. Completely cover the floor and any remaining furnishings with plastic or other drop cloths. If wall refinishing is part of the program, do the ceilings first. If door and window trim is to remain as is, drape plastic film over all doors and windows. Inswinging doors can be removed from their hinges. If the walls will not be refinished later, cover everything over with lightweight plastic drop cloths (available at hardware and paint supply stores). If there is molding at the wall/ceiling joint or the walls are covered with hard-finish paneling, you can secure the film at the ceiling line with masking tape. Do not use tape on a soft surface such as flat wall paint or wallpaper; suspend the plastic film from a series of common pins stuck into the wall at the ceiling level. Remove ceiling lighting fixtures, after turning off the electrical circuits. Or, if there are ceiling plates or trim covers on the fixtures that can be removed or lowered, do that, and wrap the fixtures in plastic bags. If necessary, arrange some temporary lighting (an old floor lamp without its shade, for example) to give you plenty of illumination at ceiling level. It is very easy to miss spots when painting a poorly-lit ceiling.

Unfinished Gypsum Wallboard

First inspect all of the taped and filled joints and fasteners dimples to make sure that they are smooth and flush and there are no surface imperfections. Do any necessary touch-up filling or sanding. If the finish will be semi-gloss or high-gloss, the surface must be as perfect as you can make it. If there are stains present, cover them with a sealer. Spots of dirt or mud can be removed with a soft rubber eraser, but take care not to abrade the paper surface and raise fibers. Immediately before applying the first coat, wipe the entire surface with a cloth to remove any accumulation of dust or cobwebs.

If the finish will be an oil-based paint, first coat the entire surface with a sealer or a primer. Use a product either made or recommended by the finish paint manufacturer, and apply it according to directions. This can be done using a brush, a roller, or a pad. Use the type recommended by either the paint manufacturer or the supplier. After allowing the sealer or primer to cure thoroughly, apply the first coat of finish paint. Depending upon the specific product and the paint color, one coat might be sufficient, especially if applied with a roller. A modacrylic roller with a $1/4$-inch or $3/8$-inch nap works well for this purpose.

If the finish is to be an emulsion paint, you can dispense with a primer or sealer coat, unless it is specifically recommended by the paint manufacturer. Top-quality latex paints, for instance, have good inherent priming/sealing properties. Often one coat is all that is necessary, especially if a special "high-hiding" variety is applied over a clean, neutral-colored surface. However, in most cases a second coat will give a better appearance. Best results are obtained by using a roller; the modacrylic type with 1/4-inch nap serves well. Use a 1 1/2- or 2-inch high-quality nylon bristle brush or a trimming type of pad applicator to first paint a band along the ceiling/wall joint or molding.

Coated Surfaces

If a ceiling surface has been previously finished, it can be recoated with either an emulsion or an oil-based paint. It is unlikely that the two will be incompatible but if there is any question in your mind, select a latex paint.

Wash the entire surface with warm water and a mild detergent. Check thoroughly for evidence of flaking, lifting, or peeling paint. Scrape such spots well but try not to abrade the undersurface. Very light sanding will help to feather out the edges of these areas. If the old coating is thick and sanding would leave depressions, skim on a thin fill coat of vinyl paste spackle and then sand smooth (two or more applications might be needed). Fill all dents, gouges, nail holes, and other imperfections with spackle, and sand them smooth. If there are stains or water spots, lightly sand away any roughness. Coat all stained areas and spackle patches with sealer, then apply the new finish. If the old and new colors are similar, one coat might be sufficient. Otherwise, assume that you will need two coats.

Ceiling Tile

Almost all kinds of ceiling tile can be successfully refinished. A latex ceiling paint is likely to give the best results and afford the easiest application. If the tile surfaces are smooth, a modacrylic or similar roller with a 3/8-inch nap length will work well. If the tile surfaces are embossed or textured, a 1/2-inch nap will do a better job. If the tiles are bevel-edged, with grooved joint lines, you might find it easiest to paint the grooves first with a small nylon bristle brush or a foam applicator. In many instances the groove edges do not have a finish coating, so two or even three coats might be necessary to achieve a good appearance.

Before starting, wash the tile faces with warm water and a mild detergent, then go back over them with clean water. Allow the tiles to dry thoroughly, then make any necessary repairs. Holes or gouges can be filled with vinyl paste spackle, then carefully and gently sanded smooth. Stains and water spots should be coated with a sealer. Use a pad or brush to paint a perimeter band around the ceiling/wall joint or molding, then do the edge grooves, if necessary. Fill in the ceiling field last. If the tiles contain many tiny holes—as most acoustic tile does—try to apply the paint with as few roller passes as possible, do not scrub the roller around, and load as little paint onto the roller as you can.

Raw Wood

If the wood is to be painted, it must first be primed. Use a primer that is recommended by the manufacturer of the paint you will use. If stains or resinous knots are present, seal those individually with one or two coats of a sealer before priming. If the spots still bleed through the primer, apply more sealer. After priming, repair any defects such as holes, cracks, or gouges. Fill them with vinyl paste spackle, then sand the patches smooth. If the primed surface is rough from dust nibs or raised wood fibers, sand the entire surface lightly with a fine-grit sandpaper. Then apply the finish in one or two coats, as required. If the wood is bevel-edged planking or if it has been sandblasted or resawn, or is a striated surface, you might prefer to paint with a 2- to 3-inch brush. Use a nylon bristle type for emulsion paints, a natural bristle type for emulsion paints. If the surface is smooth, a low-nap modacrylic roller will work well.

If the wood is to be stained, you can make minor repairs first with a compatible, stain-accepting wood filler. Run some tests on scrap wood beforehand to make sure that the patches will blend in well. You can also fill defects after the final finish coat has cured, using putty sticks. Sand the entire surface smooth before staining, in order to achieve uniform coloration; the stain will appear blotchy and darker on rough areas. Apply the stain with a cloth pad, wiping it in thoroughly as you go along. Allow the stain to cure well before applying a protective topcoat—a week is not too long. The topcoat can be a paste or semi-liquid wax, or a varnish. Apply paste or wax with a cloth pad and then buff. Varnish, which is available in flat, satin, semi-gloss, or gloss, can be applied with a brush or a roller in the same way as paint. One coat is usually sufficient for a flat finish, but two will probably be needed with satin or gloss finishes to achieve a uniform appearance.

If the wood is to have a clear or natural finish, sand the surface smooth first. A paste or semi-liquid wax works well. Apply it directly to the wood with a cloth pad and then buff it out, which will give a low-sheen, almost flat appearance. One coat is usually enough to bring out the grain in the wood, but two will give more protection and a good finish. For a hard coating, apply a clear sealer first. This can be a commercial sealer, or a half-and-half mixture of your topcoat material and mineral spirits or paint thinner. The topcoat can be any of the clear varnishes, applied with a roller or brush. Again, two coats rather than one will afford greater protection and a better, more uniform appearance, especially if the wood is soft or porous. Note that it is almost always easier to prefinish new wood ceiling material, then touch up as necessary after installation.

Papered Ceiling

A papered ceiling can be successfully painted over, provided that the paper is tight, even, and well bonded to the ceiling at all points. If the paper is smooth, the finished painted surface will likewise be smooth. If it is textured, the texture will be apparent on the paint surface. In any case, the paper joints will probably be visible, unless the edge mating is virtually perfect.

First, wash the entire surface with warm water, a mild detergent, and a damp sponge. Do not soak the paper, because this will loosen the glue. If the paper is thin, fibrous, and uncoated, washing must be done very gently so as not to raise the fibers. Allow the paper to dry for several days, then prime the surface with a top-quality oil-based primer. After the primer has cured, inspect the surface for defects like holes or scrapes. Repair them with vinyl paste spackle and sand the patches smooth with fine sandpaper. If there are areas of raised fibers or dust nibs, knock them down by sanding very, very lightly with fine sandpaper. Then apply one or two coats of finish paint, preferably with a roller. You can use either an emulsion or an oil-based paint. Load the roller with just enough paint to work with and roll the paint on in a thin, even coating.

TEXTURING CEILINGS

All of the texturing finishes that are suitable for walls are equally at home on ceilings (Fig. 5-32). Those most commonly used are: a relatively thin and smooth finish incorporating tiny metallic or mica chips that impart a "glitter" appearance; a somewhat thicker material compounded with sand that looks like coarse sandpaper; a compound that results in a nubbled surface; and a thick, pasty compound that can be worked after application into any one of dozens of patterns—such as peaks, swirls, striations, or waves. Colors of white predominate, although some manufacturers offer a small selection of muted colors, and tinting of many texturing products is possible.

Texturing finishes can be applied to new gypsum wallboard or standard gypsum sheathing ceilings, or to new wood. They can also be applied to most kinds of existing ceilings, if they are in good enough condition to accept the coating. Application on ceiling tile, however, might be problematical because of the joints, and because the tiles are not always well anchored. The acoustic properties are also nullified. This combination is not generally recommended, nor should texturing compounds be applied over wallpaper.

Fig. 5-32. A typical textured ceiling.

Preparations

Room preparation for texturing a ceiling is the same as for painting (refer to "Painting Ceilings" earlier in the chapter). If the ceiling material is new and unfinished gypsum board, the joints should be taped, filled, smoothed, and the fastener dimples filled. For thin or medium-thick texturing compounds, the surface should be nearly as smooth and flat as it would be for paint, because imperfections will appear through to the new surface. If a heavy coating of thick, workable compound is applied, this is much less crucial. Minor imperfections, fastener dimples, dips, and ridges can be readily covered over and the material used as a leveling coat as well as a finish.

Other existing ceiling surfaces, except for raw wood, should be washed with warm water and mild detergent and allowed to dry thoroughly. If previously painted, check for loose or scaling paint; sand or gently scrape away any loose material. Repair any damage that might show through the new finish. If the ceiling is papered or fabric covered, strip the material off. Wash the surface at least twice to remove all the old adhesive. In some cases, a primer might be needed before the texturing material can be applied; follow the manufacturer's instructions. If the ceiling is suspended acoustic panels, remove the panels from the gridwork, texture them individually, and replace them only after the coating has had several days to cure. Remember that the acoustic properties will be lost. While the panels are out, you might also wish to repaint the gridwork. This is most easily done with a small foam brush applicator and a relatively thin semi-gloss emulsion paint. Be sure to wash the surfaces first.

Other details of ceiling texturing are the same as for walls; refer to "Texturing Walls" in Chapter 3 for more complete information.

PAPERING CEILINGS

Covering a ceiling with wallpaper or a fabric wallcovering is more difficult than covering a wall only in that you must work overhead and have a little more patience. The details and the procedures are the same as for wallpapering, so read through "Papering Wall" in Chapter 3 before proceeding. Think of the ceiling as one big wall in an awkward position.

The same materials that are used as wallcoverings can be applied to ceilings. However, bear in mind as you make your selection that the heavier and/or stiffer the material, the more difficult it will be to put up, and the more reluctant it will be to stay there until the adhesive sets. A medium-weight, fairly tough and supple paper of good grade will probably afford the best results, especially if you have never attempted this sort of project before.

Minimizing Difficulties

There are a few tactics you can use to minimize the difficulties of working over your head and fighting the law of gravity. First, make everything as easy on yourself as possible. Enlist the aid of a helper; this is really a two-person job. Second, clear everything possible out of the room to provide maximum working space. Set up the worktable in the same room, or at least close by. Plan plenty of time for the job. If the ceiling is a large one, don't try to do the whole job in one session. Work on one full-length strip at a time and spread adhesive only for that strip. That way, when you get tired you can stop with no adverse consequences. And when you start to get tired, *do* stop—that is when mistakes start to happen.

Before you begin, set up a solid, substantial platform from which to work, big enough for two people. The platform should be at least 2 feet wide, and at just the right height that you can comfortably lay the paper or fabric against the ceiling and work it with relative ease. Heavy planks set on solid supports like milk crates might do the job, or better, rent a set of low scaffolding with adjustable legs, and preferably locking casters as well. Adjust the platform height as necessary. Also, arrange for one or two temporary light sources that you can shift around to brightly illuminate the ceiling area on which you are working. A few floor lamps minus shades will do a good job.

Preparations

Do not mix your own wallpaper paste from a package; use a prepared super-sticking variety that has strong early holding power and is applied to the ceiling instead of the wallcovering. Although the process is a little messier and takes more cleanup, putting the adhesive on with a roller is faster and requires less energy. If you can, discuss the project with a wallcovering supplier who not only retails the supplies but has experience in this sort of work, and perhaps subcontracts it as well. An expert's recommendations concerning particular wallcoverings and both brands and types of adhesives that are locally available can be invaluable.

Wallcoverings can be applied to any smooth, sound ceiling surface, finished or unfinished. If the surface is unfinished, it must first be sealed or primed with a product recommended for that particular sheathing material. If painted, the finish must be in good condition. Any minor defects must be repaired first; vinyl paste spackle will generally do the job. Loose or flaking paint must be scraped or sanded away. If this condition exists, check the entire surface to determine how serious the problem is. If the paint bond is not good, it might simply peel off, taking the new wallcovering with it. If there is any question, you should re-cover the ceiling with a new sheathing (see "Re-covering Ceilings" earlier in the chapter).

If the finish is semi-gloss or high-gloss paint or varnish, cut the gloss with a deglosser or by light sanding. Wash the surface with warm water and a mild detergent to remove dust, grime, and grease. If the finish is a wallcovering, strip it away. Then wash with plain water or a commercial glue solvent to remove the old adhesive. In some cases, applying a primer or sealer coat to a previously finished surface is advisable; follow the recommendation of your wallcovering supplier. In all cases, the surface should be thoroughly coated with wall size and allowed to dry before you begin.

One further note: The atmospheric conditions can play an important part in a ceiling papering job. Arrange to do the job when the room temperature is between 65° and 80°

Fahrenheit and the relative humidity is in the middle range—40 to 60 percent. Very high humidity lengthens the set-up time of the adhesive, and the covering will tend to peel off of its own weight during this time. Low humidity hastens the process and shortens your working time too much, and the adhesive can start to cure before you are ready, resulting in bonding that is uneven or weak. High temperature speeds up the adhesive tacking and set-up times, and also makes a miserable working condition, especially at ceiling level. Low temperatures impede the curing process and lead to improper bonding. A happy medium is needed, for both good results and good working conditions.

PLANK CEILINGS

Wood plank ceilings are a popular alternative to conventional painted plaster or gypsum wallboard. The planking is not especially difficult to put up, and it can be prefinished; some planking is factory prefinished. Any species of wood can be used; thicknesses from 1/4 to 3/4 inch are most appropriate. Wall planking products work well, and are available in several species. They are packed in convenient bundles, end matched and tongue and grooved. General-purpose planking is also available in several species, unpackaged and in various lengths and widths, tongue and grooved but not end matched. There are also many species in numerous square-edged sizes, available through specialty wood dealers.

Planking from 1/2-inch thickness up can be attached directly to open ceiling joists or to furring strips, if their bottom edges are reasonably level and in line with one another. If they are not, the framework should be properly leveled first (see "Leveling Ceilings" later in the chapter). Tongue-and-groove planking is best for this purpose. If square-edged planking is used, first install a barrier across the framework to prevent dust and debris from dropping down through the cracks between planks. Use polyethylene sheeting only if the placement is correct for a vapor barrier. Otherwise, put up a vapor-permeable material such as building paper, or install thin gypsum sheathing.

All thicknesses and types of planking can be installed against an existing ceiling to cover it over. Planking less than 1/2 inch thick is best applied against an existing ceiling or a new ceiling sheathing, either 3/8 or 1/2 inch thick. The material most commonly used for this purpose is gypsum wallboard or gypsum sheathing, but plywood could also be used (see "Replacing Ceiling Sheathing" earlier in the chapter).

Preparations

The first step in installing a plank ceiling is to remove any ceiling lighting fixtures, after turning off the feeding electrical circuits. If an old ceiling is to remain in place, take down the fixture outlet boxes too, and any wall/ceiling joint molding. If the old ceiling or an open ceiling framework is off level, this must be made level. If the planking you will install is 3/4 inch thick, minor variations in planeness will be ironed out and unnoticeable (especially if it is tongue and grooved). The thinner the planking, the more it will follow the existing ceiling or framework contours, so the closer to planeness it should be.

Planks $1/2$ inch thick or more should be nailed solidly to joists or furring strips, whether applied over an existing ceiling or not. In the case of an existing ceiling, find the centerlines of the joists or furring strips with a stud locator and snap chalklines on the ceiling to use as nailing guidelines. Thick planks are usually set at right angles to the joists or furring strips and nailed at each crossing point. Trim the first plank so that one end snugs into the wall corner and the other lies at the centerline of a joist or furring strip (or trim it to fit wall to wall). Face nail square-edged planks with finish nails that are long enough to extend at least $3/4$ inch into the joists or furring strips. Use two nails at each point for nominal 4-inch-wide planks, three for wider ones. If the planks are tongue and grooved, set the first plank with the groove toward the wall. Face nail at the grooved edge of the plank and about $1/2$ inch in from the edge, and toenail the tongued edge (Fig. 5-33). Finish out the first run, then set the second plank, and so on across the ceiling.

The planks can also be run diagonally, but count on some waste. Each end next to a wall must be backed with a solid nailing point, and each end joint must fall over the centerline of a joist or furring strip where the two plank ends can be well anchored. Start with a triangular piece fitted in a corner. If the planking is tongue and grooved, begin with the tongue toward you. A herringbone pattern or various geometrics can also be installed if

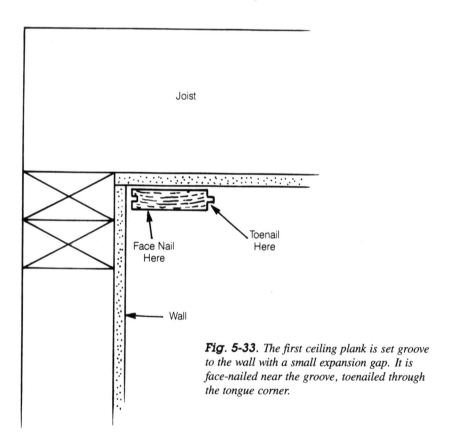

Joist

Face Nail
Here

Toenail
Here

Wall

Fig. 5-33. *The first ceiling plank is set groove to the wall with a small expansion gap. It is face-nailed near the groove, toenailed through the tongue corner.*

you plan ahead and work the design to fit the joist or furring strip spacing. For some patterns the starting point is in the center of the ceiling. This means that you will have to install a sizeable piece of plywood flush with the joist or furring strip surface, to give you a solid surface to which you can nail the relatively small starting pieces.

Planking thinner than $1/2$ inch, particularly packaged wall planking which may be as thin as $1/4$ inch, is best secured with construction or panel adhesive. Often a combination is handiest: Use glue on every piece plus a few small panel nails here and there to hold everything together and aligned. Adhesive will bond very well to new gypsum sheathing or wallboard. If the backing is an existing ceiling, wash it first with warm water and a mild detergent, and allow plenty of drying time before putting up the planking. If the old ceiling is painted, make sure the paint is well bonded and scrape or sand any loose spots. The glue bond to the planking will only be as strong as the paint bond to the ceiling. Use nails along with the glue here. If the ceiling is papered, strip the paper first, for the same reason.

There are several advantages to using the packaged wall planking products. Although relatively expensive, the planking is light and easy to work with, it is tongue and grooved as well as end matched for easy installation and alignment, and you can lay it up in any pattern or direction you desire. Snapping chalklines along joist or furring strip centerlines for nailing guides is still a good idea if you are working on an existing ceiling, however. Then you can easily drive nails wherever convenient. If the ceiling is plaster over wood lath, you can nail into the lath.

Procedure

Start the installation in a corner or along a wall for diagonal or parallel planking, or at centerpoints as required for patterns. Work with the grooves away from you so that you can nail into tongues as necessary. Apply fairly liberal beads or zigzags of adhesive to each piece, fit the groove onto the previous tongue, and press the piece against the ceiling. Nail wherever you can with finish nails or long ring-shank panel nails. You might have to drill small pilot holes for the nails if you are using wood that is hard, like oak or ash. Have two or three lengths of 1-×-2 furring on hand, cut to fit between the floor and the new ceiling level. If a plank doesn't want to fit tightly against the ceiling while the adhesive cures, you can jam a prop under it to hold it in place. At some points you might find that face nailing a piece is easier than trying to toenail through a tongue. That creates no problem: Set the nail slightly and fill the hole with a putty stick, and it will be unnoticeable.

After every few pieces, check the alignment of the planks to make sure they are not becoming misaligned. Use the end-matched joints in the ceiling field as much as possible. When you have to cut pieces, abut the cut ends against a wall where they will be covered later with a molding. To cut down on waste, you can reshape square cut ends into similar matching ends by rabbeting them on a table saw. Then shape any bevels or rounded edges with a wood file or a plane.

If the planking is factory prefinished, no further work must be done after installation, except perhaps nail hole filling. If it is not prefinished, consider finishing the individual planks before installing them. This is much easier than finishing after installation, and probably only minor touchup at cut joints will be needed later, if at all.

Usually a wood molding is installed at the wall/ceiling joint to cover the raw and often somewhat uneven edges or ends of the planking. Any molding design can be used. The molding is usually painted to match other trim in the room. This can be done beforehand and then touched up as necessary after installation. If the planking is painted, the molding can be painted to match. If the planking is stained or has a natural or clear finish, occasionally the molding is made up of the same wood and finished to match the planking.

There are many similarities between planking ceilings and planking walls. For more details, refer to "Paneled and Planked Walls" in Chapter 3.

ACOUSTIC TILE CEILINGS

Acoustic ceiling tile, usually simply referred to as ceiling tile, is available in a wide variety of patterns and is suitable for almost every room in a house (Fig. 5-34). It is not recommended for use in high moisture areas such as saunas, steam rooms, or bathrooms, but can be used in moderate moisture areas such as kitchens, laundries, or lavatories. Most ceiling tile is made in interlocking 12-inch squares $1/2$ inch thick. However, plank style tile in 4-foot lengths, packaged in assorted widths is also available (Fig. 5-35). Each tile is formed with a $1/2$-inch-wide mounting flange on two adjacent sides and a mating slot on the other two adjacent sides. There are two types: wood fiber composition, and fire-retardent mineral fiber composition. Ceiling tile can be installed in three different ways; stapled to furring strips, glued to an existing ceiling, or clipped onto special metal tracks made for the purpose. The tracks themselves are usually attached at right angles to either joists or furring strips. Sometimes they are attached to and parallel with furring strips, either directly or through an existing ceiling sheathing.

The first consideration is the relative planeness and levelness of the existing ceiling or ceiling framing. If this is satisfactory, tiling can begin. If not, the ceiling must first be leveled. See "Leveling Ceilings" earlier in this chapter.

Estimating Material Requirements

The next step is to calculate your tile needs. Usual practice is to center the tile field in the ceiling area for a balanced appearance. This means, in most cases, that the border tiles—those next to all of the walls—must be trimmed to other than full size. Measure the distance between two opposite walls, then the other two opposite walls. If the measurements in both directions are exactly to the foot, whether odd or even numbers, plus zero or minus about $1/2$ inch maximum, you are in luck. The border tiles will be either full or half size; just multiply to find the area in square feet of the ceiling, and that will be the number of tiles you will need. Then, because ceiling tile is most often sold by the case, you will have to round to that amount.

It is likely, however, that the wall-to-wall measurements will not be the same in both directions. Suppose the distance from wall A to wall B is 10 feet $5^{1}/2$ inches. Ignore the

Fig. 5-34. *A typical ceiling tile installation. (Courtesy Armstrong World Industries, Inc.)*

Fig. 5-35. *A typical plank ceiling tile installation. (Courtesy Armstrong World Industries, Inc.)*

feet, take the 5^1/$_2$ inches, and add 12 inches—a total of 17^1/$_2$ inches. Divide that by 2, for 8^3/$_4$ inches, which is the width of the border tiles along walls A and B. If the measurement between walls C and D is 8 feet 10 inches, the border tiles along those walls will be 11 inches wide (10 plus 12 equals 22, divided by 2 equals 11). Now take the overall room measurement of 10 feet 5^1/$_2$ inches by 8 feet 10 inches and round up to the next highest foot, which is 11 feet by 9 feet. Multiply these to arrive at the ceiling area—99 square feet—and translate that to the number of cases of tile required.

Layout

In the case of an irregularly shaped room, you will have to decide whether the tile field should be centered, offset, or simply based against one wall. For example, one long wall might be mostly covered by built-in floor-to-ceiling bookshelves, leaving an inset jog at each end. You could center the tile field between the outer edge of the shelving and the opposite wall, or between the two walls. If there are relatively small structural niches such as a bay window, or protrusions such as a masonry fireplace, the tile field is usually centered between the principal wall lines and the tiles trimmed or filled in to suit the odd shapes. If the room is oddly shaped or has several walls running off at different angles, the best bet might be to base a row of full border tiles along the longest wall and trim all the other border tiles to fit. Some will be oddly shaped and/or less than half a tile.

Installation Considerations

The next steps depend upon the method you will use to mount the tiles and the surface to which they will be attached. However, consider the following items that apply to all installations:

- If you purchase your tile well before you can get on with the installation, store the unopened cases in a dry place away from sunlight. The temperature should not fall below about 60° Fahrenheit or go above whatever the outdoor temperature dictates—preferably about 80°.
- About two days before you plan to install the tile, move it to the room where it will be installed, take the tiles out of the boxes and stack them flat, and let them adjust to the atmospheric conditions.
- When you install the tiles, make sure your hands are clean and dry; the tiles take smudges and fingerprints readily.
- To trim the tiles, use a very sharp razor knife or utility knife and change the blades often. Cut with repeated, even strokes against a straightedge, with the tile face up. The thick steel of a try square works very well for this.
- Inspect each tile face to make sure it is undamaged and dust it with a soft brush before you put it up. Also, remove any loose bits of material from the inner mating edges.
- Remove pencil marks or smudges with a gum eraser, but do not scrub hard.

- Wear safety goggles or a face shield as you work, because bits of fuzz or mineral particles do drift down on you.

Note: Some people find that the tile material, especially the fire-retardant mineral type, causes mild skin irritation. If you run into that problem, wash, then use a good hand cream. Wear thin plastic throwaway gloves to prevent irritation.

Gluing Tile to Existing Ceiling

First, square the room by the 3-4-5 method, as follows. This is necessary because the walls of few rooms are square to one another, so the tiles are likely to go off line if you try to align them to the walls. The border tiles must be measured and cut individually to prevent this. Select one of the long walls of the room and measure out from it at each end the width of the border tiles, plus $1/2$ inch. Taking wall C from the example given earlier in this section, you would measure out 11 inches plus $1/2$ inch for the tile flange. Make a mark at the two points and snap a chalkline on the ceiling between them, from wall A to wall B. Now measure back along the chalkline from wall A a distance equal to the width of the wall A border tiles—8 $3/4$ inches, as in the earlier example—plus $1/2$ inch. Make a mark X. From mark X, measure back along the chalkline 3 feet and make a mark Y. Again from mark X, measure alongside wall A 4 feet and strike a small arc. Then from mark Y, measure toward that arc 5 feet and strike a second arc that crosses the first. The intersecting point of the two arcs is mark Z. Snap a chalkline from mark X across mark Z, from wall C to wall D. The two chalklines are in square, and will be your guidelines for setting the tile (Fig. 5-36).

Unlike floor tile, ceiling tile installation starts in a corner, and this must be whichever corner you square up. In some rooms you can select any corner to square up, in others, however, there might only be one or two 90-degree corners to work from. Select the most convenient corner.

Measure and trim one tile to fit in the starting corner within the intersecting chalklines. The dimensions for this tile, according to the previous example, would be 11 inches along the edge abutting wall A and 8$3/4$ inches along the wall C edge. This was an estimating measurement, however, not an actual dimension, nor does it account for the $1/2$-inch flange. Take the actual measurements from the chalklines to the walls. Transfer those measurements to the tile and trim it so that you discard the grooved edges of the tile and leave the flanged edges. The flanged edges will face outward, the trimmed edges against the wall. Include the flange width, but allow about $1/4$ inch of clearance at the walls. Following the instructions included with the ceiling tile adhesive, daub the adhesive on the tile back and press the tile into place. The flanges should exactly align with the chalklines, with about a $1/4$-inch space at the walls (Fig. 5-37).

Next, trim a border tile to fit along wall C. Again (and as a rule of thumb), the flanges should face outward. Apply the adhesive and slip this tile in place while holding the first one steady and in position. Make sure the joint is snug. Then trim a third tile to fit next to the first one along wall A, and set it in place. The fourth tile is a full one, mating with the

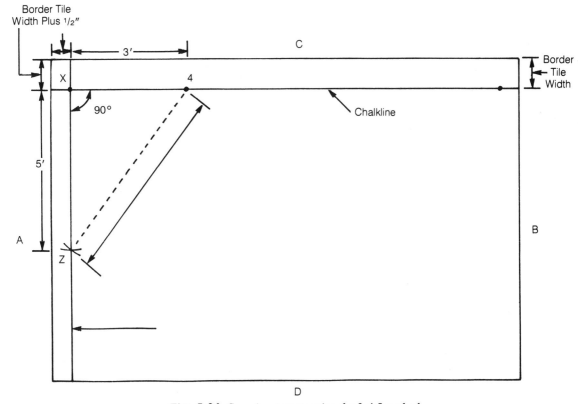

Fig. 5-36. *Squaring a room using the 3-4-5 method.*

three previously set. Continue by setting a border tile or two, filling in with full tiles, and so on (Fig. 5-38). When you come to the end of each row, measure the border tile from both corners of the previous tile to the wall. If the wall runs off-square to the others and to the tiles, you might have to trim the border tile to that same angle rather than square across.

Stapling Tiles to Furring Strips

Do not use a pneumatic stapler for this procedure; it is too powerful and will drive the staples right through the tile flanges. Many electric staplers will do the same. A heavy-duty manual stapler that fits your hand comfortably and ejects the staple very close to the snout of a flat-faced head is the best bet. Follow the tile manufacturer's recommendation as to staple length. This is usually ¹/₂- or ⁹/₁₆-inch leg length, either the narrow or wide crown type.

The furring strips should be kiln-dried 1-×-3 or 1-×-4 stock. They can be fastened at right angles to joist bottoms or leveling boards, or to other furring strips that have been installed for leveling purposes, or through an existing ceiling into joists or furring strips.

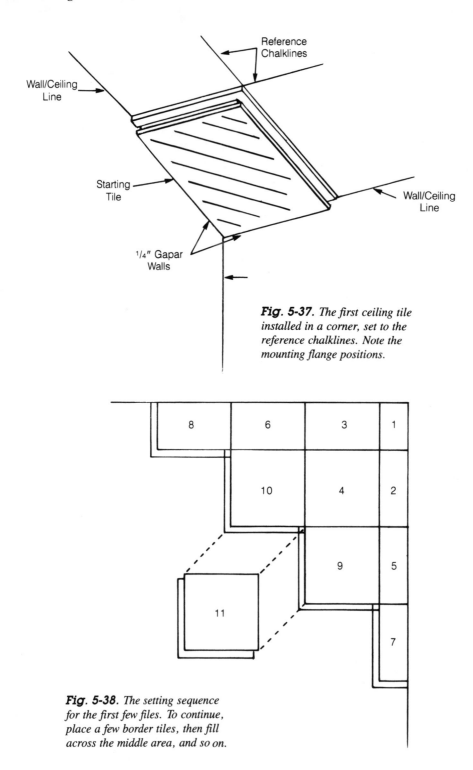

Fig. 5-37. The first ceiling tile installed in a corner, set to the reference chalklines. Note the mounting flange positions.

Fig. 5-38. The setting sequence for the first few files. To continue, place a few border tiles, then fill across the middle area, and so on.

Use a pair of nails at each nailing point, long enough to penetrate at least 1 inch. End joints of strips must fall over the centerline of a joist or furring strip for solid mounting, and should be staggered so no two are in line.

If the old ceiling is still in place and needs a leveling operation, refer to "Leveling Ceilings" earlier in the chapter and relate that information to the following details on proper installation and spacing of furring for ceiling tiles. If the old ceiling is suitably plane and level but needs to be furred because the surface is unsatisfactory for installing ceiling tiles with adhesive, the first step is to find the joists or furring strips that lie above it. Do this with a stud locater or by tapping with a hammer (a solid thud indicates a joist or strip, a hollow bonk means nothing). Drive nails partway through the ceiling to find the outer edges of each joist or strip at each end, and mark their centerlines. Then snap chalklines on the ceiling to indicate each strip or joist centerline. In most cases they will be 16 inches apart, but you might find some variations.

Begin by installing the first furring strip snugly against a wall at right angles to the joists or old furring strips. Then, measure out from the wall at each end along a joist or chalkline a distance equal to the width of the border tiles that will go along that wall, plus $1/2$ inch for the tile flange. If this were wall C of the previous example, this would be $11^1/2$ inches. Mount the second furring strip so that its centerline is this distance from the wall. Mount the third and subsequent strips exactly parallel with the first two and 12 inches apart center to center (Fig. 5-39). The next-to-last strip at the opposite side of the room should fall with its center the same distance from the wall as the width of the border tile, minus $1/2$ inch. The last furring strip is mounted snugly against that wall.

The next step is to square the room, using the 3-4-5 method as explained earlier. The only difference is that instead of snapping the chalklines on the ceiling, you will make the

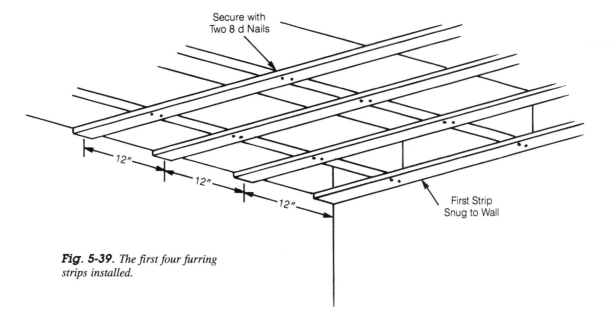

Secure with
Two 8 d Nails

12"

12"

12"

First Strip
Snug to Wall

Fig. 5-39. *The first four furring strips installed.*

first line and marks X and Y on the centerline of the second furring strip out from the wall. The arcs and mark Z will fall on the sixth strip (Fig. 5-40).

To install the tiles, follow the same procedures and setting pattern as outlined in "Gluing Tiles to Existing Ceiling." Instead of glue, use two staples on the flange of trimmed tiles that seat fully against a furring strip, and use one or two staples through flanges that straddle two furring strips. Drive two or three staples through the seated flange of full tiles, plus one through the far end of the straddling flange. At the far ends of the tile rows—especially where you have to fit around a protrusion or into a niche—there might not be any (or very little) tile flange present. In that case, face-nail up through the tile close to the edge with a pair of 1-inch wire nails (not brads, which have no appreciable heads). Drive the nails until the heads just pull the tile up snugly but before they crush through the tile surface.

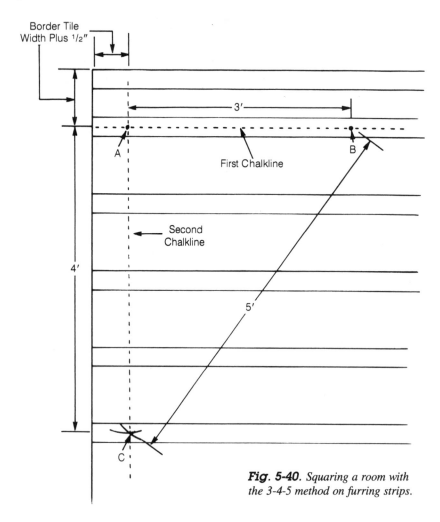

Fig. 5-40. Squaring a room with the 3-4-5 method on furring strips.

Runner or Track Installation

Setting up the track or runner system for ceiling tiles (Fig. 5-41) is a fairly complicated bit of business, but once that is done and the first few tiles are set, the remainder proceeds quickly and easily. The installation details of this system vary among products. The easiest approach is to first determine whether or not such a system is appropriate for your installation, and if so, which particular one. Read the manufacturers' literature. If you decide to use one, determine from the product estimating chart how many installation kits you will need for the job. Then spread some of the bits and pieces out on the floor in approximately the same way they will go on the ceiling. Mull over the installation instructions that are included to get an idea of how the system all fits together. When you get everything sorted out, proceed with the installation. The key point to keep in mind is that if you install the tracks or runners just right, you will end up with a perfect ceiling. If you don't, the results will be less satisfactory.

Tiling Juts and Jogs

Most rooms have at least one jut-out or jog-in that you will have to fill in with tile or cut out around. This chore is a little bit easier if you can arrange your starting corner and wall to be opposite the one with a jut or jog. Regardless of their positions, however, if you are gluing the tiles to an existing ceiling, all you need to do is measure, trim, and fit each tile successively to the wall contours. The rule of thumb, as mentioned earlier, is to always keep the mounting flanges of the tile you are installing toward you and exposed, locking it over the flanges of the previous tile. However, there are occasions when you can do the opposite and work from the backside, especially when putting small pieces into awkward spots. Position each piece first, without adhesive, to make sure you can get it properly into place.

When furring a ceiling, set the principal furring strips across the main field of the ceiling first. If the strips are at right angles to a jog or jut, just run them fully into or against it. If parallel, bypass it with the full-length strips, then go back and install the necessary number of parallel strips, properly centered within the jog area. Or install a single strip snugly against the face of a jut. Then measure, trim, and fit the tiles into or around the jog or jut in logical succession.

It can happen that the only convenient starting corner is itself in a jog, or is crowded by a jut-out, leaving insufficient room to square up using the 3-4-5 method. In that case, add the width of the border tiles along the starting wall to the combined width of as many full tiles as are necessary to move out onto clear ceiling area. Snap the first chalkline here. Suppose that the corner is inset 2 feet, for instance, and the border tiles for a properly centered tile field are to be $8^1/2$ inches wide. Measure out $32^1/2$ inches, plus $^1/2$ inch for the tile flange, and snap the chalkline there. Then square up in the usual fashion from that line. Once that is done, you can measure back toward the wall to determine the correct tile line positions, and the actual size and positioning of the starter tile in the corner. When that is set, all else will follow.

Fig. 5-41. Ceiling tiles can be installed on a special steel runner system that saves a considerable amount of work. (Courtesy Armstrong World Industries, Inc.)

Tiling Odd Shapes

Rooms or areas can be oddly shaped, or might have only one or two right-angle corners. This creates options and perhaps problems in squaring up and setting the tile field for a uniform appearance. In such a case, you must simply make whatever choice seems most reasonable. In Fig. 5-42, for example, you might center the field between walls A and B and square up from either main back corner. That, however, will mean that the border tiles along walls C and D probably will not be alike; one row might be narrow and the other wide, depending upon the specific room dimensions. Or, you could take the centerline of wall E and balance the tile field on that. Thus, the border tiles along walls C and D would be alike, but those along walls A and B would probably not be alike. The same situation is true in the opposite direction. You could center the field between walls E and F, or between walls G and F. You could not center between walls A and C and also B and D because that would give you two mismatching fields.

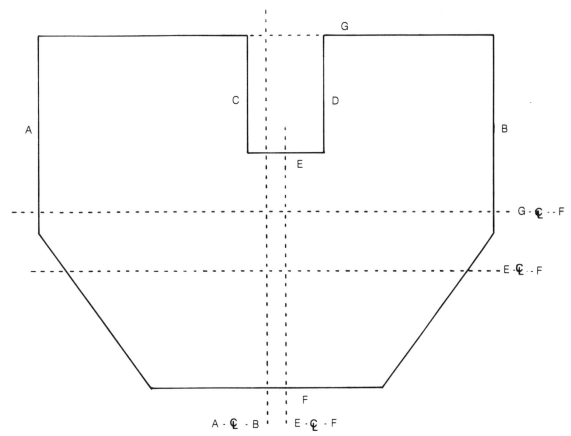

Fig. 5-42. *In many odd-shapped rooms there are several possibilities for establishing a base reference centerline, all of which result in different tile field alignments. This is an actual example.*

There are also rooms or areas that have no right-angle corners, as might be found in a dome or a round house. In that case, you must make an artificial right-angle corner to square up the tile field and to work from. Figure 5-43 shows examples of this process.

SUSPENDED CEILINGS

A suspended ceiling (Fig. 5-44) is a complete system consisting of an interlocking metal grid framework, ceiling tiles to fit within the grid, and a series of hanger wires. The grid is made up of wall molding, main runners, and crosstees. The visible portions of the grid are usually painted white, but black and simulated wood grain are also available, and the gridwork faces can be repainted in any color. Different interlocking methods are used, and the gridwork pieces come in standard lengths that must be trimmed to fit as necessary.

The tiles, usually called *panels*, are available in wood fiber and fire-retardent mineral fiber types, in numerous patterns and textures, and in both 2- × -2 foot and 2- × -4 foot sizes. Most are both acoustic and washable. There are two styles: regular and recessed. Regular or standard tile lies within the flanges of the gridwork; the thin faces of the gridwork hide the joints between tiles. Regular, or recessed, tile simulates the appearance

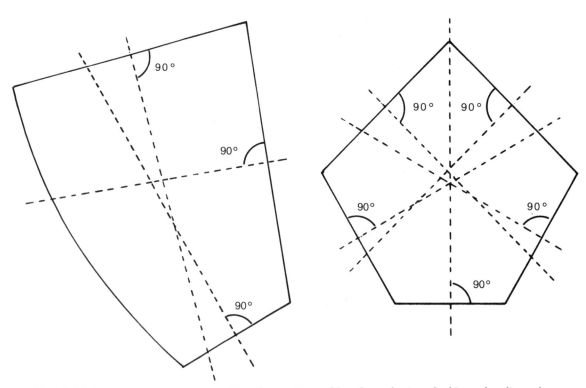

Fig. 5-43. *In some rooms no true centerline reference is possible only a selection of arbitrary baselines where a convenient right-angle starting point can be established.*

Fig. 5-44. *A typical suspended ceiling installation. Recessed panels are used here.*
(Courtesy Armstrong World Industries, Inc.)

of large ceramic tile; the tile faces lie below the gridwork to create wide, deep joint lines. The hanger wire used in residential installations is ordinary 16-gauge steel wire. The hangers can be attached to common screw eyes or screw hooks, or even nails in some instances. Also available are special fluorescent troffer lighting fixtures that fit directly into the gridwork. Translucent plastic panels are set into the gridwork beneath the lighting fixtures; several different patterns are available.

Although this system sounds complex, it actually goes together easily and makes for a fine appearance at a competitive cost. A suspended ceiling can be installed in any room except where high humidity or steam is present. It is an ideal solution for lowering the ceiling level in a high-posted room; or for easily covering over an old ceiling in poor condition; or for covering a maze of pipes, wires, and joists in a basement ceiling. It can be installed below any existing ceiling, or where no ceiling exists.

The only criterion for a suspended ceiling is that there be sufficient space between the gridwork and the ceiling or joists above, to allow working space and room enough to slide the tiles into place. About 6 inches is a practical minimum, but much depends upon the lie of the grid members with relation to exposed joists, the presence and position of pipes and electrical wiring, and whether or not troffer or other lighting fixtures will be installed. If space is tight, discuss your proposed installation with your supplier before starting. The minimum height for a finished ceiling is considered to be 7 feet 6 inches above the floor, except for kitchens, hallways, and bathrooms, where 7 feet is acceptable. The maximum distance between the suspended ceiling and the existing ceiling or structural framework above is dictated by practical considerations.

Layout

The first step is to diagram the ceiling layout to scale on graph paper. Decide upon the new ceiling level, keeping in mind that there should be a clear line all around the room with no interference from anything that cannot be moved or worked around. Measure the length of each wall, including any jogs or jut-outs, and transfer them to the paper in the form of an outline of the room as it is shaped at the new ceiling level. Make a note of each dimension on the drawing. Also, note the greatest width and length of the area, which is simply wall to wall in both directions in most rooms. And, determine the direction in which the ceiling joists lie and note that.

Next, decide whether you will install 2-×-2-foot or 2-×-4-foot tiles. In small or long, narrow rooms, most folks feel that 2-×-2 tiles look better; either size is appropriate for larger rooms.

The gridwork is formed on a 2-×-4-foot module, and the initial layout is made on that basis. If 2-×-2-foot tiles are used, additional crosstees are added after the basic modular framework is installed. Your next task is to decide which way the basic grid should be installed—parallel with the long axis of the room, or at right angles to it. If you will install 2-×-4 tiles, do you want the tiles to lie lengthwise along the long walls, or parallel to the short walls? If you will install 2-×-2 tiles, this orientation makes no difference, with two possible exceptions: First, one direction or the other might affect the ease of installation or the amount of grid material required. Crosstees come in standard 2- and 4-foot lengths, and the main runners in standard 12-foot lengths. Second, if the ceiling is long and narrow and you intend to install 4-foot troffer lighting fixtures (in a corridor kitchen, for example), the fixtures might best be centered and positioned parallel with the long axis of the ceiling. Thus, the grid layout must also run that way.

The grid system must be centered in the room, so the border tiles will almost never be full units. To find the side border tile size with 4-foot panels running parallel with the short axis, determine the total maximum length of that axis in inches. Divide that number by 48. To the remainder of your answer, add 48. Round this answer up to the nearest whole number, then divide by 2. That answer is the length of the border panels at each end of the long-axis runs. Example: The long axis measures 12 feet 3½ inches, or 147½ inches. Dividing by 48 equals 3 with a remainder of 16½. Adding this to 48 equals 64½. Round this off to 65, which divided by 2 equals 32½ inches—the length of the border tiles. Now repeat the process with the length of the long axis to find the width of the end border panels.

If the panels will parallel the long axis of the room, or if 2-×-2-foot panels will be installed, follow the same procedure. Use the short-axis measurement, but divide by 24 instead of 48, and add 24 to the remainder instead of 48.

Now, lightly draw in the border panel lines on your drawing—as close to scale as you can manage—the required distance from the long and the short wall lines. Then heavily shade in the two border panel lines that lie at right angles to the joist direction. These lines represent the two outside main runners. Between them, draw in more heavy parallel lines at 4-foot intervals, as required, to represent intermediate main runners. Note that in a narrow room where the panels are parallel with the long axis, there might be only two main runners, and they might only be 2 feet apart. Or in a hallway that is 4 feet or less in width, there could be one main runner down the centerline—or no main runner, just crosstees.

Next, using a different color, draw in the crosstees at right angles to the main runners, and at 2-foot intervals. There will be 4-foot lengths between main runners, less between border main runners and the walls. For 2-×-2 foot panels, add more 2-foot crosstees between the midpoints of the 4-foot crosstees, again using a different color. Then, if lighting fixtures are to be installed, note the locations on the grid for ordinary ceiling fixtures that will be cut into or mounted on the panels. Also, shade in the 2-×-4-foot spaces where troffers will be placed in the gridwork. Figure 5-45 shows a completed plan.

Estimating Material Requirements

Material estimates come next. Add up all the wall measurements noted on your drawing, and round to the nearest foot. Then divide that by the length of a standard section of wall molding, as will be used in the system you have selected. Then round up to the number of full pieces required. If the wall perimeter is 487 inches, for example, that converts to 40 feet 7 inches and rounds up to 41 feet. If each wall molding section is 10 feet long, you will need 5 sections. Note, however, that if there are any poured concrete walls in the

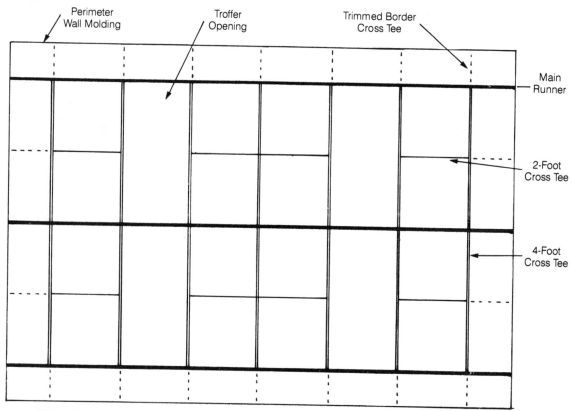

Fig. 5-45. *An example of a finished plan for the gridwork of a suspended ceiling installation.*

room, you will not use regular wall molding there. Instead, calculate the number of main runner sections that will be needed to go along those walls.

Next, figure the number of main runner sections you will need, other than for use as substitute wall molding. There is a locking mechanism at each end of each section so they can be joined, which means that you can cut no more than two pieces from one section when piecing together a long run. If a standard section is 12 feet long, for instance, and you need two 9-foot main runners, you will need three sections: Cut 3 feet off the ends of one section, add them to the other two, and throw the middle piece away. For three 15-foot main runners, you would need four sections.

Count up the required number of full-length 4-foot and 2-foot crosstees noted on the drawing. Then note the number of crosstees that will be needed at the borders, determine whether you can use trimmed 4-foot or 2-foot pieces, and add those. Count up the number of ceiling panels, making whatever allowance is needed for cutting border panels from full ones, and for fitting small pieces into jogs or odd angles. You will need enough hanger wire to suspend the main runners every 4 feet, plus six lengths for every troffer lighting fixture, plus a few spares. Each length should span the distance from the gridwork to the point of attachment above, plus 6 inches at each end for wrapping. And, you will need a screw eye or hook for every attachment point.

Installation

Having accomplished all this, you can now start the easy part—the installation. Begin at any corner at the new ceiling height. Use a 2-foot—or preferably 4-foot—spirit level as a straightedge, and lay it end over end along the wall. Draw a level line across all the walls except the last one. Snap a matching chalkline across the last wall, which should be one of the longest ones. This line will be the guideline for mounting the wall molding. Hint: If the molding has a slightly rounded bottom corner, you will probably find it easier to set the top edge of the molding against the line rather than the bottom corner. If so, position the line higher than the desired finish ceiling level by the height of the wall molding.

If the wall is wood framed, find the wall studs all along the molding line. Do this by using a stud locater, measuring for 16- or 24-inch centers, and tapping a nail through the wall sheathing. Make a mark where each stud crosses the molding line. Cut the first piece of wall molding as required to fit along one wall. Outside corners should be cut to a tight miter, and inside corners are butt jointed (Fig. 5-46). Hold the piece in place along the line and mark the back flange at each stud location. Remove the piece and cut holes for the mounting nails at each mark. You can punch them, but drilling helps prevent damage to the molding. You can set as many nails as you wish, but they should be no more than 24

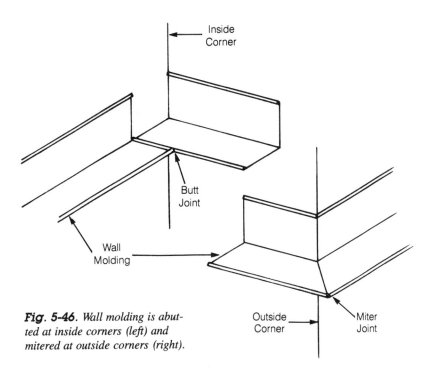

Fig. 5-46. *Wall molding is abutted at inside corners (left) and mitered at outside corners (right).*

inches apart at any point, and there should be one nail within at least 2 inches of each end. Nail the piece up—6d box nails are usually sufficient—and continue in this way all around the room (Fig. 5-47).

If the wall is concrete block, marking the molding and predrilling holes is not necessary. Drive short masonry nails directly through the molding. To avoid cracking a block, set the nails into the mortar joints between the block and the mortar.

If the wall is poured concrete, no wall molding will be nailed to it. Instead, suspend lengths of main runner from above so that they rest in proper position against the wall (Fig. 5-48). Cut the pieces to fit as necessary. Set screw eyes or other anchors above and in line with each piece, and wrap lengths of wire onto them with three or four twists. Run the wires through the slots in the pieces, adjust the runner height to coincide with the level line on the wall, and wrap those ends. If the runners hang a bit cockeyed, don't worry— the ceiling panels will straighten them out. However, the runners must be level.

Next, if the ceiling joists are not exposed, locate them and snap chalklines on the ceiling to indicate their centerlines. Refer to your drawing, and snap a chalkline across the ceiling or joists, at right angles to the joists, to indicate the position of the first main runner. Then snap exactly parallel chalklines at 4-foot intervals to indicate the remaining main runners. At every point where a main runner chalkline crosses a joist, drive a screw eye or other anchor. Thread a premeasured length of hanger wire through each and secure it with at least a triple wrap around itself (Fig. 5-49).

The alignment strings are next. Accuracy here is essential because the entire gridwork follows these alignment strings. Pick your starting corner and measure out along the back face of the wall molding a distance equal to the border tile width for that wall. Repeat along the adjacent wall. Go to the opposite ends of those walls and measure out the appropriate border tile widths. Then stretch a pair of taut lines between those respective points

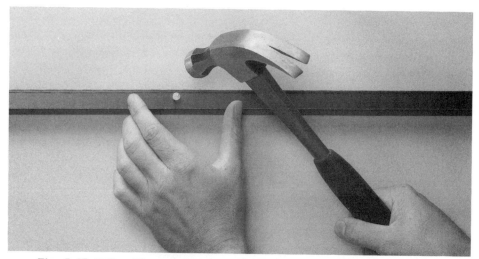

Fig. 5-47. *Wall molding nailed in place. (Courtesy Armstrong World Industries, Inc.)*

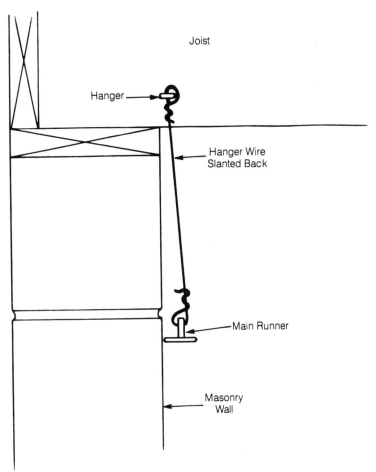

Fig. 5-48. *Instead of attaching wall molding to poured concrete or block walls, main runner lengths can be suspended.*

(Fig. 5-50), set to lie just below the wall molding. The best way to do this is to slide the string up behind the molding and anchor it at any handy point above—drive a nail if necessary—and let it lie against the bottom of the molding.

With a framing square, check the angle at the starting corner where the taut lines cross. They must be exactly at right angles to one another. If they are not, move the far ends of the lines until they are. One of these lines indicates the position of the first main runner. The other indicates the first set of crosstees. Which is which depends upon your orientation of the 4-foot panel module you selected earlier. For explanatory purposes: The walls parallel to the 4-foot module are the sidewalls; those at right angles are end walls.

Now you can put up the first main runner. Measure carefully, and trim off the wall end of the first main runner section. This way, when the end is set on the wall molding and snug against the end wall, the first locking slot of the main runner lies directly above the

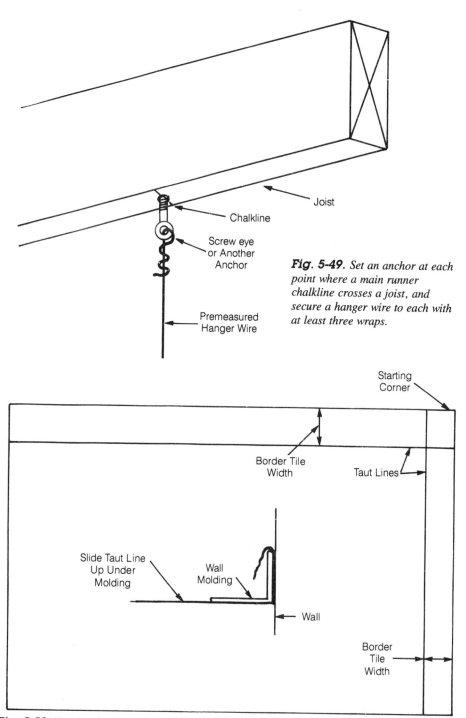

Fig. 5-49. *Set an anchor at each point where a main runner chalkline crosses a joist, and secure a hanger wire to each with at least three wraps.*

Fig. 5-50. *Taut line layout for establishing the position of the first main runner. Note how the lines are tucked up under the wall molding and drawn tight (center inset).*

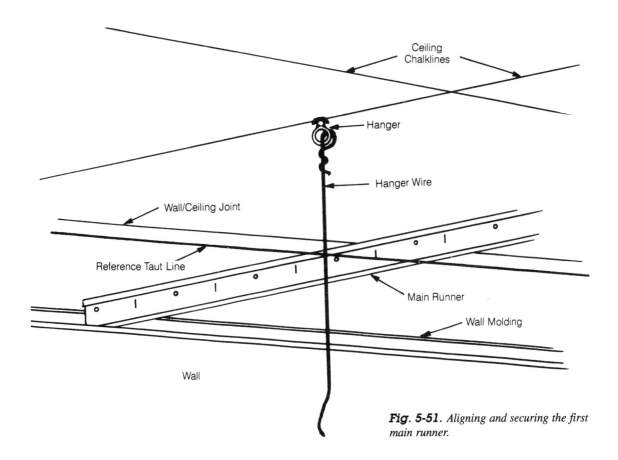

Fig. 5-51. *Aligning and securing the first main runner.*

border crosstee taut line that is parallel with the end wall (Fig. 5-51). If the total length of the main runner is less than that of this trimmed section, just measure from end wall to end wall, then transfer that measurement to the section you have trimmed, minus about $1/16$ inch for clearance, and trim off the other end. Set the runner in place on the end wall moldings, aligned with the sidewall taut line.

At each wire hanger, locate the nearest hanger support hole (do not use the locking slots). You probably will have to tilt the wire forward or backward along the runner. At the point where the wire crosses the hole, use a pair of pliers to crimp a sharp right-angle bend in the wire. Slip the end of the wire through the hole, bend the wire straight up, and twist it around itself three or four times. Meanwhile, you have to make sure that the runner does not lift up off the wall moldings. Hint: Use a small C-clamp and a protective piece of scrap wood to lock the runner to the wall molding while you work. Then install the remaining main runners in the same way, positioning them 4 feet apart (this need not be exact at this point). Measure each main runner individually for its particular placement; do not use the first one as a pattern for the others.

If the main runner total length is greater than one standard section, there are several different ways to hang them level. If the length is no more than about a section and a half, trim the first end wall section as just explained, to properly align the crosstee slot. Then measure the distance from the free end of the first main runner section to the opposite end wall. Trim a second section of main runner to that length, minus $1/16$ inch for clearance, and lock the two sections together. Place a rigid, straight length of wood or strap iron across the joint and against the vertical web of the runner. Clamp it tightly on both sides of the joint with small C-clamps, so that the pieces are aligned and straight. Set the runner on the wall moldings and attach the hanger wires, remove the clamps, and go on to the remaining runners.

If the total length of the main runners is longer than a standard section and a half, here are two other useful approaches: First, trim one end of the first section as explained earlier, to set against the end wall with the first crosstee slot properly aligned. Set the wall end of this runner in place on the wall molding and hold the other end up approximately level. Find the hanger support hole that is farthest from the wall but closest to the farthest hanger wire. In other words, the wire must be tilted away from the starting end wall to reach the hole (Fig. 5-52). Note the hanger wire, set the runner down, and tie a piece of string to the hanger anchor.

Temporarily secure a small spirit level into the angle of the runner section, toward the free end where you will be able to see it. Reposition the runner section and have someone hold the wall end in place on the wall molding (or clamp it). Thread the free end of the string through the hanger support hole, raise or lower the runner on the string until it is dead level, and tie off the string. Go back to the next hanger wire in line, crimp it, and secure it with a few wraps (Fig. 5-53). Then fit the next section of main runner to the first one. If that section is a trimmed one and fits against the opposite end wall, just set it in

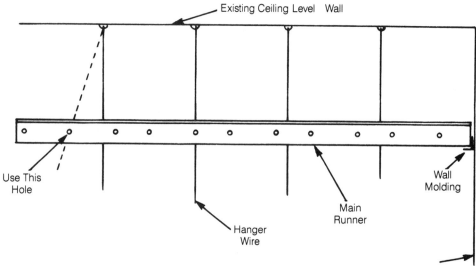

Fig. 5-52. *The end hanger wire must tilt forward so that the main runner is forced back against the wall molding and held in place while the other hanger wires are attached.*

place on the wall molding and go back down the line, attaching all the hanger wires. Remove the level and the string. If still another section is needed, full or trimmed, repeat the performance with the string and level.

The other approach is a bit more accurate but takes longer. Trim and set the first runner section and find the farthest hanger support hole, as just explained. Instead of noting the hanger, mark the hole on the runner. Take the runner to the sidewall and set it on the wall molding, with the trimmed end abutting the end wall. Mark the sidewall through the marked hole. Then do the same at the opposite wall. Drive a nail in the wall at each mark (down at a sharp angle if there is no stud there). Stretch a string from nail to nail, but at the level of the marks. Swing each of the nearby hanger wires to just meet the string, and at the point of intersection, crimp the wire to a 90-degree angle (Fig. 5-54). Remove the string. Set the end of the main runner section on the wall molding, slip the prebent end of the hanger wire through the support hole, and wrap it tightly (Fig. 5-55). Then attach the remaining hanger wires. Follow the same procedure for the first sections of the remaining main runners. If the next sections required are also full lengths, repeat the whole process. If they are trimmed sections, simply lock the sections in place from the first sections to the wall molding and attach the wire hangers.

When all the main runners are in place, you can install all the full-length 4-foot crosstees between them (Fig. 5-56). Ignore the border crosses for the moment. If you are installing 2-×-2-foot panels, put in the full-length 2-foot crosstees as well, locking them at the midpoints of the 4-footers. If you plan to install troffer lights, be sure not to inadver-

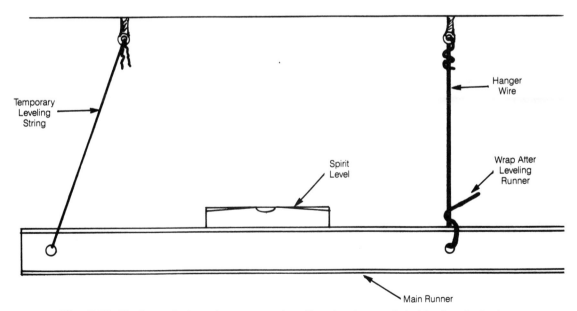

Fig. 5-53. *The free end of a main runner can be adjusted and suspended with a length of string to hold it in place while the hanger wires are attached.*

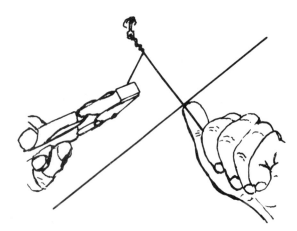

Fig. 5-54. *Crimping a hanger wire to a right angle where it intersects with a taut line.*

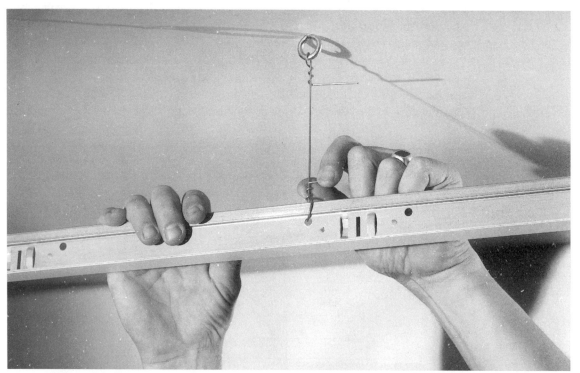

Fig. 5-55. *Attaching the hanger wire to a main runner. (Courtesy Armstrong World Industries, Inc.)*

tently lock 2-foot crosstees into those openings. Once all the full crosstees are in place, drop a few ceiling panels into the grid at random places to help to stabilize the whole affair (Fig. 5-57).

Now cut and fit the border crosstees along the sidewall parallel with the first main runner. Use the taut line as a guide. Shift the main runner (and the rest of the grid) as

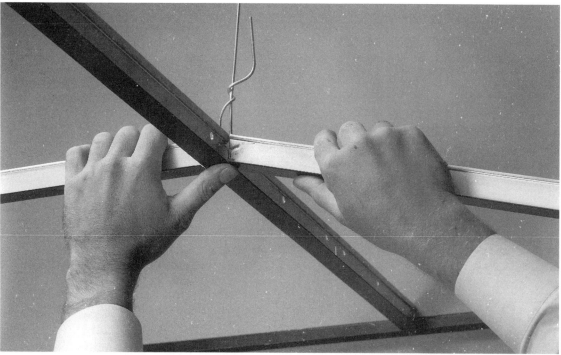

Fig. 5-56. Installing the crosstees. (Courtesy of Armstrong World Industries, Inc.)

Fig. 5-57. Dropping the ceiling panels in place. (Courtesy Armstrong World Industries, Inc.)

necessary until it is centered on the taut line. Then measure from the wall to the nearest edge of the main runner at the crosstee location closest to your starting corner. Transfer the measurement to a crosstee and trim it to fit with sharp tin snips or a hacksaw. Lock it to the main runner with the cut end resting upon the wall molding and snug against the wall. Continue this process, fitting each piece individually, all along that sidewall. Then go back to the end wall at the starting corner, and cut and fit those border crosstees. Continue along the opposite sidewall, then across the opposite endwall.

Next, cut each border tile to fit its individual space. Measure carefully. For regular panels, subtract about $1/4$ inch maximum in each direction for fitting leeway. Mark and cut them with the panel on a flat surface, face up. Use a thick straightedge (a scrap of runner or crosstee works well), and make repeated strokes with a very sharp utility or razor knife until you cut through.

For regular panels, follow the same procedure and cut each piece to a fairly snug fit. Then drop the piece into the grid, position it, and with a pencil trace lightly around the edge of the gridwork from below, on the face of the panel (Fig. 5-58). Hold the piece firmly to keep it from shifting in the grid. Take the panel out, and with a straightedge and knife cut halfway through the panel along all the lines. If the exposed portions of the factory edges are beveled, make these cuts at approximately the same angle. The blade must be very sharp for this, or the edges of the cut will tear and the ragged edges will be visible. Now slice into the edge of the panel at right angles to the first cuts (Fig. 5-59); you can use

Fig. 5-58. *Tracing the trim outline on a regular panel.*

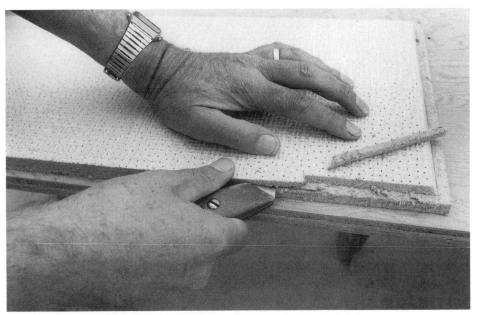

Fig. 5-59. *Cutting the relief edge in a regular panel.*

another panel to judge the proper position of the cut. Make the two cuts meet, then lift off the waste strip, brush off the fuzz, and set the panel in place. Continue fitting the border panels until all are in place.

With the border panels and a random few full panels in the grid, this is a good time to double check to see if all the gridwork is hanging level—no droops or high spots. You can sight along the runners and crosstees by eye, or check here and there with a spirit level. If minor adjustments are needed, you can probably accomplish these by turning in or backing out the hanger wire screw eyes a half turn or so. If more than a slight adjustment is needed, you will have to redo some of the hanger wires. If you will install troffer lights, go to the grid corners where the lights will rest and add an extra hanger wire and anchor at each one. Then position and connect the fixtures and install the translucent plastic panels. Finally, drop in the remaining panels, and the job is done.

Not all rooms are neatly arranged with two long walls and two short ones and no awkward spots. Some suspended ceiling installations require a bit of ingenuity, the process is not always as straightforward as just explained. But as you can see from the plans of an actual installation of a regular system shown in Fig. 5-60, almost any shape of ceiling can be installed. The key is simply to set the main runners first, then make sure the principal crosstees—and thus the main portion of the gridwork, wherever it lies in relation to the ceiling perimeter—is in square. Then secondary and usually shorter main runners and their crosstees can be installed, working outward from and squared to the main field. Then, working again outward from and squared to them, the remaining bits and pieces around the edges can be individually fitted into place.

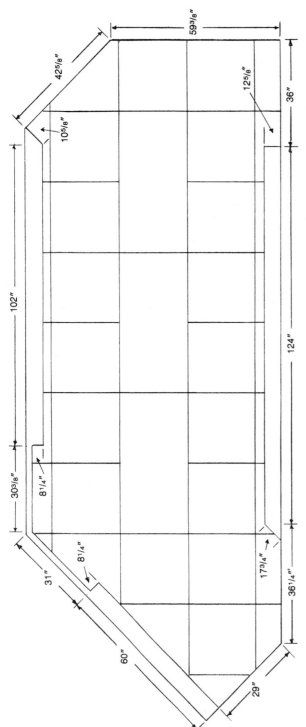

Fig. 5-60. As this plan of an actual installation shows, a suspended ceiling can be hung in a room of just about any configuration.

DROPPED CEILINGS

A dropped ceiling can hide an existing ceiling. It accomplishes the same purpose as a suspended ceiling. The differences are that a dropped ceiling is built up from scratch from conventional building materials and looks like a conventional ceiling when completed. Also, there is no easy access into the above-ceiling area as there is with suspended panels. A dropped ceiling framework consists of a series of wood members arranged in the same manner as that of any conventionally framed ceiling system. The framework is not an integral part of the building's structure, but instead is attached to the perimeter walls and carries only its own weight.

Note that ceiling insulation should not be placed directly on top of a dropped ceiling, but rather at the true ceiling level—if it is required at all. Also, electrical wiring, as for lighting fixtures, cannot be simply extended from an upper ceiling using the old outlet boxes. Correct fixture installation and accessibility of all connections must be considered; consult an electrician.

There are two approaches to dropped ceiling construction, depending upon the circumstances. In a small narrow room such as a bathroom, corridor kitchen, or hallway—where acoustic ceiling tile will be installed—the framework can be made up of nominal 1-×-4 stringers. Furring strips are fastened to these to accept the ceiling tile. This is a lightweight construction. The other method, which is suitable for any kind of room, employs nominal 2-inch-thick dimension stock for the framework. This is the common construction. The width of the members depends upon the span of the room and the weight of the ceiling sheathing and finish. The ceiling itself may be attached directly to the stringers, or to furring strips attached to the stringers. (Ceiling tile is always mounted to furring strips).

Lightweight Construction

To construct the lightweight version, first determine the desired height from the floor to the finished ceiling surface. Then determine the thickness of the complete ceiling system. For example, the 1-×-4 framing will stand about $3^1/2$ inches high. Furring strips add another $3/4$ inch and ceiling tiles $1/2$ inch, for a total of $4^3/4$ inches. Add this to the finish ceiling height above the floor. At that height, start in a corner using a 2- or 4-foot spirit level laid end over end as a straightedge and draw a level line along one wall. Repeat along all the walls except the last one; snap a chalkline across this last one to join the pencil lines.

Use a stud locater, or tap on the walls with a hammer and drive a finish nail through the wall sheathing, to find all of the wall studs along the line. Mark each one. Then trim and fit lengths of 1 × 4 as necessary and mount them with their top edges flush with the line, in a band all the way around the room. Fasten them with nails long enough to penetrate at least 1 inch into the studs—or better, with long wallboard screws and a power driver. Use two fasteners at each stud.

The stringers go up next, full length across the narrowest dimension of the room. Cut two pieces of $14^7/8$-inch-long 1 × 4 for nailing blocks. Mount one directly on one long-

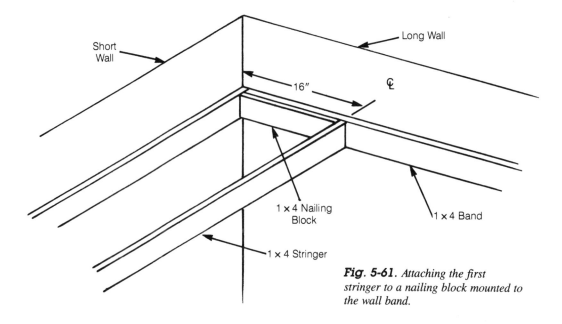

Fig. 5-61. *Attaching the first stringer to a nailing block mounted to the wall band.*

wall band, tucked into the corner of a short wall. Mount the other one on the opposite long-wall band. You can do this with either ring-shank 6d nails or wallboard screws; three or four staggered fasteners are sufficient. Then trim the first 1-×-4 stringer to fit between the long-wall bands, set it in place against the free ends of the two short nailing blocks (Fig. 5-61), and nail or screw each end to the ends of the nailing blocks. The centerline of the stringer should be 16 inches from the wall.

The next two nailing blocks are 15¹/₄ inches long, to maintain the 16-inch stringer centering. Mount these to the long-wall bands, snug against the first stringer, and fasten the second stringer to the open ends of the blocks. Continue until you have installed all of the stringers. The band length between the last stringer and the short-wall band need not be fitted with a nailing block.

The last step is to install the 1-×-4 furring strips for the ceiling tiles, at right angles to the stringers. The installation is made just as though the stringers were ordinary ceiling joists.

Common Construction

The more common construction for a dropped ceiling frame is a series of dimension stock stringers in a joist-like arrangement. The stringers are set across the narrow dimension of the room. You can install 2 × 4s on 16-inch centers for spans up to 8 feet with ceiling tile attached, or 6 feet for other materials. Use 2 × 6s for spans up to 11 feet, 2 × 8s up to 14 feet, 2 × 10s to 21 feet, and 2 × 12s to 26 feet. These are conservative numbers; for special situations or if your spans are longer, consult with a builder or engineer to make a final determination. The usual choice for the stock quality is ordinary kiln-dried construction grade or framing lumber, also referred to as common construction lumber. If possible, select the stock yourself at the lumberyard so that you can get the straightest and flattest pieces available. Make your purchase immediately before starting the job, and if possible, store the material for one or two days in the same environment where it will be installed. Do not let it sit out in the sun or rain.

The initial construction steps are the same as for the lightweight framework just described. Mount a perimeter band on the wall, using the same size of stock as the stringers. Fasten the band to the studs with two or three 16d nails at each point, depending upon the band width. To avoid a lot of pounding and vibration, which might cause damage to the wall finish—especially on the opposite side of the wall—consider using a pneumatic nail gun (rentable). A tedious alternative is to drill clear holes in the band and drive long $1/4$- or $5/16$-inch lag screws.

With the band in place, mark off 16-inch centers along both long walls. Start 16 inches from a short wall, not at the centerline of the short-wall band. At each of these points, center and mount a steel joist hanger of appropriate size for the stringers (Fig. 5-62). Nail them in place with the recommended size and type of nails. Note: If the hanger positions do not coincide with the wall stud locations, they can be attached before the band pieces are put up, again avoiding a lot of pounding against the walls.

Cut and fit each stringer individually (do not use the first one for a pattern), and fasten it into the anchors. As you do, check each one to make sure that it has no crown or bow. Stretch a length of string lengthwise along an edge from one corner to the other. If there is a crown, snap a chalkline along the side of the piece opposite the crown and plane that

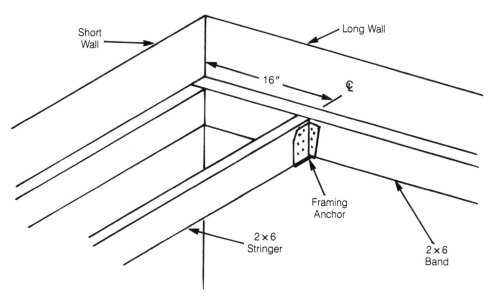

Fig. 5-62. *For heavier constructions, stringers are best secured to the wall bands with metal framing anchors.*

edge to the line (Fig. 5-63). When you install the piece, place it with the crown upward. If each piece has a flat, level bottom edge and the perimeter band around the room is level, the finished ceiling surface will also be plane and level.

When all the stringers have been installed, complete any necessary electrical work and put up the ceiling itself. If furring strips for ceiling tiles are required, refer to "Acoustic Tile Ceilings" earlier in the chapter. If furring will not be installed and the stringer span is more than about 10 feet, installing a run of solid blocking between the stringers at the lengthwise midpoint of the span is advisable. This will stiffen the framework and help to keep the stringers from twisting.

COVES, CORNICES, AND VALANCES

Many conventional ceilings are simply broad, horizontal expanses that join the walls at a bare and unobtrusive right-angle joint. However, there are numerous and often overlooked methods of adding character and emphasis to a ceiling area that do not entail any great amount of work or expense. Chief among these are coves, cornices, and valances.

A *cove* is a concave surface that makes a transition from a vertical wall to a ceiling, whether horizontal or angled upward. It eliminates the interior angled break between the wall and ceiling. A cove can also be a trough of almost any shape mounted on a wall just below ceiling level to house concealed lighting; in this case it is not necessarily concave in cross section.

A *cornice* is a continuous horizontally projecting feature that may be located at the top of a wall, or at some other high point on a wall, to divide it for decorative or aesthetic reasons. If at a wall top and also joined to a ceiling, it is simply a cornice. If it projects only from a wall, it can be called a *floating cornice*. Although usually of solid or completely enclosed construction, a cornice can be open at the bottom, the top, or both.

A *valance*, strictly speaking, is a short piece of drapery placed across the top of a window as a decorative heading, usually to hide the tops of curtains or shades and their hardware. However, the term has come to include any kind of wood or metal cover, built-in or not, that is in the same position. It might or might not actually conceal draperies, and

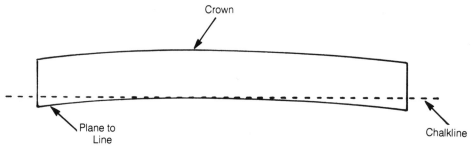

Fig. 5-63. *Crowned joists should be planed flat opposite the crown for ceiling applications (along the crown for floors).*

it can be open top and bottom, closed at the top to form a shelf, or continue upward to join a ceiling.

Coves

A true cove can be built in a number of ways, depending largely upon the desired size and radius of the concavity. In a room with a low, horizontal ceiling, a small-radius cove is the usual choice. The higher a horizontal ceiling, the larger the cove radius can be without seeming out of proportion. If the ceiling is angled upward, cathedral style, the cove radius is usually very large—but because of the widely divergent wall/ceiling lines, the cove is not overbearing.

One method of making a cove at an existing 90-degree joint is to install a feather-edged wood cove molding (Fig. 5-64). The molding would have to be custom milled; select a fine-grained, nonporous wood such as ponderosa pine for a smooth finish. Coat the molding face with a wood sealer, then nail it snugly into place so that the feathered edges lie flat against the wall and the ceiling. If necessary, hide the edges wherever necessary with patching plaster or vinyl paste spackle, and sand smooth. Fill the nail holes, sand smooth, and apply a finish.

A similar arrangement can be used when both the ceiling and the wall sheathing is to be replaced. Use a custom-milled molding that has a coved face and squared-off top and bottom edges, and is the same thickness as the wall and ceiling sheathing material. After applying a wood sealer, nail the molding to the top wall plate, snug against the ceiling joists or furring strips. Fit the wall and ceiling sheathing to the squared-off edges of the molding, leaving a slight gap (Fig. 5-65). Cover the joints with joint tape and compound and sand smooth, spreading the compound out onto both the ceiling and the walls.

A third method is to mold the cove in place with either slow setting regular plaster or a molding plaster that sets more quickly. Trowel the plaster into place along the wall/ceiling joint, then smooth it into the desired concavity. There are no available tools for this purpose, so you will have to make your own. For example, a short length of pipe drawn along the joint at a shallow angle will give you the radius of the outside diameter of the

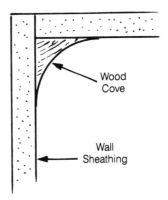

Fig. 5-64. *A feather-edged cove.*

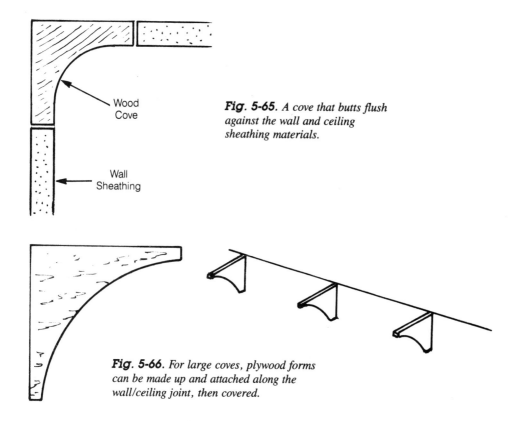

Fig. 5-65. *A cove that butts flush against the wall and ceiling sheathing materials.*

Fig. 5-66. *For large coves, plywood forms can be made up and attached along the wall/ceiling joint, then covered.*

pipe. Or, you can have a piece of heavy-gauge sheet metal of required dimension rolled to a half-circle at a local sheet metal shop. The process is similar to tooling the joints in a concrete block or brick wall. The outer extremities of the cove can be smoothed along the ceiling and the walls with a joint knife and/or a damp sponge. If the cove, and consequently the amount of plaster in the joint, is small, it can be directly applied. If the radius is large and the joint must hold a considerable amount of plaster, install a row of closely spaced nails or a strip of screening, hardware cloth, or metal lath for the plaster to cling to. If the radius is large enough to allow working room, the plaster can be applied with a trowel in the same manner as plastering a wall.

Large-radius coves are built differently. Establish the desired radius, then make a template of plywood as shown in Fig. 5-66. Using this as a guide, make up enough forms of $1/2$-inch (or thicker) plywood to install every 16 inches around the perimeter of the ceiling. Cover the forms with a thin, stiff material that will bend to that radius, such as $1/8$-inch hardboard, 5-millimeter or $1/4$-inch plywood, or $1/4$-inch gypsum wallboard. If the material is a bit too stiff to conform, score the backside lightly along parallel lines, about 1 inch apart. The depth of the cut depends upon the resistance of the material to the necessary bend, but should not be more than halfway through. If the material is plywood, cut parallel with the face grain. Once this skin is nailed in place, cover the joints along the wall and ceiling with fiberglass joint tape. Then apply joint compound in several applications, and feather the edges out onto the wall and ceiling surfaces and into the cove. Fill over the nail heads as well, sand smooth, and apply a finish.

Note that these coves eliminate all angular joints, and so there is no natural break point to separate two different finishes on the wall and ceiling. If you want such a demarcation, you can introduce one in either of two ways: First, form the cove with either the upper or lower edge squared or rounded off, and protruding beyond its mating surface (Fig. 5-67). Second, install a separate, surface-mounted cornice molding (any of numerous standard or custom configurations can be used) on the wall just below the cove (Fig. 5-68).

A cove of the sort used to contain lighting fixtures is easy to build, and might actually contain fixtures—or not, as you wish. The construction is basically an internally bracketed and supported shelf fitted with a fascia across the front edge. The fascia may be vertical or set at an outward angle, and the whole assembly normally extends only about 6 to 8 inches from the wall (Fig. 5-69). It can be built in place without disturbing the existing wall finish. If no lighting is involved, the cove can be positioned at any point beneath the ceiling, leaving an open space of whatever height is desired between the top of the fascia and the ceiling. Likewise, the height of the fascia can be whatever seems suitable. When indirect lighting is installed, the bottom of the cove is placed anywhere from about 10 to 18 inches below the ceiling. The outward angle of the fascia and the distance between its top edge and the ceiling depend upon the sight lines of the room and the angle of light. The height of the fascia depends upon its outward angle and the fixture type and placement.

To build a fixtureless cove, first decide upon the materials. The bracketing can be plain pine 1-×-4 stock. The bottom, or soffit, and the fascia can be made up of top grade, kiln-dried, nominal 1-inch pine stock; plywood of any sort from $1/4$ inch thick up; or even particle or chip board. Determine the height from the floor of the bottom surface of the

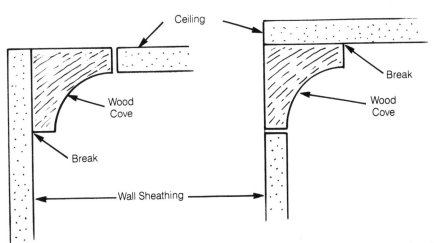

Fig. 5-67. *To establish a break line, a cove can be set flush at one edge and squared off at the other.*

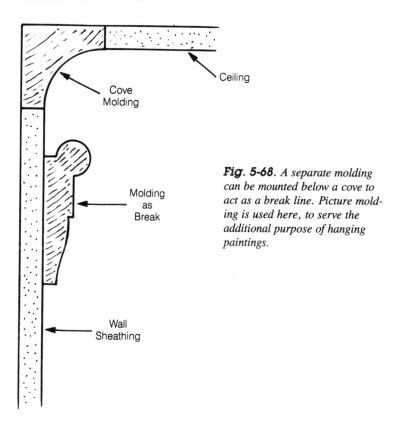

Fig. 5-68. *A separate molding can be mounted below a cove to act as a break line. Picture molding is used here, to serve the additional purpose of hanging paintings.*

Ceiling

Cove
Molding

Molding
as
Break

Wall
Sheathing

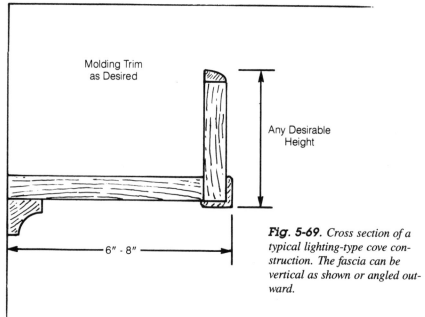

Molding Trim
as Desired

Any Desirable
Height

6" - 8"

Fig. 5-69. *Cross section of a typical lighting-type cove construction. The fascia can be vertical as shown or angled outward.*

cove on the wall, add the thickness of the soffit material, and make a mark at a wall corner. Draw a level line across the wall, using a flip-flopped 2- or 4-foot spirit level as a straightedge. Find the wall studs with a stud locater, or by tapping and/or driving finish nails through the sheathing, and mark each at the level line.

Cut a piece of 1 × 4 to length for a band. Attach the brackets—cut to the width of the soffit—to the face of the band in a T-joint. Drive two long wallboard screws through the back of the band and into the ends of the brackets. Fasten the band/bracket assembly to the wall with the bottom edge of the band on the level line. Use a pair of 8d or 10d nails or long bugle-head wallboard screws at each stud. Then cut and fit the soffit. Attach it at right angles to the bottom edge of the band with countersunk wallboard screws every 12 inches. Drive two more screws up through the soffit into the bottom edge of each bracket, one near the front and the other near the back. Finally, fit the fascia across the front. Secure it with a pair of screws set into the end of each bracket, plus at least one every 2 feet along the front edge of the soffit. Attach corner and/or cap moldings if you wish. Putty over the fastener heads and apply a finish. Figure 5-70 shows a typical assembly.

There are two further points: First, if the cove is made of lightweight materials, you can assemble it with glue and/or finish nails. This is simpler and also requires less hole filling. Second, if the cove is relatively small and light—and depending upon how it must fit into place—you might find it easier to build the whole assembly on the floor. First remove the fascia to give yourself working room, then mount the assembly on the wall and

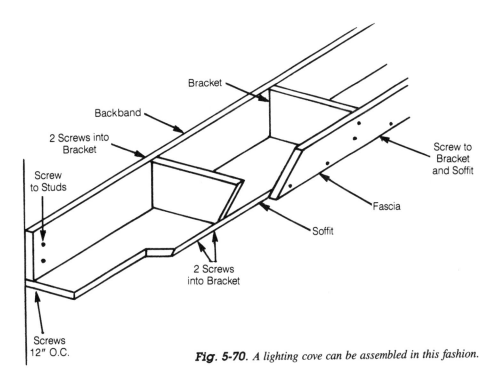

Fig. 5-70. *A lighting cove can be assembled in this fashion.*

replace the fascia. With proper planning, you might even be able to build and finish it on the floor and mount it as a completed unit.

Cornices

Ancient cornice designs are elaborate and bulky, intended for huge rooms with high or vaulted ceilings. In modern residential design, however, a cornice is typically a boxed-out construction, at the wall/ceiling joint. Some are wide enough that they could be called perimeter dropped ceilings.

Construction of this sort of cornice is basically the same as for the lighting cove just described (which is itself actually a cornice). The dimensions are variable to any reasonable degree, but typically the cornice drops from the ceiling about 6 to 12 or 14 inches, and extends from the walls about 12 to 24 inches. The proportions are horizontally rectangular rather than square or vertically rectangular. Construction details for a relatively small, lightweight cornice follow those of the lighting cove, with a few added points.

The fascia, whether vertical or angled, joins the ceiling at the top instead of being open. This means that a nailing strip must be positioned on the ceiling just behind the fascia edge, to which it can be secured. When the cornice lies at right angles to the ceiling joists, a length of 1 × 4 nailed to the ceiling joists is ample for this purpose. When the cornice lies parallel with the ceiling joists, there are several options. One is to calculate the width of the cornice so that the top edge of the fascia runs just to the outside of a joist. Then a parallel nailing strip can be attached to the joist, and the fascia to the strip (Fig. 5-71). If the fascia top edge lies beyond the first joist out from the wall, you can nail short

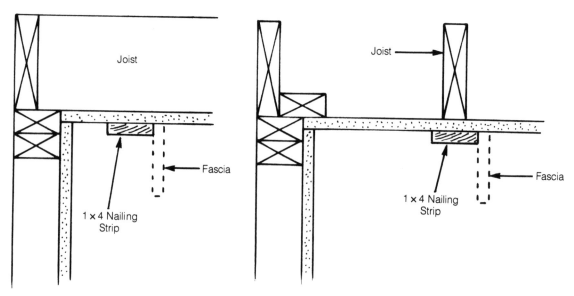

Fig. 5-71. For a box cornice running at right angles to the ceiling joists, a fascia nailing strip is nailed to the joist bottoms (left). If the cornice runs parallel with the joists, the nailing strip can be secured lengthwise to an appropriate joist bottom (right).

lengths of 1 × 4 at right angles to the joists. They are spaced about 12 inches apart and toenailed to the wall plate at the inner end, and face-nailed to the first joist near the outer end (Fig. 5-72). Then secure the fascia by nailing it to the free ends of these strips. If the fascia top lies between the wall and the first joist out, secure lengths of 1 × 4 flat to the ceiling with hollow-wall anchors staggered on 12-inch centers. Then fasten the fascia to the strips with screws (not nails).

Larger cornices, especially if faced with thick and heavy materials, are more easily constructed by other methods. One is to make up a series of rectangular box frames with 1-×-4 stock nailed—or preferably screwed and glued together (Fig. 5-73). Fasten one frame to each wall stud with nails or screws, flush against the ceiling. Secure each frame top to a joist or to the ceiling with hollow wall anchors. Then sheath the box frames with ³/₈-inch or thicker plywood. If you prefer to sheath with thin plywood, hardboard, or gypsum wallboard, add four nailing strips of 1-×-2 (or larger) stock across the edges of the box frames. Set two strips across the bottom edges, one flush against the wall and the other flush with the outside corners. Set two more across the outer edges, one flush with the ceiling and the other flush with the bottom edge. The whole affair resembles a box kite frame (Fig. 5-74). Use screws to attach the sheathing to these strips, at about 8-inch intervals along the strips.

Cornice edges and joints can be fitted together with butt or miter joints, as required, and the assembly finished as is. Or you can apply molding in any of a number of patterns to conceal the joints and ornament the assembly. For example, the lower outside corner of a cornice with a vertical fascia might be capped with corner molding. If the fascia is angled outward, two meeting moldings such as plain lattice could be mounted, one along

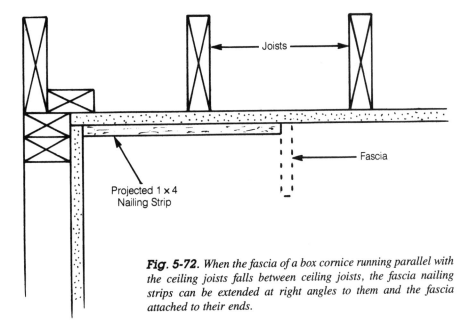

Joists

Fascia

Projected 1 × 4
Nailing Strip

Fig. 5-72. *When the fascia of a box cornice running parallel with the ceiling joists falls between ceiling joists, the fascia nailing strips can be extended at right angles to them and the fascia attached to their ends.*

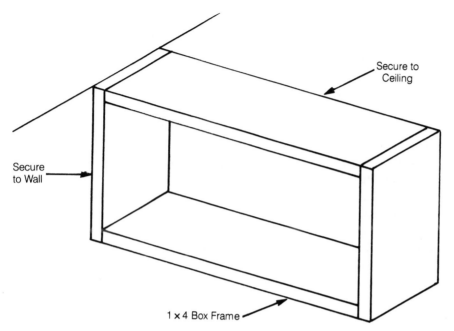

Secure to
Ceiling

Secure
to Wall

1 × 4 Box Frame

Fig. 5-73. *Simple frame assembly for a box cornice; one frame is attached to each wall stud.*

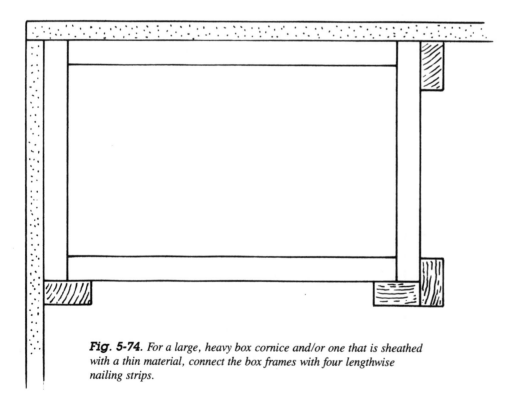

Fig. 5-74. *For a large, heavy box cornice and/or one that is sheathed with a thin material, connect the box frames with four lengthwise nailing strips.*

the soffit edge and the other along the fascia bottom edge. The wall/soffit joint might be coved, or fitted with a cove, crown, or bed molding. The same treatment can be used at the ceiling fascia joint.

A cornice can also be made with an integral lighting cove. Extend the soffit of the cornice beyond a vertical fascia by 6 to 8 inches or so. Then attach a second fascia—angled as required if lighting fixtures will be installed—to the outer edge of the soffit (Fig. 5-75). If the soffit material is thick and rigid, such as ³/₄-inch plywood, no further support or bracing will be needed. The outer fascia can be fashioned from a relatively thin but stiff material such as ¹/₄-inch plywood that will also be self-supporting—eliminating the need for bracketing there.

Valances

Valances fit over window tops and extend to the ceiling. They serve to break up a monotonous ceiling/wall joint line, as well as hide drapery tops or roll shades and their hardware.

The easiest approach to building a valance is to take the necessary measurements and then construct it on the workbench or the floor. If the valance is very large and heavy, however, it might better be built in place. For materials, either ³/₄-inch cabinet-grade plywood or top-grade nominal 1-inch pine boards work best.

Determine how far down from the top of the window the bottom edge of the valance must be to do its hiding job and provide the desired appearance. Measure from that point to the ceiling to get the full valance height. Then measure across the window from casing

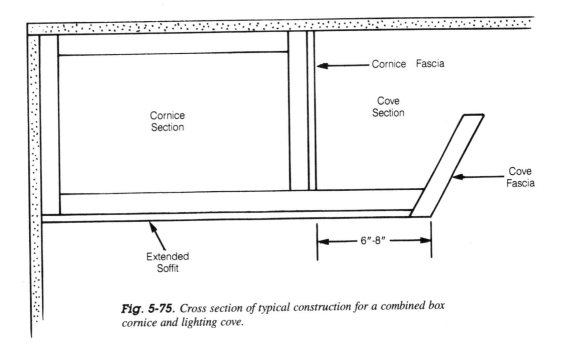

Fig. 5-75. *Cross section of typical construction for a combined box cornice and lighting cove.*

outside edge to casing outside edge to get the inside width of the valance. Construct a simple three-sided box to match the measured dimensions, with glued and nailed miter joints at the corners. If the valance is small—2 to 3 feet wide and 6 inches or so deep—that is all that is needed. If it is larger and fairly heavy, or if the valance does not extend to the ceiling and a shelf top is desired, fit a top piece within the three sides (Fig. 5-76). This will stiffen the assembly and also give you some ceiling attachment points, if necessary.

The bottom, exposed edge of the valance can be left squared off, or molded to some other contour with a router or shaper before assembly. You can also cap it with a 3/4-inch-wide molding, such as base shoe for a curved edge or half-round for a rounded edge. Or, you might cut a decorative pattern across the front edge—scallops, for example. In any case, sand the valance and apply a finish before installing it. To install, fit the valance over the window casing and nail it to the casing edges with 6d finish nails. If a top plate lies against the ceiling, nail up through it and the ceiling into a previously located joist or furring strip. If there is nothing to nail it to, set two or three hollow-wall fasteners. Fill the nail holes with glazing compound, let the compound harden for a few days, and touch up the finish with an art brush.

To construct a valance in place, mount the top plate to the ceiling or the top edge of the head casing first, its ends aligned with the outer edges of the window side casings. Fasten the valance ends to the top plate ends and the window side casing, then attach the valance fascia to the top plate front edge and the valance ends. Apply trim as necessary, take care of any hole plugging and touch-up sanding, and apply a finish.

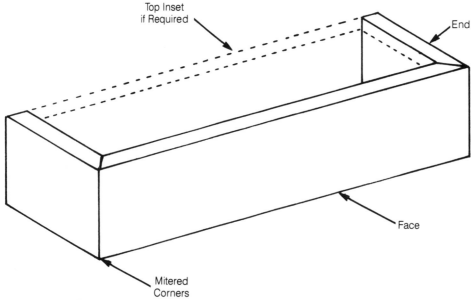

Fig. 5-76. *A simple valance assembly; the inset top panel is optional.*

The joint between the top edge of the valance and the ceiling can be left alone if the fit is good. But good or not, it can also be concealed with a molding. If there is no wall/ceiling joint molding, any type that appeals can be installed around the valance. If there is an existing molding, install the same kind around the valance. Use mitered corners on the valance and coped ends to fit against the existing molding at the wall. To save time and trouble, prefinish the molding and fill and touch up the nail holes after it has been put up.

In some situations the valance must extend beyond the window casings, perhaps to clear a drapery rod that is wall mounted, or just as a matter of balance and perspective. In that case, build the valance with a full-width top plate. Attach the valance ends to the top plate ends, and let them hang free; the fascia will anchor the ends at the front. The entire valance assembly must be fastened to the top edge of the window head casing with nails or screws, or to the ceiling with nails or screws into joists or furring, or with hollow-wall fasteners through the ceiling.

CEILING TRIMWORK

Many ceilings are characterless, blank expanses—but they don't have to be. A little imagination and the addition of some inexpensive trimwork can change a plain ceiling into one that adds elegance or decorative emphasis to a room. The principal aspects to consider—apart from the color or texture of the ceiling field itself—are edge (or perimeter, or cornice) moldings, decorative field moldings, and beams.

Edge Moldings

Edge moldings are the most widely used trimwork, installed at the wall/ceiling joint all around the room. Their purpose can be simply decorative, but they usually serve to hide the unfinished joint as well. To keep the molding size and configuration in proportion to the other elements of the room, the rule of thumb is that small rooms and/or low ceilings should be fitted with small, plain moldings, while the larger the room and/or the higher the ceiling, the larger and/or more ornate the molding can be without seeming overbearing.

The most commonly used molding is a small cove with a right-angled back (also sometimes called *scotia*.) The $3/4$-$\times$-$3/4$-inch size is most readily available, but there are other standard sizes as well. Larger cove moldings are open-backed and create a cove as large as 3 inches, sometimes more. Crown moldings are designed for wall/ceiling joint applications. The face pattern, of which there are several common variations, is more ornate than a cove; numerous sizes are available. Bed moldings are also used for this purpose; but the choice of sizes here is smaller. There are other options as well: quarter-round, base shoe (installed either upside down or with the narrow dimension against the wall), or glass bead. Stop molding, made in a wide range of sizes and with several different edge patterns, can be used, as can shingle, panel, or brick moldings. Picture molding, which is designed to support picture hanging hooks, is sometimes installed just below ceiling level. The profile of the molding obscures the wall-ceiling joint, which in this case should be a finished one.

Though there are many standard patterns of wood moldings, only a few are usually available at any given time from local suppliers. However, others can be special ordered and virtually any can be made up by a good woodworking shop. Most stock moldings are pine, although some are oak and mahogany. Again, moldings can be custom made from nearly any suitable wood. Note also that you can build up an edge molding from two or more stock moldings, or use plain square-edged stock and one or more moldings. Thus, you can combine locally available stock patterns into a unique configuration. This is also a useful course to follow when a large and complex pattern is desired (Fig. 5-77).

To install wood edge moldings, you will need a top-quality miter box and saw, or a power miter (chop) saw that will allow you to make accurate angle cuts. Alternatively, use a radial arm saw fitted with a very fine-toothed blade. A table saw will do the job but with

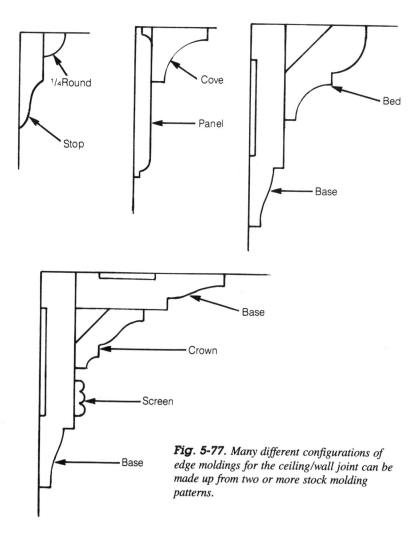

Fig. 5-77. *Many different configurations of edge moldings for the ceiling/wall joint can be made up from two or more stock molding patterns.*

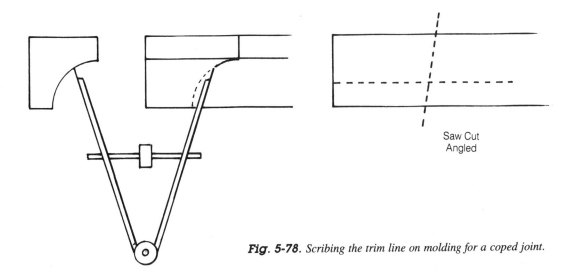

Saw Cut
Angled

Fig. 5-78. *Scribing the trim line on molding for a coped joint.*

more difficulty. Outside corners must be mitered to fit. Inside corners can be either mitered or coped; the latter is preferred by most professionals because it gives a tighter, cleaner joint. To make coped joints you will need a coping saw fitted with a fine-toothed blade and a half-round wood file. If there is an outside corner in the room, start the installation there and proceed around the room. If not, start in a corner and work continuously left or right. Do not start at one corner, then go the next corner and work back to the first piece, because you will have matching problems.

To make a miter joint, split the total angle of the corner in half: A 90-degree corner takes a 45-degree cut on each mating molding face, a 45-degree corner takes two $22^1/_2$-degree cuts, and so on. Mount the two pieces cut face to face. In a coped joint, such as at an inside corner, the first piece is butted directly into the corner. The end of the mating piece must be shaped to lie flat against the face of the first piece. There are two ways to do this. One is to set the two pieces of molding on a flat surface in the same relative position that they will be mounted, with a gap of about 1 inch between them. Hold or clamp the pieces firmly, and with a compass scribe the profile of one onto the end of the other (Fig. 5-78). Then cut along the line with a fine-toothed coping saw, the blade cocked at an angle about 95 degrees to the back of the molding.

The other method is to place the piece to be coped in a miter box, set against the back-stop of the box in its mounting position. Make a miter cut (half of the particular total angle involved) on the molding, with the face of the cut outward (Fig. 5-79). Follow the profile line (the outer edge) of the miter cut with a coping saw. Hold the saw at an angle to the back of the molding that is a bit sharper than the total angle of the joint (Fig. 5-80). Trim the edges of the cut with a fine-toothed wood file to make a final, tight fit.

Prime, sand, and paint the uncut molding lengths before installing them. Use brads or finish nails of appropriate size to install the molding. Some types must be nailed at the top, others at the bottom, and a few through the center at a 45-degree angle. Much depends

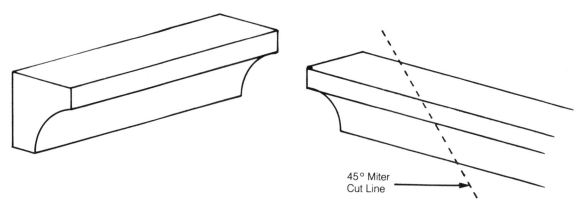

Fig. 5-79. *The first step in the alternative method of cutting a coped joint is to position the molding piece as it will lie, then make a miter cut down through it.*

45° Miter Cut Line →

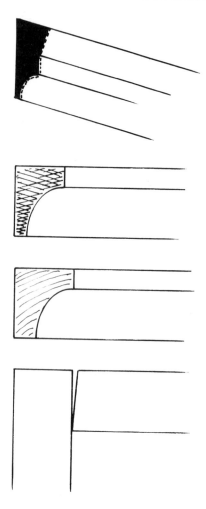

Fig. 5-80. *The miter cut leaves a contour outline along the face edge of the cut; the shaded portion must be cut away (top). Viewed straight on, the cross-hatched portion is cut away with a coping saw. The cut end should fit cleanly against the second molding piece. Shown from the top, the coped molding is back-cut so that only its outer edge actually fits tightly against the second molding (bottom).*

upon the mounting surface and the size and shape of the molding, so use your own best judgement. Drilling pilot holes is advisable when nailing close to ends, or if the wood is hard. Set the nails, fill the holes with putty, carefully smoothed off, and touch up with paint.

Ordinary wood molding patterns might be described as plain or smooth faced. There is another class of edge moldings that is altogether different; architectural cornice moldings (Fig. 5-81). These are substantial in size and carved or patterned in forms ranging from relatively plain and simple to very ornate. Although they once were built up and hand carved from wood stock or made from plaster, they now are mostly machine-molded, typically from polymers, in one-piece standard lengths. Most follow the classical patterns. These moldings too can be built up by adding accessory moldings, also in classical patterns. They are suitable for use in any kind of house, but they are best used in large rooms of classical proportions with ceilings at least 8 feet high. Other accessory pieces such as domes, dome rims, and medallions are available to use with these moldings.

Decorative Field Moldings

Decorative field moldings can be installed on any flat ceiling, regardless of size or height, and the number of patterns and arrangements is virtually limitless. You can use plain wood moldings, stock or custom-milled patterns. For narrow stock, shelf edge or

Fig. 5-81. A typical architectural cornice molding; this is the classical Georgian pattern. Many other patterns are available. (Courtesy Focal Point Inc., Atlanta, Georgia.)

screen mold are ideal. Half-round gives a fine, smooth face, while astragal presents a distinct form. Lattice can be used in dozens of ways. It is thin, so it does not look bulky or heavy when installed. For wider, plain-faced patterns, edge-to-edge stops or half-rounds work well; for a more involved pattern, use edge-to-edge base cap coves, or shingle molds. For a very wide and sculptured effect, install edge-to-edge casings or bases (Fig. 5-82).

Installation of decorative field molding is not difficult. The key is to lay out the pattern first and make all the miter cuts with precision. Accurate measurements for placing the components of the pattern on the ceiling are imperative; double check everything. Start a pattern with the shorter pieces, and line the longer ones up on them. If the molding is small and limber, make sure that the pieces line up straight as you secure them, and do not bow out of line. Prepaint the uncut molding lengths if they are of a different color than the ceiling, or if the whole affair is not to be repainted at once. Trim and fit to your pattern and secure the pieces with a moderate amount of construction or panel adhesive, spread thin so it does not ooze out onto the ceiling. Lightweight moldings will stay in place with glue alone. Where heavier pieces are installed, drive a finish nail into the furring or joists at whatever points you can. If the moldings were prepainted, go back after the installation is complete and fill nail holes with putty, then touch up the patches and the cut joints using paint applied with a small art brush.

False Beams

False beams—because of their bulk and appearance of heaviness, as well as the way they tend to make a ceiling appear lower than it really is—are generally installed only in rooms of substantial size, with ceilings at least 8 feet high. False beams are usually set across the narrow dimension of the room at regular intervals of at least 4 feet, more if the room is very large. If the room is wide enough, other beams of equal or somewhat smaller size can be installed at right angles to the first. If too many beams are installed, however, the effect is lost and the ceiling appears crowded and cluttered. A ceiling 20 feet long and

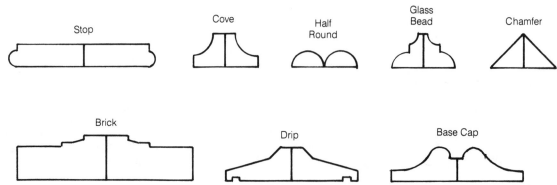

Fig. 5-82. *Some moldings can be installed edge to edge to form unusual configurations in decorative field trimwork.*

16 feet wide, for example, might be outfitted with three beams across the narrow dimension only, or with one or two more intersecting along the long axis to form a grid. False beams are usually at least 6 inches wide, but more often 8 to 12 inches. They may extend downward from the ceiling field as little as 2 or as much as 12 inches.

There are three different approaches to beaming a ceiling. One is to use actual beams of the desired dimensions. They must be full length, fitted into and supported by framed pockets within the walls (Fig. 5-83). This is a project of major proportions, because the wall finish must be removed and replaced and heavy beams boosted into place. However, the end result—especially if you are lucky enough to find some nice old hand-adzed cherry or walnut beams—can be worth the effort.

Another method is to have full beams, antique or otherwise, rip-sawn in half at a mill. If light and thin enough, these beam sections can sometimes be anchored to the ceiling joists with lag screws set in counterbored holes. The screw heads can be covered with plugs fashioned from the same beam material so they blend in, or with plugs made from a different, contrasting wood, set to protrude slightly and resemble pegs.

The most common procedure is to build up box beams from ordinary construction-grade dimension stock and finish-grade boards or plywood. They may be trimmed out with wood moldings, depending upon the way the beams are made. There is a lot of flexibility in design, but a popular arrangement consists of a 2 × 6 or 2 × 8 lagged flat to the ceiling to form the beam base. To each side attach a 1 × 6 or 1 × 8. Fit another length of 1-inch stock at the bottom between the two sides, or across the bottom and overlapping them. The full length of the beam can be made up from short pieces by butting them randomly along the beam length. You can add wood moldings at the top or bottom—or both—in any configuration that appeals (Fig. 5-84).

Fig. 5-83. *Two possible methods of framing beam pockets into a wall.*

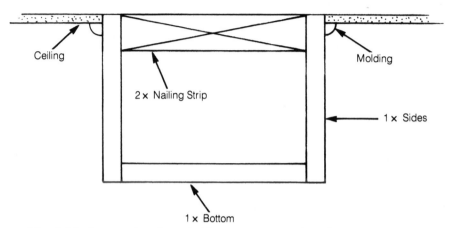

Fig. 5-84. *Cross section of a typical beam construction. Details vary considerably.*

False beams are frequently stained a medium to dark color and topcoated with varnish. If the characteristics of the wood are appealing, a natural or clear finish might be the last choice. However, beams are also sometimes painted to match or just slightly contrast the ceiling color, or to match a contrasting color applied to other trimwork in the room. Usually the finish work is done after installation, except for moldings applied at the ceiling joint. Newly made false beams can also be "antiqued" in various ways: denting with a hammer, scarring and scratching with tools, chipping off edges, and making simulated adze marks with a hatchet. Even a box beam can be made to look like an authentic old barn beam if it is carefully assembled so the joints do not show.

OPEN CEILINGS

Depending upon the design and construction of a house, a flat, horizontal ceiling can sometimes be partially or wholly opened up into a second floor. It can also be converted to a cathedral ceiling, where the underside of the roof system becomes the effective ceiling. This tends to make a room or area feel much more airy and expansive. The process can also open the way for a sunspace or solar room conversion, the installation of large skylights, or, in a two-story house, an interior balcony or sleeping loft arrangement. Conversely, an existing cathedral ceiling design can be closed off by constructing a conventional flat, horizontal ceiling across the open area. This results in a smaller, more snug environment. In a single-story house, this decreases the volume of interior air that must be heated and/or cooled, and in a larger house it might allow the construction of one or more additional rooms above the new ceiling system.

If you would like to do away with a conventional flat ceiling, the first step is to bring in an architect or a builder to see if, and how, the job can be done. It involves removing the ceiling, its supporting members, and perhaps a floor above. Some or all of these members might be an integral part of the house skeleton, and therefore not practicably removable. For example, if the roof structure is composed of trusses, the bottom leg of each truss supports the ceiling as well. They cannot be removed, nor can the web of intermediate truss members above them, so the ceiling cannot be opened up. However, if the roof framing consists of rafters and the ceiling is only attached to joists, removal of the ceiling and joists is often feasible. The interior walls of the room can be extended to the roof as necessary, and the exposed underside of the roof system covered to make a finished ceiling. If a small amount of structural integrity is lost during the process, usually it is possible to replace it in any of a number of ways. This is dependent upon the configuration of the structure and the details of the new installation.

Closing off an existing cathedral ceiling is usually less problematical, because the existing structural skeleton remains intact. Even so, consulting with a professional to make sure that your plans are both workable and safe is a wise idea. Basically, the process involves constructing a support system across the open area. If only a ceiling will be installed, this might consist of a relatively lightweight frame of 2 × 6s or 2 × 8s (depending upon the span and the weight of the ceiling to be installed), secured to headers at each end, which in turn are anchored to the wall studs at an appropriate height. If flooring at the second level is part of the plan, the floor/ceiling framework system must be engineered to be capable of sustaining the floor loads as well as the ceiling. If the span is wide, this in turn might require some sort of additional support from the first-floor level, such as columns beneath a girder.

Unless a cathedral ceiling area is to be closed off completely and ignored, access of some sort must be provided. This might be done by installing a doorway from adjacent second-level living quarters, by constructing a stairway from the first floor, or by installing a hatch or pull-down folding stairway in the new ceiling.

Sometimes the plans for closing off or opening up a ceiling include other remodeling work, such as the addition of a room or two or a loft, a sunspace or greenhouse conversion, or revamping a roof line. Any such work constitutes a major project, and professional help is definitely advised, especially in the planning stages and wherever structural remodeling is involved.

Index

Other Bestsellers of Related Interest

INCREASE ITS WORTH: 101 Ways To Maximize The Value of Your Home—Jonathan Erickson

"...an idea book, filled with sensible advice on what makes a home valuable."—San Francisco Examiner

The author profiles the three basic types of home buyers, defines the factors that affect resale value, explains two basic methods of determining your home's resale value, and shows you what rooms play the biggest role in deciding the value of a home. 208 pages, 105 illustrations. Book No. 3073, $14.95 paperback, $23.95 hardcover

KITCHEN REMODELING—A Do-It-Yourselfer's Guide—Paul Bianchina

"...offers all the know-how you need to remodel a kitchen economically and attractively."—Country Accents

Create a kitchen that meets the demands of your life-style. With this guide you can attractively and economically remodel your kitchen yourself. All the know-how you need is supplied in this complete step-by-step reference, from planning and measuring to installation and finishing. 206 pages, 187 illustrations. Book No. 3011, $14.95 paperback, $23.95 hardcover

FENCES, DECKS AND OTHER BACKYARD PROJECTS—2nd Edition—Dan Ramsey

Do-it-yourself—design, build, and landscape fences and other outdoor structures. The most complete guide available for choosing, installing, and properly maintaining every kind of fence imaginable. Plus, there are how-tos for a variety of outdoor structures, from sheds and decks to greenhouses and gazebos. Easy-to-follow instructions, work-in progress diagrams, tables, and hundreds of illustrations. 304 pages, Illustrated. Book No. 2778, $14.95 paperback, $22.95 hardcover

ATTIC, BASEMENT AND GARAGE CONVERSION: A Do-It-Yourselfer's Guide—Paul Bianchina

Achieve the space, appearance, and functional practicality you want in your home using the space that already exists in your home. This book combined with your own creative imagination will produce professional results. Information on tools and techniques is featured along with complete step-by-step instructions for converting underutilized basements, garages, and attics into spacious, attractive living spaces. 208 pages, Illustrated, with 8-color photo section. Book No. 3271, $16.95 paperback, $24.95 hardcover

KITCHEN AND BATHROOM CABINETS—Percy W. Blandford

Kitchen and Bathroom Cabinets is a collection of wooden cabinet projects that will help you organize and modernize your kitchen and bathroom and make them more attractive at the same time. Clear step-by-step instructions and detailed drawings enable you to build wall and floor cabinets and counters, corner cupboards, island units, built-in tables, worktables, breakfast bars, vanities, and more. 300 pages, 195 illustrations. Book No. 3244, $16.95 paperback, $26.95 hardcover

GARAGES: Complete Step-by-Step Building Plans—Ernie Bryant

A tremendous savings if you elect to contract out the project are the five building plans included in this book for garages in cape, colonial, and contemporary styles. One-, two-, and three-car garages, with or without living quarters above, are featured. These plans used with Bryant's explicit, illustrated, step-by-step instructions make it possible for you to build an attractive garage without the added cost of hiring professional help. 192 paperback, 127 illustrations. Book No. 3314, $14.95 paperback, $22.95 hardcover

MAKE YOUR HOUSE RADON FREE—Drs. Carl and Barbara Giles

Safeguard your home and family from the dangers of radon using this practical guide. What radon is, what it does, how it enters your home or workplace, how to remove it, and how to prevent it from recurring are covered in detail. Specific brands of radon-measuring and radon-deterring equipment, products, and materials are recommended. Tips on building a radon-resistant home are also included. 144 pages, Illustrated. Book No. 3291, $9.95 paperback, $15.95 hardcover

PROFESSIONAL PLUMBING TECHNIQUES— Illustrated and Simplified—Smith

"...useful for the experienced handyperson or the homeowner who wishes to verify an estimate of work needed."—Booklist

This plumber's companion includes literally everything from changing a washer to installing new fixtures: installing water heaters, water softners, dishwashers, gas stoves, gas dryers, grease traps, clean outs, and more. Includes helpful piping diagrams, tables and charts. 294 pages, 222 illustrations. Book No. 1763, $11.95 paperback, $16.95 hardcover

WHAT'S IT WORTH: A Home Inspection and Appraisal Manual—Joseph V. Scaduto

"...replete with diagrams and written in language that can be understood by even the most novice house seeker...a must for anyone looking at older houses."—**The Boston Globe**

"...a truly no-nonsense manual for home buyers to use in a self-inspection process."—**New York Public Library, New Technical Books**

This book is packed with practical advice that could save you hundreds, even thousands of dollars in unexpected home maintenance and repair costs! 288 pages, 299 illustrations. Book No. 3301, $16.95 paperback, $24.95 hardcover

BASIC BLUEPRINTING READING—John A. Nelson

With the knowledge gained from this book, you will become expert at reading not only mechanical drawings, but construction, electrical, and plumbing drawings as well. Using a step-by-step approach, John Nelson incorporates the latest ANSI drafting standards as he covers all aspects of blueprint reading. Through straightforward language and excellent example illustrations, Nelson shows you how to identify and understand one-view, multi-view, sectional-view, and auxiliary-view drawings. 256 pages, 235 illustrations. Book No. 3273, $18.95 paperback, $27.95 hardcover

---